From Message to Mind

FROM MESSAGE
TO MIND

Edited by

Directions in Developmental Neurobiology

Stephen S. Easter, Jr.

Kate F. Barald

Bruce M. Carlson

The University of Michigan

SINAUER ASSOCIATES, INC. • PUBLISHERS
Sunderland, Massachusetts 01375

THE COVER

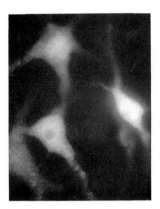

Light micrograph of some of the spinal cord motoneurons
that innervate the forelimb muscle (medial triceps) of
the bullfrog. Cells were labeled with tetramethyl-rhodamine
conjugated to horseradish peroxidase by retrograde transport
of the label from a lesion of their axons in the muscle
nerve. They were viewed with fluorescence optics and
photographed at 320× magnification. This procedure is an
example of the type used to specifically label neurons for
studies of synaptic connectivity during development or
regeneration. Micrograph courtesy of M. Westerfield,
Institute of Neuroscience, University of Oregon.

FROM MESSAGE TO MIND

Library of Congress Cataloging-in-Publication Data

From message to mind.

    Bibliography: p.
    Includes index.
    1. Developmental neurology. 2. Nervous system—Growth.
3. Brain—Growth. 4. Molecular biology. I. Easter, Stephen S.
II. Barald, Kate F. III. Carlson, Bruce M.
QP363.5.F76 1987 612′.8 87-16647
ISBN 0-87893-157-0
ISBN 0-87893-158-9 (pbk.)

Printed on paper that meets the guidelines for permanence
and durability of the Committee on Production Guidelines
for Book Longevity of the Council on Library Resources.

Printed in USA

3 2 1

Contents

Preface

When the idea was raised that we should hold a meeting in Ann Arbor devoted to developmental neurobiology, the question of scope came up immediately. The development of the nervous system is studied at so many levels, from the molecular genetic to the behavioral: Which should we emphasize? After relatively little debate, we concluded that we should attempt to cover a wide sampling of approaches, in an effort to get people with widely divergent approaches talking and sharing ideas in a "small meeting" setting rather than in the smorgasbord atmosphere of a national meeting where one tends to talk only to others with similar interests. The book's title is meant to convey this breadth—from the message at the molecular level to the emergent property of mind. Not surprisingly, the center of gravity (and most chapters) is pitched at the level of the cell.

The next question was, What should the book be? Like all symposium organizers, we wanted to avoid producing "just another symposium book." We wanted to present up-to-date material, but also to make the material accessible, both to specialists from other areas, and, above all, to students. To this end, we have added introductions to groups of related chapters. These section introductions, all written by developmental neurobiologists from the University of Michigan, are intended to set their chapters in context, to provide a preliminary source of reading for the complete novice, and to highlight the significance of each chapter. The specialist may find parts of these section introductions unnecessary, at least in his or her own field, but we suspect (and hope) that they will prove useful for those chapters distant from one's own particular specialty.

We asked the authors not only to present their current ideas to an audience of neurobiologists whose interests are wide-ranging, but to try to speculate on the directions their individual fields are taking and why. As you will see, they have succeeded. Most chapters contain new material, previously unpublished, and all contain broad reviews that go beyond the author's own personal contributions. We thank the authors for achieving this breadth, and we thank Sinauer Associates for their thoughtful and careful support in the publication of this book.

Stephen S. Easter, Jr.
Kate Barald
Bruce Carlson
Ann Arbor

Contributors

KATE F. BARALD
Department of Anatomy and Cell Biology, University of Michigan School of Medicine, Ann Arbor, Michigan

MICHAEL J. BASTIANI
Department of Biological Sciences, Stanford University, Stanford, California

JIM BOULTER
Department of Molecular Neurobiology, The Salk Institute for Biological Studies, San Diego, California

J. P. BROCKES
MRC Cell Biophysics Unit, King's College, London, England

MARY ANN BROW
Department of Molecular Biology, Research Institute of Scripps Clinic, La Jolla, California

STEVEN J. BURDEN
Department of Biology, Massachusetts Institute of Technology, Cambridge, Massachusetts

BRUCE M. CARLSON
Department of Anatomy and Cell Biology, Medical School, and Department of Biology, University of Michigan, Ann Arbor, Michigan

DONA M. CHIKARAISHI
Neuroscience Program, Tufts University School of Medicine, Medford, Massachusetts

KATHRYN L. CROSSIN
The Rockefeller University, New York, New York

PATRIA DANIELSON
Department of Molecular Biology, Research Institute of Scripps Clinic, La Jolla, California

STEPHEN S. EASTER, JR.
Department of Biology, University of Michigan, Ann Arbor, Michigan

GERALD M. EDELMAN
The Rockefeller University, New York, New York

JAMES C. EDMONDSON
Department of Pharmacology, New York University Medical School, New York, New York

JUDITH S. EISEN
Institute of Neuroscience, University of Oregon, Eugene, Oregon

D.M. FEKETE
MCR Cell Biophysics Unit, King's College, London, England

P. FERRETTI
MCR Cell Biophysics Unit, King's College, London, England

CHRISTINE T. FISCHETTE
The Rockefeller University, New York, New York

ERIC FRANK
Department of Neurobiology, Anatomy and Cell Science, University of Pittsburgh School of Medicine, Pittsburgh, Pennsylvania

PAUL GARDNER
Department of Molecular Neurobiology, The Salk Institute for Biological Studies, San Diego, California

MARTIN GODBOUT
Department of Molecular Biology, Research Institute of Scripps Clinic, La Jolla, California

DANIEL GOLDMAN
Department of Molecular Neurobiology, The Salk Institute for Biological Studies, San Diego, California

COREY S. GOODMAN
Department of Biological Sciences, Stanford University, Stanford, California

H. GORDON
MRC Cell Biophysics Unit, King's College, London, England

ROGER A. GORSKI
Department of Anatomy and Laboratory of Neuroendocrinology, Brain Research Institute, University of California School of Medicine, Los Angeles, California

WILLIAM T. GREENOUGH
Departments of Psychology and Anatomical Science, Neural and Behavioral Biology Program, University of Illinois, Urbana-Champaign, Illinois

ALLEN L. HARRELSON
Department of Biological Sciences, Stanford University, Stanford, California

MARY E. HATTEN
Department of Pharmacology, New York University Medical School, New York, New York

STEVE HEINEMANN
Department of Molecular Neurobiology, The Salk Institute for Biological Studies, San Diego, California

KARL HERRUP
Department of Human Genetics, Yale Medical School, New Haven, Connecticut

RICHARD I. HUME
Department of Biology, University of Michigan, Ann Arbor, Michigan

MICHAEL V. JOHNSTON
Departments of Pediatrics and Neurology, University of Michigan School of Medicine, Ann Arbor, Michigan

KATHERINE KALIL
Department of Anatomy, University of Wisconsin, Madison, Wisconsin

ERIC I. KNUDSEN
Department of Neurobiology, Stanford University School of Medicine, Stanford, California

LYNN LANDMESSER
Department of Physiology and Neurobiology, University of Connecticut, Storrs, Connecticut

JEFF W. LICHTMAN
Department of Anatomy and Neurobiology, Washington University Medical School, St. Louis, Missouri

VICTORIA N. LUINE
The Rockefeller University, New York, New York

MICHAEL F. MARUSICH
Institute of Neuroscience, University of Oregon, Eugene, Oregon

BRUCE S. McEWEN
The Rockefeller University, New York, New York

RANDALL D. McKINNON
Department of Molecular Biology, Research Institute of Scripps Clinic, La Jolla, California

BRUCE MENDELSON
Department of Neurobiology, Anatomy and Cell Science, University of Pittsburgh School of Medicine, Pittsburgh, Pennsylvania

BRUCE OAKLEY
Department of Biology, University of Michigan, Ann Arbor, Michigan

JIM PATRICK
Department of Neurobiology, The Salk Institute for Biological Studies, San Diego, California

MARK M. RICH
Department of Anatomy and Neurobiology, Washington University Medical School, St. Louis, Missouri

CAROLYN SMITH
Department of Neurobiology, Anatomy and Cell Science, University of Pittsburgh School of Medicine, Pittsburgh, Pennsylvania

PETER M. SNOW
Department of Biological Science, Stanford University, Stanford, California

J. GREGOR SUTCLIFFE
Department of Molecular Biology, Research Institute of Scripps Clinic, La Jolla, California

KATHRYN TOSNEY
Department of Biology, University of Michigan, Ann Arbor, Michigan

KRISTINE S. VOGEL
Institute of Neuroscience, University of Oregon, Eugene, Oregon

MICHAEL W. VOGEL
Department of Human Genetics, Yale Medical School, New Haven, Connecticut

JOSEPH B. WATSON
Department of Molecular Biology, Research Institute of Scripps Clinic, La Jolla, California

DAVID A. WEISBLAT
Department of Zoology, University of California, Berkeley, California

MONTE WESTERFIELD
Institute of Neuroscience, University of Oregon, Eugene, Oregon

JAMES A. WESTON
Institute of Neuroscience, University of Oregon, Eugene, Oregon

ROBERT S. WILKERSON
Department of Anatomy and Neurobiology, Washington University Medical School, St. Louis, Missouri

I

MOLECULAR APPROACHES TO NEURAL DEVELOPMENT

BRUCE OAKLEY

The emergence and maintenance of the differentiated state of cells is certifiably a central question in biology. How is it, for instance, that various discrete parts of a cell's membrane come to differ radically in their morphology and biochemistry? Not only is the origin of membrane differentiation during development an issue, but it remains an issue thereafter, for it has been clear since the publication of the remarkable book *The Dynamic State of Body Constituents* (Schoenheimer, 1942) that underlying the apparent stability of mature tissues is a quiet but relentless turnover of virtually all constituent molecules. Thus, the long-lived cells of adults must contain programs that assist in maintaining membrane specializations. All of the chapters in this section deal with molecules in membranes.

The intricate neuronal connections that comprise the complex circuits of the brain are unlikely to be established chaotically. Rather, one assumes that precisely detailed neuronal connections arise reliably during development. How are these circuits put together? Developmental neurobiologists like to point out that there are many more synapses in the brain than genes in the genome—hence the intricacies of neuronal circuits cannot be based on a one gene–one synapse formula. It is much more reasonable to argue, as Edelman and Crossin do in Chapter 1, for the importance of epigenetic controls in which the continuous interplay among the cells and the environment regulate gene expression of a given cell. Indeed, they argue that it may take only a few genes so modulated by cell interactions to develop the principal form of the nervous system. Thus, the general morphology of a tissue or organ may be developmentally regulated by cell surface molecules that adhere to other cells or to the extracellular substrate. Adhesive interactions at the border between cell groups or collectives may be particularly important for development of normal morphology. Edelman and Crossin have correlated morphological development with the distribution in time and space of CAMs (cell adhesive molecules) and SAMs (substrate adhesive molecules) in the chick. They have further shown that treatment with an antibody to CAM can disrupt normal events in development, such as feather formation, alignment of axons into bundles, the migration of brain neurons, or the formation of layers in the retina. One would like to know how many details of the circuitry can be correlated with differences in CAM expression. What additional informational factors might there be which are necessary to establish the most detailed nuances of circuit connections, such as termination of axonal endings on selected subsets of dendritic spines? By acting as general purpose adhesives that make cells stick together, CAMs appear to be required for the gross orderliness of the development of neural tissue. If the finer grain of neuronal interconnections is to be explicable in terms of cell adhesion molecules, one would expect to find families of adhesion glycoproteins whose members are expressed by cells at different times under varying circumstances. Chapter 1 explores these issues.

Burden (Chapter 2) addresses the problem of the stability of membrane specializations by seeking out proteins that are closely affiliated with the acetylcholine receptor of muscle. He describes a 43-kd protein that may have a role in stabilizing the clusters of acetylcholine receptors to constrain them to the synapse, and concludes his presentation by examining the feasibility of similar studies on the mechanism of spatial stabilization of synaptic transmitter receptors in the central nervous system.

Chemical or surface interactions among cells are believed to be important in controlling the selection and fate of targets contacted by axons. As an axon elongates, gene expression of both the neuron and its target may be altered by the cell–cell interactions that occur. The alteration of the target is nowhere more dramatic than in the instances of urodele limbs (Sicard, 1985) and mammalian taste buds, whose regeneration is entirely dependent upon reinnervation. The extent to which a target's development and regeneration are controlled by common principles may vary with the particular model system. Recent experimental demonstrations that mammalian taste buds are neurally induced during a sensitive period in development have suggested that some processes, such as the induction of stem cells, are unique to development and do not reoccur in taste bud regeneration (Hosley et al., 1987a,b). In Chapter 3, Fekete et al. have used a monoclonal antibody to reveal differences in the development and neurally mediated regeneration of urodele limbs. The initial developmental responsiveness of progenitor cells to growth factors may be supplanted by an emerging dependence upon chemical factors derived from neurons (or from Schwann cells). This approach, utilizing antibodies as cell markers, opens up new vistas on the origin and time of appearance of blastema cells involved in limb formation.

CHAPTER ONE

The Molecular Regulation of Neural Morphogenesis

GERALD M. EDELMAN AND
KATHRYN L. CROSSIN

Introduction

Recent studies of cell adhesion molecules (CAMs) and substrate adhesion molecules (SAMs) suggest that many key events related to neural pattern formation depend upon cell surface modulation. This chapter provides a short review of the evidence supporting this conclusion.

Although morphology is under genetic control, morphogenesis occurs largely by epigenetic mechanisms involving local interactions of cells, some of which are stochastic. The reconciliation of these facts is a large challenge, particularly in attempting to understand the development of neural structures in which the apparent specificity of patterns is so dominant. One possible solution is to consider that certain gene products mediate mechanochemical interactions among cells and that the expression of such protein molecules depends in turn upon the results of these interactions, particularly of cells in collectives. Adhesion molecules are obvious candidates for this role, for the following reasons: (1) they link cells into epithelia; (2) their regulation governs conversion of epithelia to mesenchyme (epithelial-mesenchymal conversion), (3) they permissively regulate cell migration; and (4) differences in their specificity can account for border formation between neighboring cell collectives that exchange signals for embryonic induction and cytodifferentiation.

A number of CAMs have been isolated and characterized in terms of specificity, binding properties, expression sequences, and regulation at a variety of embryonic sites including the nervous system (for review, Edelman, 1984a, 1985, 1986a,b). In addition, a new SAM called cytotactin has been found (Grumet et al., 1985) that shows unusual expression sequences in the embryo and the developing nervous system (Crossin et al., 1986). Some of the characteristics of these molecules are summarized in Table 1.

Two general observations support the notion that gene control of these molecules may control the dynamics of morphogenesis as a result of their mechanochemical functions. The first is that the expression of two primary CAMs, N-CAM (the neural cell adhesion molecule) and L-CAM (the liver cell adhesion molecule), follows a defined set of rules in bordering cell collectives at all sites of embryonic induction (Crossin et al., 1985; Edelman, 1985, 1986a,b). The second is that a mounting body of evidence (much of it from studies of the

TABLE 1. Some Characteristics of CAMs and Cytotactin

	Type	Molecular Weight	Carbohydrate Characteristics	Binding Mechanism	Ion Dependence
CAM					
N-CAM (neural CAM)	1°	180–250 kd (E form) 180, 140, 120 kd (A forms)	glycoprotein with unusual polysialic acid	homophilic	none
L-CAM (liver CAM)	1°	124 kd	glycoprotein	homophilic	Ca^{2+}
Ng-CAM (neuron-glia CAM)	2°	200 kd 135 kd 80 kd	glycoprotein	heterophilic	none
Cytotactin		190 kd 200 kd 220 kd	glycoprotein	higher order	none

nervous system) suggests that perturbation of CAM function affects pattern formation and that perturbation of morphological continuity alters CAM expression in both embryogenesis and regeneration (Edelman, 1984a, 1985).

In this chapter, we first briefly review induction events, stressing an instance (the developing feather) in which primary CAM expression rules lead to border formation during induction and cytodifferentiation. Perturbation of CAM action in the early stages of feather induction and morphogenesis has been shown to lead to large alterations of pattern and fate. We then turn to adhesion molecules in the nervous system and provide evidence for their morphogenetic roles at a number of sites and processes: neural plate formation, neural crest cell migration, development of the otic placode, cell migration in cerebellar morphogenesis, mapping in the retinotectal projection, and regenerative events in the periphery. In the course of this survey, we provide evidence that abrogation of CAM function results in altered morphogenesis and that disruption of morphology affects

the course of expression of the CAMs that are important in establishing and maintaining neural structures. We also extend the idea of regulatory control of expression sequences as they influence morphogenesis to SAMs, using cytotactin as an example.

Expression Rules in Neural and Secondary Induction

Two primary CAMs, N-CAM and L-CAM, appear very early in development (prior to gastrulation) in a definite temporal sequence establishing boundaries between adjoining tissues involved in embryonic induction. After organogenesis, a projection of CAM distributions onto a fate map of the blastoderm shows that each of these CAMs crosses borders assigned to organ structures (Figure 1). This indicates that CAMs are not tissue specific per se but rather are involved in early interactions of cells, leading to borders that govern later histogenetic events.

In the chick blastoderm, the two primary CAMs, N-CAM and L-CAM, appear together

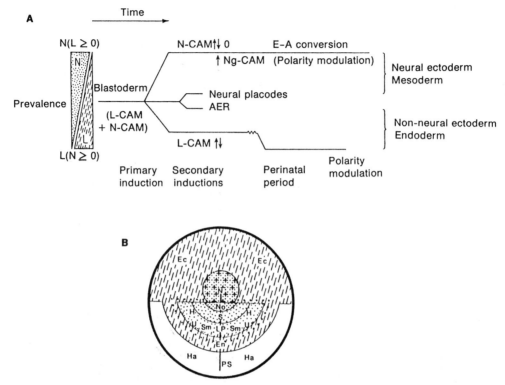

FIGURE 1. Major CAM expression sequence and composite CAM fate map in the chick. (A) Schematic diagram showing the temporal sequence of expression of CAMs during embryogenesis, starting from the stage of the blastoderm and proceeding through neural and secondary inductions. Germ layer derivatives are indicated at right. Vertical wedges at the left refer to approximate relative amounts of each CAM in the different layers or parts of the embryo; e.g., the line referring to blastoderm has relatively large amounts of each CAM, whereas that for neural ectoderm has major amounts of N-CAM but little or no L-CAM. After they diverge in cellular distribution, the CAMs are then modulated in prevalence within various regions of inductions, or actually decrease greatly when mesenchyme appears or cell migration occurs. Note that placodes, which have both CAMs, echo the events seen for neural induction. As neuronal development proceeds, a secondary CAM (Ng-CAM) emerges; unlike the other two CAMs, this CAM would not be found in the fate map shown in B before 3.5 days. In the perinatal period, a series of modulations occurs: embryonic to adult (E → A) conversion for N-CAM, and polar redistribution at the surface of individual cells for L-CAM. (B) Composite CAM fate map in the chick according to Vakaet (1985). The distribution of N-CAM (stipple), L-CAM (slashes), and Ng-CAM (crosses) on tissues of 5-to 14-day embryos is mapped back onto the tissue precursor cells in the blastoderm. Additional regions of N-CAM staining in the early embryo (5 days) are shown by larger dots. In the early embryo, the borders of CAM expression overlap the borders of the germ layers, i.e., derivatives of all three germ layers express both CAMs. At later times overlap is more restricted: N-CAM disappears from endoderm, except for a population of cells in the lung. L-CAM is expressed on all ectodermal and endodermal epithelia but remains restricted in the mesoderm to epithelial derivatives of the urogenital system. The vertical bar represents the primitive streak (PS); Ha, hemangioblastic area; Ec, intraembryonic and extraembryonic ectoderm; En, endoderm; H, heart; LP, lateral plate mesoderm; N, nervous system; No, prechordal and chordamesoderm; S, somite; Sm, smooth muscle; U, urogenital system.

on all cells (Edelman et al., 1983); at gastrulation, however, ingressing cells stain much less with anti–N-CAM and anti–L-CAM. Cells in the chordamesoderm then re-express N-CAM at neural induction, and endodermal cells later re-express L-CAM. During neural induction, a remarkable transition occurs in the epiblast: after completion of induction, the presumptive neural plate shows only N-CAM and the surrounding somatic ectoderm shows mainly L-CAM. Thereafter, at all sites of secondary embryonic induction, N-CAM or L-CAM or both undergo a series of prevalence modulations following two main rules (Crossin et al., 1985; Edelman, 1985, 1986a,b): all conversions from epithelia to mesenchyme to condensed mesenchyme show a modal transition N→O→N, where O means either low or undetectable levels of the CAM (rule I), and epithelial cells show other modal transitions in CAM expression, either NL→L or NL→N (rule II). Table 2 is a list of these modulations

and their corresponding rules. As already mentioned, the expression of CAMs in epigenetic sequences following these rules is map-restricted: particular CAMs appear in a particular ordered pattern in a fate map of the chick blastoderm (see Figure 1B).

One of the striking aspects of these epigenetic sequences is the finding that at many sites of induction, cell collectives consisting of mesenchymal condensations following rule I are found in proximity to epithelial cell collectives following rule II. This reflects two facts: (1) with the exception of urogenital structures, once mesodermal mesenchyme is formed, it never re-expresses L-CAM, and (2) this mesenchyme plays an important inductive role at many sites. Borders at such sites are between mesenchymally derived collectives expressing N-CAM (rule I) and epithelial collectives expressing either N-CAM, or N-CAM and L-CAM together (rule II).

Perhaps the most dramatic example of the

TABLE 2. CAM Expression in Mesenchymal Conversions and Epithelia[a]

	Ectodermal	Mesodermal	Endodermal
Mesenchymal conversions (rule I)	N → O → N neural crest cell/ganglia	N → O → N somite/skeletal muscle N → O → N (→ *)[b] somite/chondrocytes	
Epithelia (rule II)[c]	NL → N (→ *) neural tube N → L (→ *) somatic ectoderm	N → NL → L kidney	NL → L gastrointestinal tract epithelium

[a]These rules reiterate the earliest modulations of CAM expression seen in the blastoderm. The epithelial blastoderm expresses both N-CAM and L-CAM; at gastrulation both CAMs are down-regulated as cells become mesenchymal and ingress to form the middle layer. The chordamesoderm thus formed condenses, re-expresses N-CAM (NL → O → N), and subsequently induces the overlying epithelium at neurulation to form neural ectoderm (expressing N-CAM, NL → N) and somatic ectoderm (expressing L-CAM, NL → L).
[b](*) represents differentiation products (e.g., keratin, crystallin) with disappearance of the CAM.
[c]Epithelia (rule II) show replacement of one CAM by another or disappearance.
For detailed examples of each entry, see Crossin et al., 1985.

recursive application of these epigenetic rules for expression of these two primary CAMs occurs in the feather. Examination of this periodic and hierarchically organized structure provides an opportunity to analyze the coupling of cell collectives in detail and to relate their interactions to cytodifferentiation events within a dimensionally well-organized appendage. Feathers are induced through the formation of dermal condensations of mesodermally derived mesenchyme, which act upon ectodermal cells to form placodes (Sengel, 1976). As feather induction proceeds in rows from medial to lateral aspects of the chicken skin, such placodes and condensations are eventually hexagonally close-packed. Within each induced placode, a dermal papilla is subsequently formed as a result of a repeated cycle of inductive interactions between mesoderm and ectoderm. Subsequently, the cellular proliferation of barb ridges (with fusion to form a rachis) followed by barbule plate formation yields the basis for three levels of branching: rachis, ramis, and barbule.

An extraordinary series of events involving L-CAM–linked collectives of cells adjoining N-CAM–linked collectives, either by movement and adhesion or by cell division and adhesion, is seen at each of these levels (Chuong and Edelman, 1985a,b). Initially, L-CAM–linked ectodermal cells are approached by CAM-negative mesenchyme cells that become N-CAM positive in the ectodermal vicinity (rule I); as these cells accumulate to form condensations, placodes are induced in the L-CAM–linked cells. A similar couple, consisting of L-CAM–linked ectodermal cell collectives adjacent to N-CAM–linked mesenchymal collectives, is seen in the papilla that forms subsequently. After N-CAM–positive mesenchyme cells are excluded by a basement membrane containing laminin and fibronectin, the collar cells derived from the L-CAM–positive papillar ectoderm express both N-CAM and L-CAM (rule II).

After this event, site-restricted applications of rule II lead to a remarkable periodicity of borders (Figure 2): cells derived from papillar ectoderm by division form barb ridges and express L-CAM. In the valleys between the neighboring ridges, the basilar cells lose L-CAM and express N-CAM to form the marginal plate. A similar process occurs as the ridge cells organize to form the axial plate. The end result is a series of cellular patterns following rule II, in which cell collectives linked by L-CAM alternate with those expressing N-CAM at both the secondary barb level and the tertiary barbule level. After further extension of the barb ridges into rami, the L-CAM–positive cells keratinize and the N-CAM–positive cells die without keratinization, leaving spaces between barbules and yielding the characteristic feather morphology.

In this histogenetic CAM expression sequence, one observes periodic CAM modulation, periodic and successive formation of L-CAM–linked and N-CAM–linked cell collectives (CAM couples), and the definite association of gene expression events during cytodifferentiation with particular kinds of CAMs. The most dramatic example is the association of the gene expression of keratins with L-CAM–containing cells only. Throughout this histogenetic sequence, the regulatory process of adhesion is intimately connected with epigenetic sequences consisting of the different primary processes that act as driving forces: morphogenetic movement for the original mesenchymal induction, mitosis for the formation of papillar ectoderm and barb ridges, and death for the N-CAM–linked collectives in the terminal period of feather formation. These findings have several important implications, since they raise the possibility that CAM function is causally important in inductive sequences and suggest that a series of local signals must be responsible for particular sequences of CAM expression.

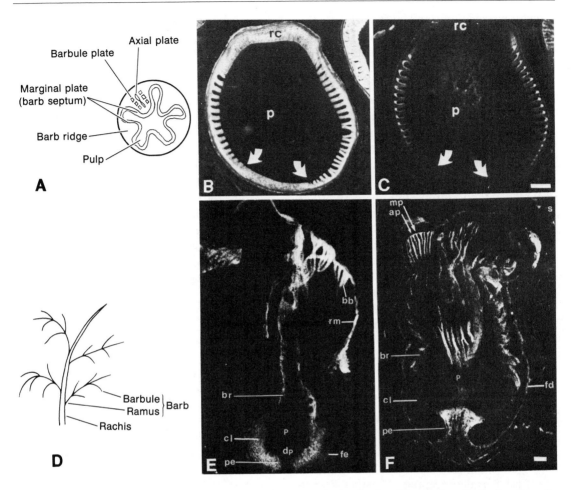

FIGURE 2. Schematic diagram of a transverse section of a developing feather (A) and a longitudinal section of a whole adult feather (D), showing the terminology used. Transverse (B, C) and longitudinal (E, F) sections of developing feather follicles from a newly hatched chicken wing, summarizing the entire sequence of expression of L-CAM (B, E) and N-CAM (C, F) during adult feather histogenesis. (B, C) The formation of the bar ridges starts from the dorsal side (the side with the rachis) and progresses bilaterally toward the ventral side, making a dorsoventral maturation gradient. Positions of the last formed ridges are marked by curved arrows. L-CAM (B) stains all of the cells of the barb ridge epithelium. Bright N-CAM staining (C) starts to appear in the valleys between pairs of the barb ridges. N-CAM appearance starts about eight ridges away from the last formed ridge and increases in staining intensity and distribution dorsally until it reaches the rachis. (E, F) Proceeding from the dermal papilla (dp), the epithelial cells develop from papillar ectoderm (pe) to collar (cl) to barb ridges (br) and then to rami (rm) and barbules (bb), reflecting temporal developmental stages on a spatial array ranging from base to tip. ap, axial plate; fd, follicular dermis; fe, follicular epidermis; mp, marginal plate; p, pulp cavity; s, skin. Bar = 100 µm. (From Chuong and Edelman, 1985b.)

This view is supported by experiments suggesting that CAMs play a key role in causal sequences during induction and histogenesis of the feather in vitro (Gallin et al., 1986). Anti-L-CAM antibodies added to chick skin explants resulted in the alteration of the pattern of N-CAM–linked dermal condensations from a hexagonal symmetry to a pattern with a tendency to fusion of feather placodes mediolaterally into stripes. Longer-term cultures treated with anti-L-CAM showed formation of scalelike plates rather than the feather-like filament patterns that were seen in unperturbed controls. Thus, antibodies to one CAM in an inducing cell collective can alter patterns formed by cells in the other induced cell collective. A computer model based on the notion that the antibody alters the signaling sent to dermal condensations by epidermal cells reproduces the pattern of symmetry breaking (Gallin et al., 1986). This example in a non-neural tissue provides a background for examination of neural tissues as they express neuronal CAMs. The nervous system shows similar but additional features, including the expression of SAMs and of secondary CAMs.

Neural Histogenesis: Neuronal CAMs

As seen in the example of feather formation, one of the key structural events in embryogenesis and histogenesis is boundary formation, an event that appears to be strongly tied to the early differentiation rules found for the two well-characterized primary CAMs (Edelman, 1985, 1986a,b; Crossin et al., 1985). The otic placode, an example of a specialized structure with neural components (Figure 3), shows the expression of these epigenetic rules (see Table 2) particularly well (Richardson et al., 1987; Crossin et al., in press). An inducing mesenchymal collective re-expressing N-CAM (rule I) induces an epithelial placode showing N-CAM and L-CAM (rule II). There is then a site-specific formation of borders of

various presumptive structures within the developing placode by a sequence of prevalence modulations of the two primary CAMs following rule II. Neural structures related to the cochlea express N-CAM; non-neural structures express L-CAM and N-CAM in various sequences.

After neural induction and the exclusion of L-CAM from neural derivatives (rule II), a secondary CAM specific for neural derivatives, Ng-CAM, is expressed on postmitotic neurons. In the central nervous system (CNS), Ng-CAM is seen on extending neurites and in very slight amounts on cell somata; this represents a striking example of polarity modulation, i.e., unequal distribution of CAMs within the membrane of a single cell. At just those sites and times at which neuronal migrations on radial glia take place, however, Ng-CAM is strongly expressed on somata and leading processes in addition to neurites (Thiery et al., 1985a; Daniloff et al., 1986a). Ng-CAM is not seen on glia in the CNS, but both Ng-CAM and N-CAM are present on Schwann cells and neurons in the peripheral nervous system (PNS). It is notable that by immunocytochemical techniques, PNS neurons do not exhibit the polarity modulation of Ng-CAM that is seen in the CNS.

As shown in Figure 4, a remarkable site-specific microsequence (Daniloff et al., 1986a) of CAM expression occurs during CNS and PNS development that results in grossly altered distribution patterns of N-CAM and Ng-CAM with time. Two particularly striking events are the prevalence modulation resulting in down-regulation of Ng-CAM in presumptive myelinating regions, and the perinatal occurrence of embryonic to adult (E→A) conversion (a loss of polysialic acid with concurrent changes in binding properties) of N-CAM in tracts. Neural crest cells, which form the PNS among other structures, show a remarkable expression of rule I as applied to ectomesenchyme: N-CAM is lost or lowered on

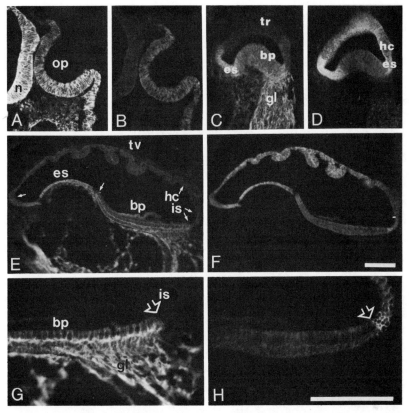

FIGURE 3. Expression of primary CAMs during development of the otic placode. The otic pit (op) in a stage-12 embryo stained strongly for both N-CAM (A) and L-CAM (B); the adjacent neural tube (n) stained only for N-CAM. N-CAM staining was most intense in the inner ventral margin of the invaginating otic pit (A), but the pit stained uniformly for L-CAM (B). By embryonic day 9, the presumptive external sulcus showed the brightest N-CAM staining (C), with presumptive basilar papilla and hyaline cells moderately stained and presumptive internal sulcus and tegmentum vasculosum faintly stained. In contrast, L-CAM (D) appeared more brightly in the presumptive external sulcus, and also began to increase in the presumptive tegmentum vasculosum (tv) and internal sulcus (is); to a lesser extent, it appeared in the basilar papilla and hyaline cell region (hc). At day 12, within the lagena, only the basilar papilla (bp) and external sulcus (es) were positive for N-CAM (E, G). The acoustic ganglion (gl) and afferent fiber tracts stained strongly with anti–N-CAM at all stages (C, E, and G). L-CAM staining intensity (F, H) was increased in the internal sulcus, external sulcus, and tegmentum vasculosum. The arrows in E represent presumptive borders within the epithelium at E12. In high-magnification micrographs it was observed that the afferent fibers and the synaptic terminal layer stained strongly for N-CAM (G). N-CAM was also found in the support cell layer below the terminals and in the hair cell layer above the terminals (G). N-CAM staining outlines the flask-shaped hair cells, but the apical surfaces of these cells do not appear to be stained. The internal sulcus was negative for N-CAM, but stained intensely for L-CAM. Arrows in G and H show the striking border between these highly N-CAM–rich and L-CAM–rich areas. L-CAM is also present in the support cells in the basilar papilla. Bar = 100 μm. (From Richardson et al., 1987.)

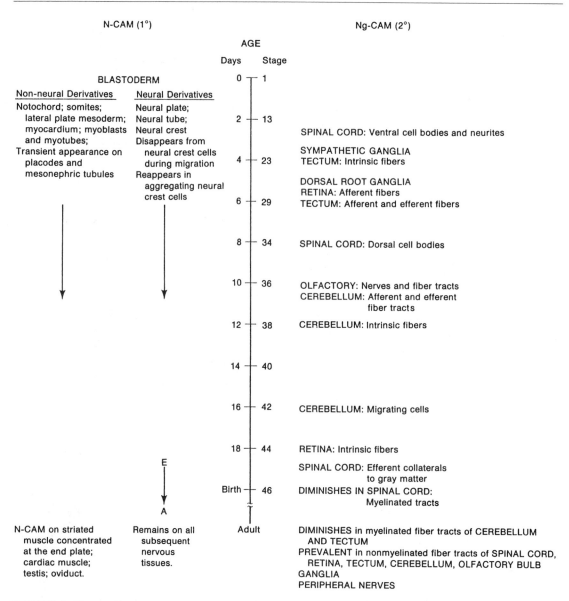

FIGURE 4. **Expression sequence of two neuronal CAMs in the developing chick nervous system. N-CAM, a primary cell adhesion molecule, appears early (blastoderm stage) in both neural and non-neural derivatives. After birth, it remains on all subsequent nervous tissues. Ng-CAM, a secondary adhesion molecule, appears later during embryogenesis and is first observed on the developing spinal cord (Thiery et al., 1985a; Daniloff et al., 1986). The sequential appearance of Ng-CAM in CNS and PNS regions is summarized. In the adult, Ng-CAM is primarily limited to the unmyelinated CNS regions (tracts or laminae) and PNS. E → A, embryonic to adult conversion of N-CAM. (From Daniloff et al., 1986a.)**

the surface of migrating crest cells and reappears at sites where ganglion formation occurs (Thiery et al., 1982, 1985b).

These sequences reveal the existence of coordinated cell surface modulation events occurring during the formation of particular neural structures: prevalence modulation, polarity modulation, and chemical modulation (E→A conversion). Recent studies on the large cytoplasmic domain (ld) polypeptide of N-CAM (Mr = 180,000) (Hemperly et al., 1986) indicate that the molecule is expressed only in the nervous system (Murray et al., 1986b). This molecule differs from the small cytoplasmic domain (sd) chain (Mr = 140,000) in having a large cytoplasmic domain that is 250 amino acids longer (Hemperly et al., 1986). It is differentially expressed in certain layers during development of the cerebellum and retina (Williams et al., 1985; Pollerberg et al., 1985; Murray et al., 1986b). Alternative RNA splicing (Murray et al., 1986a) leads to coding for the larger cytoplasmic domain in the ld polypeptide that might interact differentially with the cytoskeleton or other proteins in the cortical region of the cell (Hemperly et al., 1986). It is an attractive hypothesis that differential cell surface modulation of the ld and sd chains could lead to altered patterns of cell interaction, migration, and layering. The adequate signal for this modulation event might be one that controlled alternative splicing in a local tissue region.

SAMs in Embryogenesis and Neural Histogenesis: Cytotactin

The evidence that CAMs may play significant roles in embryogenesis and neural histogenesis by means of modulation and regulation prompts the idea that complementary functions may be exercised by substrate adhesion molecules (SAMs). A newly discovered extracellular matrix protein, cytotactin, appears to have such a role.

Cytotactin was isolated from 14-day embryonic chicken brains as structurally related polypeptides of Mr = 220,000, 200,000, and 190,000 (Grumet et al., 1985). These polypeptides were efficiently extracted in the absence of detergent and appeared to be disulfide-linked into higher polymers. Cytotactin was involved in glia-neuron adhesion in vitro by a mechanism independent from Ng-CAM, but unlike Ng-CAM, it was absent from neurons. The molecule was synthesized in culture by glia and cells from smooth muscle, lung, and kidney. It was found at the surface of cultured glia in a cell-associated fibrillar pattern.

Although cytotactin was initially identified as a molecule that mediates glia-neuron interactions, preliminary immunohistochemical localization of the molecule suggested that it was an extracellular matrix protein with a widespread but nonetheless more restricted distribution than either fibronectin or laminin. Analysis of the sequence of its expression (Crossin et al., 1986) showed that it was first present in the gastrulating chicken embryo. It appeared later in the basement membrane of the developing neural tube and notochord in a temporal sequence beginning in the cephalic regions and proceeding caudally. Between days two and three of development, the molecule was present at high levels in the early neural crest pathways (surrounding the neural tube and somites) but, in contrast to fibronectin and laminin, was not found in the lateral plate mesoderm or ectoderm. At later times, cytotactin was expressed extensively in the CNS, in lesser amounts in the PNS, and in a number of non-neural sites, most prominently in all smooth muscles and in basement membranes of lung and kidney. In the cerebellum, it appeared on glial endfeet, on Bergmann glial fibers, and in extracellular spaces.

These findings raise the possibility that certain extracellular matrix proteins such as cytotactin contribute to pattern formation in embryogenesis as a result of their restricted expression in a spatiotemporally regulated fashion at some sites but not at others. Preliminary analysis of the role of cytotactin in supporting or modulating external granule cell migration (Chuong et al., in press) supports this idea and raises the possibility that coordination and differential expression of CAMs and SAMs are major factors in regulating morphogenetic patterns.

Differential Roles of CAMs in Morphogenetic Processes: Fasciculation, Migration, and Layer Formation

The evidence that CAM modulation is a major factor in both embryogenesis and histogenesis in the nervous system is particularly compelling. It is striking that CAMs of different specificity can play complementary or differential roles in altering various morphogenetic processes at different times and places. Thus, differential cell surface modulation of CAMs having different binding mechanisms may play a major role in shaping different neural regions. In accord with this idea, different cellular systems have revealed functional differences for each CAM, reflecting its relative amount (prevalence modulation) and cellular location (polarity modulation) (Hoffman et al., 1986).

Each of three cellular processes examined in vitro was found to be preferentially inhibited only by anti–N-CAM or by anti–Ng-CAM antibodies at different sites (Hoffman et al., 1986). Both neurite fasciculation in cultured dorsal root ganglia and the migration of cerebellar granule cells were inhibited by anti–Ng-CAM antibodies. Anti–N-CAM antibodies inhibited the formation of histological layers in the retina. The relative Ng-CAM/ N-CAM ratios in comparable extracts of brain, dorsal root ganglia, and retina were 0.32, 0.81, and 0.04, respectively, as shown by quantitative immunoblotting. Thus, the relative ability of anti–Ng-CAM antibodies to inhibit cell-cell interactions in different neural

tissues was strongly correlated with the local Ng-CAM/N-CAM ratio. During culture of dorsal root ganglia in the presence of nerve growth factor (NGF), the Ng-CAM/N-CAM ratio rose to 4.95 in neurite outgrowths and 1.99 in the ganglion proper, reflecting both polarity and prevalence modulation. These findings show that the degree to which a particular CAM influences a morphogenetic process will depend upon the site, its amount relative to other CAMs, and the particular histogenetic process being mediated.

An excellent example of these principles is seen in cellular migration of external granule cells in the developing cerebellum. Recent studies (Hoffman et al., 1986; C.M. Chuong et al., 1987) indicate that antibodies to N-CAM have only a slight effect on the migration of the external granule cells in cerebellar slices in vitro. In contrast (Table 3), anti–Ng-CAM blocked movement out of the external granular layer; anticytotactin had no effect on this process but slowed migration in or exit from the molecular layer. Correlation of these effects with the spatiotemporal expression of these different molecules suggests a complex complementary scheme in which temporal expression, different binding roles, and different functional effects on interactions between neurons and Bergmann glia are all required for layer formation.

The existence of cell surface modulation in various forms indicates that individual cell shape and specialization will also constrain and alter the effects of various modulatory modes of cell adhesion in different contexts. For example, N-CAM appears enriched in neural growth cones (Chuong et al., 1982; Wallis et al., 1985; Ellis et al., 1985); the evidence is also consistent with the presence of the ld chain at this location (Pollerberg, 1985; Murray et al., 1986b). The accumulated data indicate that CAM-mediated adhesion must, therefore, by analyzed in a context-dependent fashion related to actual tissue and cell morphology. Although this brings concomitant in-

TABLE 3. Effects of Anti–N-CAM, Anti–Ng-CAM, and Anti-Cytotactin on Cerebellar Granule Cell Migration

Antibody[a]	External Granular Layer (EGL) % Distribution[b]	% of Cells Leaving EGL[c]	Molecular Layer (ML) % Distribution	% Cells that Entered ML Leaving ML[d]	Internal Granular Layer (IGL) % Distribution
Day 0					
Nonimmune (NI)	86 ± 2		3 ± 1		11 ± 1
Day 3					
Nonimmune (NI)	12 ± 9	86	15 ± 9	81	73 ± 13
Anti–N-CAM	24 ± 8	72	8 ± 5	88	68 ± 9
Anti–Ng-CAM	64 ± 6	26	13 ± 7	48	23 ± 5
Anti-cytotactin	9 ± 6	90	55 ± 11	30	35 ± 12
Mab 1D8	25 ± 6	71	44 ± 3	30	31 ± 4
Mab 6G10	37 ± 6	57	40 ± 6	25	24 ± 3

[a]For polyclonal antibodies, 3.8 mg/ml Fab' fragments were used. For monoclonal antibodies, culture supernatant was dialyzed and used.

[b]The cultures were labeled for 1 hr with [^{3}H] thymidine and either fixed immediately (day 0 value) or incubated with the indicated antibodies. At the end of day 3, the distribution of [^{3}H]-thymidine-labeled cells (mean ± s.d.) among the three cerebellar layers was calculated by counting all labeled cells from the pia to the IGL in a rectangular layer about 200 μm × 100 μm. Five to ten fields from each of four explants were counted in each experiment.

[c]The percentage of cells leaving the EGL was calculated as follows:

$$\frac{\text{\% cells, day 0} - \text{\% cells, day 3}}{\text{\% cells, day 0}}$$

[d]The percentage of cells that enter the ML and subsequently leave the ML was calculated according to the formula:

$$\frac{\text{\% cells in IGL, day 3} - \text{\% cells in IGL, day 0}}{\text{\% cells in ML, day 0} + \text{\% cells in EGL, day 0} - \text{\% cells in EGL, day 3}}$$

From Chuong et al., 1987.

creases in difficulty of analysis, the combinations of mechanisms having important effects on morphogenesis that result from such context-dependency are virtually limitless and free one from having to specify definite molecular cell addresses as the basis for neural morphogenesis (Sperry, 1963; Edelman, 1984c).

All of the experiments described above were done on in vitro preparations. By means of agarose implants that contained anti–N-CAM antibodies placed in the tectum of *Xenopus laevis*, it could be shown that severe distortion of in vivo map formation and regeneration ensued (Fraser et al., 1984) (Figure 5). A sharp decrease in the density of axonal arbors

physiological alterations (S.E. Fraser et al., in preparation). Perhaps the most striking aspect of this experiment was that a more normal map pattern returned as the local concentration of anti–N-CAM antibodies decremented in the tectum.

It is quite obvious from both the in vitro and in vivo results that perturbation of CAM binding function can alter morphology. To develop a consistent notion of the molecular regulation of neural morphogenesis, it is necessary to show that disruption of morphology or of linked structures can lead to alterations of CAM expression, thus indicating that a control loop exists between maintained form

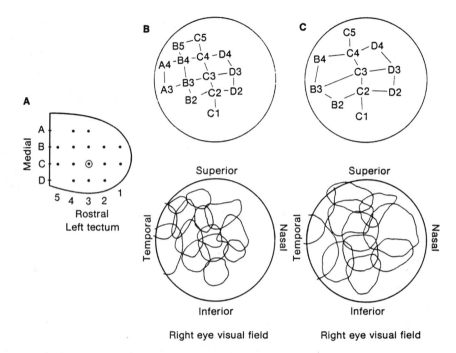

FIGURE 5. Retinotectal projection patterns following implantation of agar microspikes into *Xenopus* tecta. Agarose implants contained either normal rabbit Fab′ fragments (B) or anti–N-CAM Fab′ fragments (C). A metal microelectrode was lowered into the tectum at each of the positions indicated by dots on the representative tectum shown in A; the antibody implant was made at the circled position near the center of the tectum. Large circles in the other diagrams represent a 200° range of the field of view of the right eye, the center of which represents the fixation point of the eye. The center of the responsive region of the frog's visual field for each electrode position is marked in the upper row of circles by the code of the electrode positions shown in A. Below each of these diagrams are shown the diagrams of corresponding multiunit receptive fields. The projection patterns of control implanted animals (B, top) show the same rectilinear arrangement of receptive fields for the rectilinear set of electrode positions that is seen in normal animals. The size of the receptive fields (B, bottom) in the projection shown here was among the largest of those seen in animals with this type of pattern. In animals in which the implants contained anti–N-CAM fragments, the patterns show a distortion of the projection (C, top) and an enlargement of the receptive fields (C, bottom). This distortion is readily seen by comparing the positions of the corresponding receptive fields B3, C3, and D3 in B and C. (From Fraser et al., 1984.)

in tissues and CAM synthesis or modulation. This can be seen most clearly in analyses of the role of CAMs in the peripheral nervous system (PNS).

Pattern and Regeneration in the Periphery

After establishment of the PNS by neural crest cells, CAMs and SAMs play key roles in later pattern formation. Because of the possibility of observing regenerative effects and the occurrence of defined interaction of cell types (Schwann cells, neurons, and muscles) that may be manipulated, structures in the periphery provide an excellent opportunity to test the hypothesis that interactions of cells in different tissues regulate CAM expression. For example, N-CAM, Ng-CAM, and cytotactin are highly concentrated at nodes of Ranvier of the adult chicken and mouse (Rieger et al., 1986). In contrast, unmyelinated axonal fibers were uniformly stained at low levels by specific antibodies to both CAMs but not by antibodies to cytotactin (Figure 6). Moreover, developmental analysis suggested that the interaction between Schwann cells (which displayed N-CAM and Ng-CAM) and neurons may play a role in establishing the one-dimensional periodic pattern of nodes. At embryonic day 14, before myelination had occurred, small-caliber fibers of chick embryos showed periodic coincident accumulations of the two CAMs but not of cytotactin, with only faint labeling in the axonal regions between accumulations. Cytotactin was found on Schwann cells and in connective tissue. By embryonic day 18, nodal accumulations of CAMs were first observed in a few medium- to large-caliber fibers. These findings are consistent with the hypothesis that surface modulation of neuronal CAMs mediated by signals shared between neurons and glia may be necessary for establishing and maintaining the nodes of Ranvier.

In view of the epigenetic nature of tissue formation and the existence of defined CAM expression sequences, it is a reasonable supposition that CAM expression and modulation depend upon local signals that vary according to the state, composition, and integrity of particular interacting cell collectives. Thus, while CAMs may serve to stabilize tissue structures (along with SAMs and cell junctional molecules), their modulation and expression should in turn depend upon local cellular interactions in the stabilized structure. A number of studies have shown that disrupted morphology or altered morphogenesis can

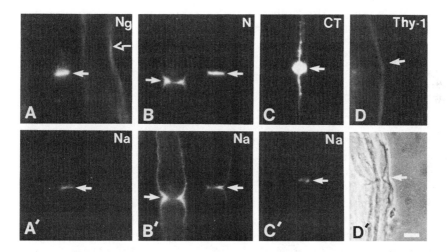

FIGURE 6. Co-localization of neuronal CAMs and cytotactin at the node of Ranvier. Ng-CAM (A), N-CAM (B), and cytotactin (C) were all seen to co-localize with the Na^+ channel (A', B', C'), a marker for the node of Ranvier (filled arrows) visualized by fluorescent α-bungarotoxin staining. In contrast, unmyelinated fibers were uniformly stained for Ng-CAM (open arrow in A) and N-CAM (not shown). Cytotactin was present in a spindle-shaped pattern seen most intensely at the node and decreasing approximately 5 to 20 μm away. Antibodies to the neuronal surface marker Thy-1 (D) uniformly stained the axon surface, indicating that increased staining at the nodes was not an artifact of the exposed membrane. (D') Phase micrograph corresponding to (D). Bar = 10 μm. (From Rieger et al., 1986.)

actually lead to changes in CAM modulation patterns. For example, in vivo perturbation of normal cell-cell interactions during degeneration and regeneration has been show to result in alteration of CAM expression and distribution (Figure 7). N-CAM is present at the neuromuscular junction of striated muscles, but is absent from the rest of the surface of the myofibril (Figure 7A) (Rieger et al., 1985). After cutting the sciatic nerve, the molecule appears diffusely at the cell surface and in the cytoplasm, but returns to normal after regeneration (Figure 7B,C) (Rieger et al., 1985; Covault and Sanes, 1985; Daniloff et al., 1986b). These experiments indicate that early events related to regeneration can be accompanied by altered CAM modulation. More recent experiments (Daniloff et al., 1986) show that both crushing and cutting a nerve have widespread effects, ranging from altered anti-CAM staining in motor neurons of the spinal cord (Figure 7G) on the affected side to modulatory changes in N-CAM and Ng-CAM within Schwann cells that are local to the lesion (Figure 7D,E).

CAM modulation has also been found to be perturbed in genetic diseases with altered morphology. As described above, N-CAM and Ng-CAM are colocalized at nodes of Ranvier in peripheral myelinated nerves (Rieger et al., 1986). In two dysmyelinating mouse mutants, trembler (*+/Tr*) and motor endplate disease (*med/med*), the distribution pattern of these molecules was found to be disrupted in the myelinated fibers. A recent study indicates that alterations in the amount of N-CAM expression are associated with a number of human myopathies (Walsh and Moore, 1985). This is in accord with the observations on alteration of CAM expression in muscle degeneration and regeneration. These failures of regulations have a counterpart in the CNS: in the mouse mutant *staggerer*, which shows connectional defects between Purkinje cells and parallel fibers in the cerebellum and extensive

granule cell death, E→A conversion of N-CAM is greatly delayed in the cerebellum (Edelman and Chuong, 1982). Data accumulated both from interventions and from examinations of disease states argue for the existence of an elaborate series of local morphology-dependent signals regulating CAM expression in various tissues.

CAM Signals: Identity of Ng-CAM and NILE

The studies of neural and non-neural tissues provide strong evidence for the existence of signaling that regulates both CAM expression and CAM function. This is clearly the case in inducing collectives as seen in the feather; it is equally striking in regenerating muscles and nerves. Only in one case, however, has a putative CAM signal been identified. The candidate substance is nerve growth factor (NGF), and its effects were understood after the demonstration in pheochromocytoma (PC12) cells that Ng-CAM and the NGF-inducible large external (NILE) glycoprotein were identical. Ng-CAM was identified in mammalian brain tissue and in PC12 cells as $Mr = 200,000$ and $230,000$ species, respectively. When PC12 cells were treated with NGF, the amount of Ng-CAM at the cell surface increased approximately three fold, while the amount of N-CAM remained unchanged (Figure 8). NILE was previously identified (McGuire et al., 1978) by its enhanced expression in NGF-treated PC12 cells. Ng-CAM and NILE are similar in molecular weight, expression during development (Thiery et al., 1985b; Stallcup et al., 1985; Daniloff et al., 1986a), and responsiveness to NGF in PC12 cells (Friedlander et al., 1986), suggesting that the two molecules are related. In addition, antibodies to Ng-CAM and NILE cross-reacted and the molecules had essentially identical peptide maps after limited proteolysis (Friedlander et al., 1986). Moreover, anti-NILE inhibited fas-

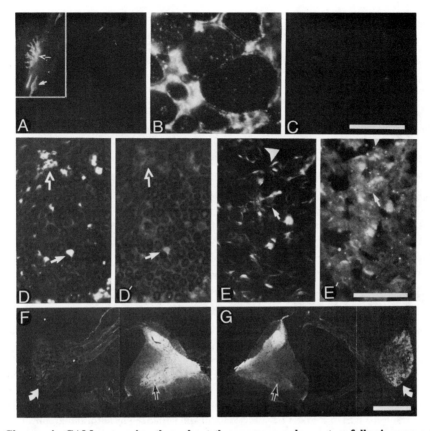

FIGURE 7. Changes in CAM expression throughout the neuromuscular system following nerve transection. Cross sections of chick gastrocnemius muscles stained with polyclonal anti–N-CAM IgG (A–C) showed that the surfaces of muscle fibers were faintly labeled in normal chicken muscles (A). The inset shows a whole-mount preparation of normal adult teased muscle fibers (magnification × 7). The muscle surface was faintly labeled with anti–N-CAM; a motor endplate (open arrow) and several mononucleated cells that are probably satellite cells (filled arrow) were intensely labeled. (B) Ten days after the sciatic nerve was cut, a dramatic increase occurred in the intensity of N-CAM; a normal pattern was restored after 150 days (C). In cross sections of normal adult nerves (D, D′), bundles of nonmyelinated axons (open arrow) and Schwann cells (filled arrow) were intensely stained for Ng-CAM (D) and were also N-CAM positive (open arrow in D′). The myelinated axons are the central CAM-positive regions within unstained donut-like structures, which represent the myelin sheath (D, D′). Only Schwann cells (filled arrow), and not fibers (open arrow), expressed S100 protein (not shown). In the distal stump 10 days after the nerve was cut, many abnormal spaces (arrowheads in E and E′) were seen in the nerve. Ng-CAM staining was most intense in Schwann cells (filled arrow in E). The overall intensity of N-CAM staining (E′) was greater than in controls (D′). Although all components of the nerve appeared to express N-CAM at this time (E′), the S100 protein staining pattern indicated that the most intense staining was associated with Schwann cells. F, G, Cross sections of the lowest lumbar segment of spinal cord and dorsal root ganglion (DRG) 20 days after cutting the sciatic nerve. On the experimental (lesioned) side (G), the intensity of Ng-CAM staining decreased in the ventral horn (black arrow) and increased within the ganglia (white curved arrow), as compared with the side contralateral to the lesion (F). (A–E) bar = 50 μm; (F–G) bar = 800 μm. (From Daniloff et al., 1986b.)

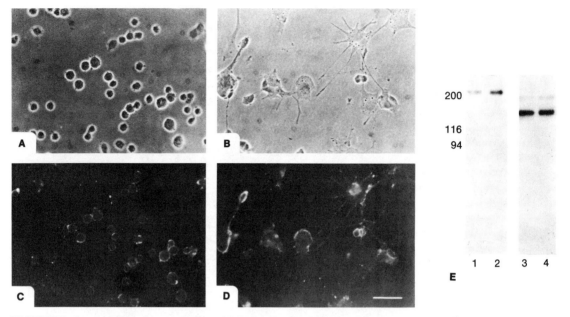

FIGURE 8. Immunofluorescent staining of PC12 cells with anti–Ng-CAM IgG. PC12 cells were grown for one week in the absence (A, C) or presence (B, D) of NGF and stained with affinity-purified anti–Ng-CAM IgG. A and B are phase-constant micrographs of the fields shown in C and D. Bar, 50 μm. E, Effects of NGF on CAM expression. PC12 cells were grown in the absence (lanes 1 and 3) or presence (lanes 2 and 4) of NGF for 1 week and were then labeled with ^{125}I. NP-40 extracts of cells containing equal amounts of counts were immunoprecipitated with specific antibodies to Ng-CAM (lanes 1 and 2) and N-CAM (lanes 3 and 4). The precipitated molecules were resolved on SDS PAGE (7.5% acrylamide) and visualized by autoradiography. (From Friedlander et al., 1986.)

ciculation of neurites in much the same fashion as anti–Ng-CAM (Stallcup and Beasley, 1985).

These data indicate that CAMs can respond selectively and differentially to growth factors and provide the first indication that signals for CAMs may include such factors, hormones, or neurotransmitters. This brings us back to the issue first confronted at the beginning of this chapter: how can CAM regulation be reconciled with both the genetic constraints on morphogenesis and the mechanochemical and epigenetic events that lead to actual form in the nervous system and in non-neural regions?

CAM Cycles and the Regulator Hypothesis

Observations indicating that there are epigenetic rules for CAM expression (Edelman, 1985, 1986a,b; Crossin et al., 1985), results of various perturbation experiments, and the demonstrated response of a secondary CAM to identified signals are all consistent with the so-called regulator hypothesis (Edelman, 1984b). This hypothesis states that *early* control of CAM expression by regulatory genes (and thus the realization of the epigenetic rules) is separate from control of expression of the molecules concerned with other cytodifferentiation events such as those governing intracellular economy (Han et al., 1986) or regulation of cell shape. According to the regulator hypothesis, one may usefully distinguish "morphoregulatory" genes regulating CAM and SAM expression from "historegulatory" genes regulating specific cytodifferentiation events. With this distinction in mind, CAM expression (Edelman, 1985b) may be viewed as taking place in a cycle (Figure 9). Traversal of the outer loop of the cycle may lead to expression or down-regulation of N-CAM genes (see Table 3, rule I), as is seen in the case of mesenchyme cells contributing to the dermal condensation of the feather (Chuong and Edelman 1985a,b) and also in the case of neural crest cells (Thiery et al., 1982); alternatively, switching off of one or the other primary CAM gene may occur in an epithe-lium expressing both CAMs (Table 3, rule II).

The action of historegulatory genes (the inner loop in Figure 9) responding to signals from the new milieux that result from CAM-mediated cell aggregation and cell motion, as well as from tissue folding and tension, may lead to the entry of altered cells into the cycle. Among other changes, this could result in the expression of gene products that alter cell shape or that govern enzymes affecting CAM action, such as those concerned in E→A conversion. The combined effects of the inner and outer loops of the cycle and the linkage of two such cycles by formation of CAM couples such as those seen at inductive sites (Crossin et al., 1985; Gallin et al., 1986) would lead to a rich set of morphogenetic structures. If proven,

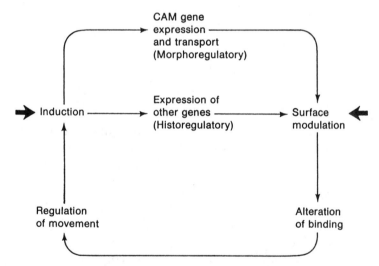

FIGURE 9. **Hypothetical CAM-regulatory cycle proposed to apply in development and regeneration. Early induction signals (heavy arrow at left) lead to CAM gene expression. Surface modulation (by prevalence changes, polar redistribution on the cell, or chemical changes such as E → A conversion) alters the binding rates of cells. This regulates morphogenetic movements that, in turn, affect embryonic induction or milieu-dependent differentiation. The inductive changes can again affect CAM gene expression as well as the expression of other genes for specific tissues; it is assumed that morphoregulatory genes affecting CAM expression are separate from historegulatory genes affecting other cytodifferentiation events. The heavy arrows refer to candidate signals for initiation of induction or surface modulation as a result of CAM binding (right) or from release of morphogens affecting induction (left). The evidence from work on the feather suggests that mechanochemical signals and morphogens may act together at key local sites to change CAM expression as seen at the origins of the marginal and axial plates.**

the CAM regulatory cycle would provide a particularly elegant solution to the problem of mechanochemical control of pattern through several levels of organization (from the gene and gene product to the cell, to tissues and organs, and back to the gene) that is posed by the occurrence of epigenetic sequences both in regulative development and in regeneration.

In the analysis of this hypothesis, molecular biological techniques will be invaluable. cDNA probes for the various CAMs are now available, and it has already been demonstrated that N-CAM is specified by a single gene on mouse chromosome 9 (D'Eustachio et al., 1985). Recent results suggest that all of the different N-CAM polypeptides are synthesized from multiple N-CAM messages generated by alternative splicing of transcripts from a single N-CAM gene (Murray et al., 1986a). Subsequent to this work, complete sequence analysis of cDNA probes corresponding to the polysialic acid–rich and cytoplasmic domains (Hemperly et al., 1986) showed that the ld polypeptide has a cytoplasmic domain that is approximately 250 amino acids larger than that of the sd polypeptide. The latter peptide arises as a result of a splicing event and, as already discussed, suppression of this event to create a message for the sd chain is nervous system specific and provides yet one more modulation mechanism for alteration of CAM function.

The fact that CAMs are specified by one or a few genes makes it feasible to test the regulator hypothesis in detail. The level at which CAM expression is controlled may be checked, and in situ hybridization experiments should allow an assessment in a given tissue of the temporal sequence of CAM messenger RNA expression as compared to that of particular tissue-specific gene products such as keratin. If the outcome of these experiments is that CAM gene expression occurs prior to that of most tissue-specific products, there would be a satisfying interpretation of the crossing of

histotypic boundaries by CAM distributions as seen in composite fate maps (see Figure 1). In this view, CAMs would first regulate movements by modulation, leading to border formation and separate milieux capable of generating subsequent signals for historegulatory genes; the expression of these genes would in turn lead to fixation of detailed histotypic boundaries.

We would expect that the nervous system should in general conform to the regulator hypothesis. Of particular interest is the possibility that events involving choice of path by neural growth cones and neurite outgrowth depend upon target-induced modulation of cell surface CAMs on pioneer neurons. Just a few such modulations in the path of choices of a single neurite could strongly alter neural pattern. The general principles of signaling for neurite choice may resemble those seen in embryonic induction systems such as the feather (Gallin et al., 1986), but on a finer scale. In view of the extraordinary local order of neural organization, the remarkable opportunity to exchange chemical signals at the synaptic level, and the continued maturation of structure after birth in the presence of behavior, it would not be surprising if additional CAMs and SAMs and new types of modulations were found to be special to neural tissue. For example, it is already clear that the ld chain of N-CAM arises by a nervous system–specific alteration of RNA splicing events. Given the number of modulation mechanisms already known and their context-specific and epigenetic character, it would not require many more mechanisms to account for the remarkable complexity of neural structure. Present knowledge of the dynamic properties of modulation mechanisms and of the genetic control of CAMs and SAMs already provides an initial basis on which to improve our understanding of both the developmental constraints on neural evolution and the somatic variability of individual neural networks.

CHAPTER TWO

Subsynaptic Proteins at Nerve–Muscle Synapses

STEVEN J. BURDEN

Introduction

A common feature of many eukaryotic cells is their ability to organize their plasma membranes into discrete domains, each of which is specialized for a particular function. Perhaps the most thoroughly studied cell type that illustrates this feature is the polarized epithelial cell. The plasma membrane of these epithelial cells has two distinct surfaces, a basal and an apical surface, each of which is distinct in its morphology, physiology, and biochemistry (Sabatini et al., 1982; Rindler et al., 1985; Simons and Fuller, 1985). Although the establishment of functionally distinct membrane surfaces is critical for normal cellular physiology, in no case is it clear how a single cell assembles and maintains distinct membrane domains within a single plasma membrane.

Specialized membrane domains are of great importance to the function of the nervous system. For example, in most neurons, voltage-activated sodium action potential channels are virtually restricted to axons and present in cell bodies and dendrites at low concentration; moreover, in myelinated axons

these channels are concentrated at 1-millimeter intervals at the nodes of Ranvier (Ritchie and Rogart, 1977). Similarly, neurons and muscle cells are capable of specializing small domains of their plasma membranes for synaptic transmission. In the case of the neuromuscular junction, less than 0.1% of the muscle cell surface area is covered by nerve terminals and is concerned with synaptic transmission; this 0.1% of the membrane is strikingly specialized. It is not clear how the muscle cell establishes this specialized postsynaptic membrane during synapse formation, nor how the specialized membrane domain is thereafter maintained.

The neuromuscular synapse is a favorable system in which to study how specialized membrane domains are established: the structure can be conveniently isolated in relatively pure form and in sufficient quantities for analysis by standard biochemical procedures (Sobel et al., 1977; Elliot et al., 1980), the synapse is easily accessible to experimental manipulation (Miledi, 1960; Brockes and Hall, 1975; Marshall et al., 1977), the mature structure forms readily in cell culture (Fischbach and Cohen, 1973; Anderson and Cohen, 1977) and there is abundant biochemical and cell biological information about the major component of the structure—the acetylcholine receptor (AChR) (Fambrough, 1979; Karlin, 1980; Numa et al., 1983).

One of the earliest events in synapse formation between nerve and muscle is the accumulation of AChRs at newly formed synapses (Fambrough, 1979; Dennis, 1981). Prior to synapse formation, AChRs are distributed throughout the membrane of myofibers (Diamond and Miledi, 1962; Burden, 1977; Bevan and Steinbach, 1977; Chow and Cohen, 1983). The distribution of AChRs is altered, however, when a muscle fiber becomes innervated by a motor neuron: innervation induces a redistribution of AChRs on the surface of the myofiber such that pre-existing AChRs migrate within the myofiber membrane and become concentrated at the newly formed synapse (Anderson and Cohen, 1977; Frank and Fischbach, 1979; Ziskind-Conhaim et al., 1984). Later during development this arrangement is made relatively permanent: AChRs remain clustered at adult synapses even after denervation. Presently, we do not understand how nerve and muscle interact to induce this redistribution of AChRs during development, nor do we have any insight into how AChRs are mobilized to the synapse and thereafter maintained at a high concentration. Nevertheless, the studies cited above and those of regenerating muscle (Burden et al., 1979; Nitkin et al., 1983) suggest that components extrinsic to the AChR polypeptides are involved in the clustering of AChRs at synaptic sites: these components include those within the synaptic extracellular matrix and the subsynaptic sarcoplasm.

Identification of Nonreceptor Postsynaptic Proteins

One approach to study the mechanism of AChR clustering at synapses is to identify macromolecular components other than the AChR itself that are required for AChR clustering. We reasoned that such components might be highly concentrated at adult synapses and might copurify with the AChR when the receptor is isolated in its membrane-bound state.

We isolated membranes from the electric organ of *Torpedo californica* that are highly enriched in the four subunits of AChR (Sobel et al., 1977; Elliot et al., 1980) and identified two nonreceptor polypeptides that copurify with the membrane-bound AChR (Burden et al., 1983). These two proteins migrate in SDS–PAGE at 43 kd and 300 kd. Neither protein is a constituent of the purified, solubilized AChR; rather, each is a peripheral membrane protein that can be solubilized from the membranes by alkaline pH or chaotropic salts (Neubig et al., 1979; Burden et al., 1983). The AChR subunits in these membranes are present at a stoichiometry of approximately 2:1:1:1 (α:β:γ:δ); the 43-kd protein is present at 1 mole per mole of AChR complex and the 300-kd protein, at 0.1 mole per mole of AChR complex (Burden et al., 1983). Analysis of AChR-rich membranes with proteases, antibodies, or surface-radiolabeling techniques demonstrates that the 43-kd protein is present on the cytoplasmic surface of the AChR-rich membranes (Wennogle and Changeux, 1980; St. John et al., 1982; Sealock et al., 1984). More recent studies demonstrate that the 300-kd protein is also present on the cytoplasmic surface of the postsynaptic membrane in both the electric organ and in skeletal muscle of higher vertebrates (Woodruff et al., 1987).

The 43-kd protein is not required for ligand-gated activation of the AChR channel, since ligand-gated activation is maintained after removal of the 43-kd protein from AChR-rich membranes (Neubig et al., 1979). It is possible, however, that the 43-kd protein has a more subtle role in regulating channel properties; for example, the 43-kd protein could be involved in controlling desensitization to acetylcholine. In this regard, there is evidence to suggest that the 43-kd protein is a protein kinase (Gordon and Milfay, 1986) and that the rate of AChR desensitization is influenced by phosphorylation (Huganir et al., in press; Middleton et al., 1986).

Immunocytochemical staining with monoclonal antibodies against the 43-kd protein demonstrates that the protein is concentrated at adult neuromuscular synapses (Froehner, 1984; Burden, 1985). Frozen sections of amphibian muscle were stained with a monoclonal antibody against the 43-kd protein and with tetramethylrhodamine-coupled α-bungarotoxin (Figure 1). All synaptic sites marked by toxin staining were stained by antibody to the 43-kd protein, and all clusters of 43-kd protein were at synaptic sites. Immunocytochemical experiments also demonstrate that the 43-kd protein appears at developing synapses virtually simultaneously with AChRs (Burden, 1985). Myocytes from *Xenopus* embryos were grown in cell culture in either the absence or the presence of spinal cord cells and stained with a monoclonal antibody against the 43-kd protein (Figures 2 and 3). Muscle cells were innervated by neurons as early as two days after plating, and synaptic sites were identified by the characteristic shape of synaptic AChR clusters in these cultures (Anderson and Cohen, 1977; Burden, 1985). Clusters of 43-kd protein were found to be coincident with all synaptic AChR clusters from the earliest stage at which synaptic AChR clusters were detected. Moreoever, the 43-kd subsynaptic protein is concentrated at AChR clusters that occur on noninnervated muscle cells in cell culture (Burden, 1985; Peng and Froehner, 1985). Thus, the 43-kd subsynaptic protein may be involved in the formation and/or the maintenance of AChR clusters. It is presently not clear, however, whether the 43-kd protein is associated only with AChR clusters or also with the more diffusely distributed AChRs present in developing muscle and denervated adult muscle.

Further support for a role of the 43-kd protein in regulating the distribution of AChRs comes from cross-linking studies demonstrating that the 43-kd protein is in close proximity to the cytoplasmic domain(s) of the AChR (Burden et al., 1983). We took advantage of the fact that the 43-kd protein contains most of the available free sulfhydryls in AChR-rich membranes (Hamilton et al., 1979; Burden et al., 1983) and used cross-linking reagents that contain an alkylating agent as one of the reactive groups. Cross-links between the 43-kd protein and AChR polypeptides were identified by

FIGURE 1. **The 43-kd protein is concentrated at adult neuromuscular synapses. Frozen sections of amphibian muscle were stained with tetramethylrhodamine-coupled α-bungarotoxin to mark synaptic sites (A) and with monoclonal-antibody to the *Torpedo* 43-kd protein followed by fluorescein-coupled goat-antimouse IgG (B), and were viewed with optics selective for either rhodamine (A) or fluorescein (B). All synaptic sites identified by α-bungarotoxin binding were stained with antibody to the 43-kd protein. This monoclonal antibody does not cross-react with synaptic sites in mammalian muscle. Bar = 25 μm. (For details, see Burden, 1985.)**

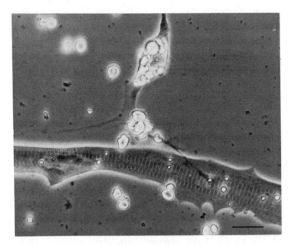

FIGURE 2. Phase-contrast micrograph of *Xenopus* nerve-muscle co-culture demonstrates the morphology of both the neurons and the muscle cells in these cultures. The bright refractile spheres both within the neuron cell body and on the culture dish are yolk granules. Bar = 20 μm.

immunoblotting and detection with antibodies against the 43-kd protein and each of the receptor subunits. These studies demonstrate that the membrane-bound 43-kd protein can be selectively cross-linked to the β-subunit of the AChR (Burden et al., 1983). Since the reactive groups on these cross-linking reagents are 1 to 1.2 nm apart, the 43-kd protein must be in close proximity to the β-subunit.

These studies have demonstrated that there is a close temporal and spatial relationship between AChR clusters and high concentrations of the 43-kd protein. Further experiments are required to determine whether the 43-kd protein is required for AChR clustering and/or for regulating other properties of the AChR.

Summary

Can these studies be extended to study subsynaptic proteins associated with postsynaptic membranes from the central nervous system? At present there is no evidence indicating that

this 43-kd subsynaptic protein is concentrated at nerve–nerve synapses. We have stained frozen sections of the frog retina, optic tectum, and cerebellum and whole mounts of the parasympathetic cardiac ganglion both with monoclonal antibodies against the 43-kd protein and with antiserum against peripheral proteins from AChR-rich membranes, and have not detected staining concentrated at synaptic sites in these tissues. It is possible, however, that the protein is present at these synapses but at a lower concentration than at the neuromuscular junction and not detectable by the relatively insensitive immunofluorescence assay.

Nevertheless, it is interesting to speculate on the possibility of studying the postsynaptic membrane of nerve–nerve synapses with the techniques that we have used to explore the postsynaptic membrane of the neuromuscular junction. The essential features of the techniques that we have used to study the neuromuscular synapse include (1) a well-characterized receptor that allows unambiguous identification of the postsynaptic membrane during isolation procedures, (2) a tissue that is homogeneous with regard to its neurotransmitter receptor and thus its postsynaptic membrane, (3) a tissue that is enriched in synaptic material and allows the identification of nonreceptor synaptic proteins by Coomassie staining of SDS–PAGE of fractionated membranes, and (4) a simple procedure to fractionate postsynaptic membrane from nonsynaptic membrane.

With regard to nerve–nerve synapses, there is no transmitter receptor that has been so thoroughly characterized and for which highly specific ligands exist. This is perhaps the key tool. Although adrenergic receptors from non-neural cells have been purified and well characterized, it is not clear that these receptors are concentrated in postsynaptic membranes (Strader et al., 1984). The elegant characterization of the glycine receptor by Betz

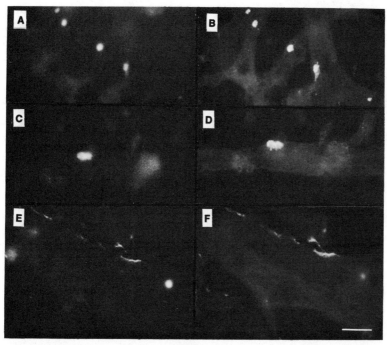

FIGURE 3. The 43-kd protein is concentrated at synapses that have formed in cell culture and at nonsynaptic AChR clusters in noninnervated muscle cells in cell culture. *Xenopus* muscle cultures (A–D) and nerve-muscle co-cultures (E, F) were stained with tetramethylrhodamine-coupled α-bungarotoxin (A, C, E) and with mono-clonal antibody to the 43-kd protein (B, D, F), as described in Figure 1. The paired micrographs (A, B), (C, D), and (E, F) demonstrate that the 43-kd protein is concentrated at both synaptic and nonsynaptic AChR clusters in developing muscle. Bar = 30 μm. (For details, see Burden, 1985.)

and colleagues (Pfeiffer et al., 1982, 1984; Triller et al., 1985), however, is likely to provide the tools necessary to characterize the glycine receptor postsynaptic membrane in some detail.

There are several reasons why fractionation of brain postsynaptic membranes and identification of nonreceptor proteins that copurify with these membranes is a formidable task. First, the brain is heterogenous with regard to neurotransmitter receptors and hence with regard to types of postsynaptic membranes. Thus, it is essential not only to fractionate postsynaptic membranes from nonsynaptic membranes, but also to fractionate one type of postsynaptic membrane

from another. Moreover, the complexity and number of different types of membranes would make it difficult to analyze the membrane protein composition by simple one-dimensional SDS-PAGE. Second, the amount of a particular type of postsynaptic membrane is small compared to the amount of postsynaptic membrane present in the electric organ. Thus, the techniques that are required to identify the postsynaptic membrane and the nonreceptor synaptic proteins would become more arduous. Third, a particular postsynaptic membrane may not be so easily fractionated from other postsynaptic membranes or even nonsynaptic membranes. Since the safety factor for synaptic transmission at synapses in the central nervous system (CNS) is unlikely to be as high as at nerve–muscle synapses, the receptor concentration at CNS synapses need not be as high as at the neuromuscular synapse. Since the high concentration of AChR in the postsynaptic membrane (high protein:lipid ratio) allows this membrane to be readily fractionated from nonreceptor membranes by equilibrium density centrifugation, more sophisticated fractionation procedures such as affinity purification methods may be necessary to isolate a particular type of brain postsynaptic membrane. Although these technical difficulties suggest that isolation and characterization of a particular postsynaptic membrane from the CNS will not be easy, methods are available to overcome most of these difficulties, and progress in this direction is likely in the near future.

CHAPTER THREE

Ontogeny of the Nerve Dependence of Amphibian Limb Regeneration

D. M. FEKETE, P. FERRETTI,
H. GORDON, AND J. P. BROCKES

Introduction

The vertebrate peripheral nervous system (PNS) plays a role in the regulation of growth. This chapter is concerned with the establishment of this role in development, particularly in relation to the phenomenon of limb regeneration in urodele amphibians such as newts or axolotls. Amputation of the limb at any proximal-distal level is followed by formation of a wound epidermis overlying a blastema—a local growth zone composed of histologically undifferentiated progenitor cells (review by Wallace, 1981). The blastemal cells divide and differentiate, giving rise to the internal tissues of the regenerate. This mode of regeneration by formation of a blastema is termed epimorphosis (Morgan, 1901) and is found in several invertebrate phyla as well as in some lower vertebrates. The phenomenon poses some difficult questions about the mechanism by which the regenerate arises from its substrate of adult tissue and the relationship of the process to the corresponding one that occurs in development.

In comparing the epimorphic regeneration of the limb with its development, it seems likely that many of the basic mechanisms of differentiation, morphogenesis, and pattern formation are identical. There is evidence that in the axolotl the specification of positional information obeys similar rules in the two circumstances, as detected by the supernumerary structures that result from confronting a forelimb blastema with a contralateral hindlimb bud (Muneoka and Bryant, 1982, 1984). On the other hand, it is likely that there are also distinctive aspects of regeneration that reflect the relationship between the adult tissue at the amputation plane and its offshoot, the blastema. This chapter is concerned with such aspects and particularly with the origin and the proliferation of the blastemal cells and the role that the PNS plays in both processes.

The importance of the PNS in regeneration of the urodele limb has been recognized since the early nineteenth century (Todd, 1823), but our current understanding is particularly associated with the experiments and proposals of Marcus Singer (1952 and 1974, for review). The forelimb is innervated principally by three spinal nerves whose axons can be cut at the brachial plexus. Singer demonstrated that regeneration depends on an adequate quantity of axons at the amputation plane, that both sensory and motor nerves are capable of meeting this requirement, and that denervation of the blastema is rapidly followed by a cessation of cell proliferation. These findings led to the hypothesis that the nerve dependence of regeneration is mediated by a neurotrophic (mitogenic) factor released by the nerve (Singer, 1952, 1974). The hypothesis has been an important stimulus to experiments that have provided more direct evidence for the proposed mitogenic effect; for example, it has allowed investigators to test the mitogenic effect on explanted blastemas of various factors isolated from newts and higher vertebrates (Globus and Vethamany-Globus, 1977; Vethamany-Globus et al., 1978; Gospodarowicz and Mescher, 1980; Mescher and Loh, 1981; Globus et al., 1983). Our work in this direction has centered on the possible relevance of glial growth factor and has been reviewed elsewhere (Brockes, 1984; Kintner et al., 1985; Brockes and Kintner, 1986). Singer's contribution illustrates the importance of focusing on one aspect of the complexities of epimorphic regeneration and defining it in sufficient detail so that hypotheses can be tested at the molecular level.

Not long after Singer proposed the neurotrophic hypothesis, results of a remarkable series of experiments were reported that seemed to challenge some aspects of it. By using the methods of experimental amphibian embryology it is possible to derive a limb that develops in the near or total absence of a nerve supply (Yntema, 1959; Thornton and Steen, 1962; Steen and Thornton, 1963; Thornton and Tassava, 1969). Such "aneurogenic limbs" are depleted in muscle (as a secondary consequence of denervation) but are otherwise relatively normal in appearance. Nonetheless the limbs were shown to regenerate when amputated, despite the absence of nerve fibers in

the resulting blastema. In a later series of experiments, the aneurogenic limbs were transplanted to a normal embryo in place of the forelimb (Thornton and Thornton, 1970). When the transplanted limbs were innervated from the host brachial nerves, they quite abruptly became dependent on an intact nerve supply for regeneration. Clearly any satisfying account of the relationship between the nerve and the limb must accommodate these observations. It is therefore necessary to understand first why an aneurogenic limb is not nerve dependent for regeneration, and second why it becomes so after innervation. It seems very likely that the interaction between the brachial nerves and the aneurogenic limb is a normal feature of limb development, and indeed most attempts to explain the observations alude to this possibility.

In this account we describe new information about epimorphic regeneration which has come from the derivation of a monoclonal antibody (termed 22/18) that identifies an early event in the formation of blastemal cells. After reviewing the distribution of the antigen, we consider the cell types that it appears in after amputation and the general nature of the events inducing its appearance. The antigen has been of considerable value in understanding the relationship between growth control in development and regeneration, and offers a new perspective on the phenomenon of the aneurogenic limb.

The 22/18 Antigen in Regeneration

Derivation of the 22/18 Antibody

After immunizing mice with various tissues of the newt (*Notophthalmus viridescens*), including the blastema, we screened the resulting monoclonal antibodies by immunofluorescence on sections of the blastema (Kintner and Brockes, 1985). Although it is relatively easy to obtain antibodies that react specifically with various differentiated tissues of the limb such as epidermis or muscle, it is difficult to derive reagents that react only with the blastemal cells. After screening approximately 3,000 antibody-secreting clones we obtained only one reagent, termed 22/18, that reacts in this way. Figure 1 shows a low-power immunofluorescence montage of a longitudinal section through an early-bud stage regenerate at eight days after amputation (stages according to Iten and Bryant, 1973). The epidermis is apparent because of autofluorescence; it does not react specifically with 22/18. About 80% of the blastemal cells react with the antibody, but

FIGURE 1. Distribution of the 22/18 antigen in a regenerating forelimb of *N. viridescens* eight days after amputation. Longitudinal sections were cut from fixed tissue and stained essentially as described in Kintner and Brockes (1985). In brief, the monoclonal IgM 22/18 antibody was detected by reaction with rabbit antimouse IgM followed by rhodamine-conjugated goat antirabbit Ig. The distal tip of the regenerate is at the top of the figure. Bar = 500 μm.

the number of positive cells falls off dramatically in the differentiated tissue of the stump. The normal limbs of most animals show little or no reactivity with 22/18, yet some animals show granular staining of a few cells scattered throughout the limb, including cells in the nerve trunk, in the dermis, and between muscle fibers. Such staining varies considerably in intensity, is qualitatively different from that of blastemal cells (see below), and has also been seen with another IgM reagent of unrelated specificity. For these reasons the staining will not be considered further in this account.

The 22/18 antigen expressed by blastemal cells is intracellular and filamentous. This is most clearly shown by staining cultured cells. Although live cells are not decorated by the antibody, loose bundles of filaments are clearly detected in fixed, permeabilized cells (Figure 2A). Cells can be stained with 22/18 (Figure 2B) and also with a monoclonal antibody that reacts with a common determinant on all intermediate filaments (Figure 2C) (Pruss et al., 1981). The correspondence in staining patterns indicates that 22/18 recognizes a determinant on an intermediate filament subunit or possibly on an intermediate filament-associated protein. Although there is increasing information about the structure and molecular diversity of these molecules, their functional role is not understood. Therefore, discussion of the antibody and its cognate antigen will be entirely in terms of their use as markers. It may be possible in the future to investigate if the antigen plays a functional role in the change in the growth control associated with its appearance (see later discussion in this chapter).

Distribution of the 22/18 Antigen

The antigen is first detected at 36 to 48 hours after amputation in the Schwann cells of the nerve sheath at the plane of amputation and in the connective tissue cells beneath the wound epidermis. As indicated in Figure 1,

A

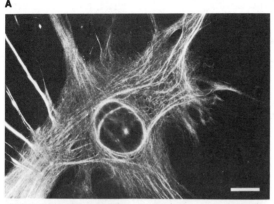

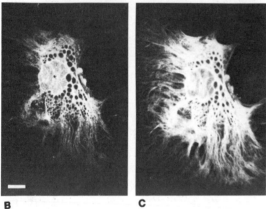

B   C

FIGURE 2. Immunostaining of acid-alcohol fixed cells derived from explants of limb tissue from *N. viridescens.* Under the circumstances of explantation, cells that migrate out of the explant are induced to express the 22/18 antigen. Antigen expression in culture is stable for months following numerous passages. (A) Filamentous staining detected with the monoclonal IgM 22/18 antibody. Double-labeling with 22/18 (B) and a monoclonal IgG anti-intermediate filament antibody (C). A similar staining pattern is observed in dissociated blastemal cells in culture. Bars = 20 μm.

most of the cells in the newt blastema react with the antibody during the first two weeks after amputation. As the blastemal population expands during the third week, there is a marked decline in the number of 22/18-positive cells. It is not clear if the decrease represents death of positive cells or, more likely, conversion from an antigen-positive to an antigen-negative phenotype. At later stages the remaining positive cells are seen where muscle differentiates, and double labeling studies with a muscle-specific marker suggest that at least some of the 22/18-positive cells give rise to muscle (Kintner and Brockes, 1984, 1985). These cells lose reactivity with the antibody as muscle differentiates, so that like the original limb, the final regenerate shows little or no reactivity. The maximal expression of 22/18 occurs during the earlier events of blastema formation and epimorphic regeneration. The reactivity of 22/18 in regeneration is discussed in more detail by Kintner and Brockes (1985).

One important aspect of 22/18 to which we will return later is that it functions as an excellent marker for those blastemal cells whose division is dependent on the nerve. This is revealed by experiments in which the regenerate is denervated and then injected with tritiated thymidine after allowing two days for the axons to degenerate (Kintner and Brockes, 1985). After a labeling period of 24 hours, the blastema is sectioned and analyzed by combining autoradiography and immunofluorescence. When the denervated side is compared to the contralateral control blastema, the labeling index of 22/18-positive cells is approximately sevenfold lower. This result is also obtained at the palette stage (25 to 30 days after amputation), when the 22/18 cells are in a minority. The 22/18-negative population is less dependent on the nerve, particularly at the later stage, when such cells constitute the majority. This finding is consistent with the classical observation that while denervation of the early blastema (majority of cells are 22/18-positive) arrests regeneration, denervation at the later stage (majority of cells are 22/18-negative) produces a limb that is well formed although smaller. These insights would be very difficult to obtain without the use of cell markers.

Figure 3 illustrates the relation of an early-bud stage blastema to the mature tissues such

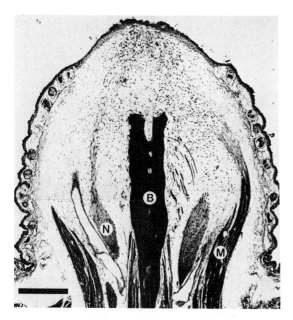

FIGURE 3. Photographic montage of an early regenerating limb (*N. viridescens*) stained by Bodian's protargol method. The original plane of amputation is marked by the boundary between mature dermis and the thickened wound epidermis at the distal tip of the limb (top of figure). The bone (B) has been trimmed back approximately 0.5 mm from the plane of amputation of soft tissue. Blastemal cells accumulate beneath the wound epidermis, while mature tissues such as nerve (N) and muscle (M) lose their differentiated characteristics as they approach the blastema. For the purposes of illustration, the right and left halves of the figure were photographed from different sections of the same half-limb and one side was photographically reversed. Bar = 500 μm.

as bone, muscle, and nerve. Between these two zones is a region where the integrity of the mature tissues (especially muscle and nerve) breaks down. It is such images that in the past led to the hypothesis that de-differentiation of mature tissues constitutes a source of blastemal cells (for review of early literature, see Thornton, 1938, 1942). Despite some experimental efforts addressing this issue (for example, Namenwirth, 1974; Maden, 1977), it has not been possible to obtain definitive proof of

cellular de-differentiation (for discussion, see Hay, 1974; see Chapter 7 in Wallace, 1981).

Double labeling with 22/18 and cell type-specific markers has been used to identify potential sources of blastemal cells in both tissue and blastema (Kintner and Brockes, 1985). This method is limited in that the cell type–specific markers are lost from the blastemal cells, and therefore no information is avialable about the fate of cells from different sources, and it does not provide a quantitative estimate of the relative contributions. These reservations notwithstanding, there is clear evidence of a contribution from muscle, Schwann cells, and interstitial cells of the connective tissue (Figure 4). The muscle contribution could be from myofibers or a myogenic reserve cell (Cameron et al., 1986) or possibly from both sources. The Schwann cells of the nerve sheath provide a clear example of de-differentiation, in that they lose expression of the myelin proteins and gain expression of the blastemal antigen. The interstitial cells do not react with the differentiation markers, and hence there is no justification at present for attributing de-differentiation to them.

Induction of the 22/18 Antigen

Some information has been obtained about the general nature of the events inducing the 22/18 antigen. It is clearly not necessary for cells to enter S phase to express the antigen. This is shown in experiments where animals are injected with tritiated thymidine immediately after amputation, that is, about five to six days before nerve-dependent proliferation begins. During this early period (days three to five) most of the 22/18-positive cells are not labeled with tritiated thymidine. Furthermore, expression of the antigen does not require a wound epidermis. If the end of the stump is covered with a skin flap, thus preventing the formation of a wound epidermis, 22/18-positive cells arise, although their subsequent proliferation is arrested. This finding is

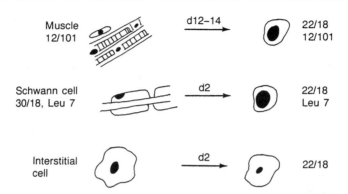

FIGURE 4. Schematic illustration of the cell or tissue types that have been demonstrated to give rise to 22/18-positive blastemal cells. The antibodies used to identify mature tissue types are listed on the left. The earliest expression of 22/18 in each tissue type is indicated above the arrows as the number of days after amputation. In the case of muscle and Schwann cells, this was determined by double-labeling cells with 22/18 and the antibodies listed on the right.

consistent with experiments in which a small number of blastemal cells (identified by standard histological stains) were observed after a similar manipulation (Mescher, 1976; Tassava and Loyd, 1977).

There are several contexts in which 22/18 is induced by injury in the absence of amputation. First, if the spinal nerves are cut at the level of the brachial plexus, the axons in the limb degenerate and the denervated Schwann cells develop reactivity to 22/18 (Kintner and Brockes, 1985). Second, if the arm is locally bruised, reactivity develops in both Schwann cells and interstitial cells. Third, if a limb muscle is removed, minced into cubic millimeter fragments, and replaced, the subsequent repair myogenesis is accompanied by expression of 22/18 in denervated Schwann cells and in numerous mononucleated cells (presumed to be myogenic precursors because some can be double-labeled with a muscle marker) (K. Griffin et al., in preparation). One possible interpretation of the induction of the antigen is that it identifies a transient, quasipathological response to the trauma of injury akin, for example, to the heat shock response (Schlesinger et al., 1982). However, several lines of

evidence suggest that the appearance of 22/18 is a potentially stable transition. If the progress of regeneration is arrested by early denervation or by replacing the wound epidermis with mature dermis, it is possible to detect cells reactive with 22/18 at least six weeks after the original amputation. Furthermore, Schwann cells that are stimulated by denervation at the brachial plexus continue to express the antigen in the distal stumps for at least six weeks, as long as the axons are prevented from regrowth. Finally, if limb or blastemal cells are transferred to cell culture, they continue to divide and express the antigen even after six months in vitro. Thus the transition is potentially stable, even though during the normal, unarrested course of regeneration the expression progressively declines after two weeks.

In summary, 22/18 identifies an early event in the formation of blastemal cells from Schwann cells, myogenic cells, and interstitial cells. The event is provoked by injury, but does not require the presence of the wound epidermis or that cells enter S phase. It is a potentially stable transition, and cells reactive with the antibody are dependent on the nerve for division.

The 28/18 Antigen in Development

The Normal Limb Bud

It is of obvious interest to determine if 22/18 is expressed during normal development.[1] We have investigated this issue with embryos of both *N. viridescens* and *Pleurodeles waltl*. During development the antigen is expressed detectably in glial cells and transiently in the epidermis and aorta, providing an assurance that the fixation conditions are satisfactory. Very few mesenchymal cells stain with 22/18. Figure 5 is a section through an early limb bud stained to detect nuclei (Figure 5A) and several different antigens, including 22/18 (Kintner and Brockes, 1985). It is clear that no cells in this section are staining with 22/18 (Figure 5D) or 12/101, a muscle marker (Figure 5C), although they do stain with 22/31 (Figure 5B), a more general marker of limb mesenchymal cells. In surveying sections of different embryos, occasionally limb mesenchymal cells do react with 22/18; these appear to be more common in the developing hindlimb than in the forelimb. Nonetheless these cells usually constitute fewer than 1% of the cells in a section, and thus the distinction between the developing bud and the regeneration blastema is a very clear one (Fekete and Brockes, 1987).

Response on Amputating the Limb Bud

Although only a subset of organisms have the ability to regenerate appendages as adults, almost all developing systems exhibit the phenomenon of regulation to a greater or lesser degree. That is, if part of a developing embryo is removed, there is a strong tendency to mount a local growth response and replace the part. In animals capable of epimorphic regeneration as adults, it is of obvious interest to examine the relationship between regulation

[1]In this discussion we will refer to developing limbs as buds and to regenerating ones as blastemas.

and regeneration. To do so, we amputated the developing bud approximately midway along the proximal-distal axis at various stages. After two to eight days, the animal was fixed, embedded in sucrose-gelatin, frozen sectioned, and processed for immunocytochemical localization of 22/18. The presence of the nerve was assessed by using an antiserum to the 70-kd subunit of the neurofilament triplet (Jacob et al., 1982). It should be noted that at all stages the severed buds healed and showed morphological regeneration.

Figure 6 summarizes the results of these experiments on the ontogeny of 22/18 expression after amputation (for a more detailed description, see Fekete and Brockes, 1987). During the early stages of limb outgrowth, amputation does not provoke 22/18 expression, although the transected bud regrows a normal limb. By stage 42 there is a detectable response on the amputated side after amputation through the humerus. Reactivity is centered around the distalmost nerve bundles, although the distribution of nerves versus 22/18 staining is not absolutely identical (Figure 7). Although the issue requires further investigation, we suspect that the induction of the 22/18 antigen at the earliest stages reflects the arrival and response of Schwann cells at the amputation plane. Unfortunately, our markers for Schwann cells in the adult newt are not specific to Schwann cells in the limb bud. By stage 45, many of the cells in the blastema are positive for 22/18. Figure 6 illustrates that the development of the hindlimb is arrested relative to that of the forelimb. In fact, it is possible to amputate both forelimb and hindlimb on one side and yet observe a response only in the forelimb (Figure 8). This rules out the possibility of any circulating factor controlling the time of appearance of 22/18, although such a factor might obviously play an important permissive role in the process.

Despite uncertainties about the origin of 22/18-positive cells at the onset of the re-

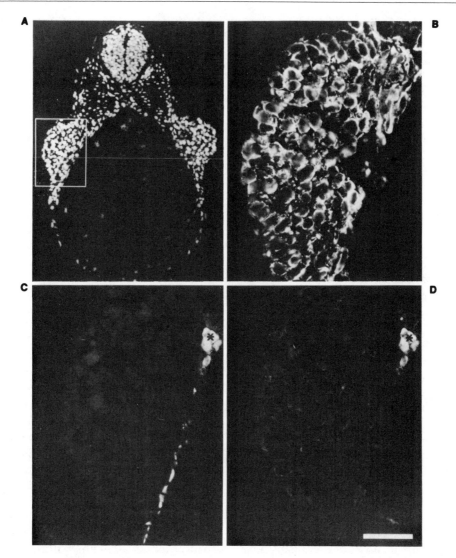

FIGURE 5. Immunostaining of a stage 32-larvae of *N. viridescens*. (A) Low-power photomontage of a transverse section through the level of the limb buds, stained with Hoechst nuclear dye (bis benzamide). The left limb bud is outlined and the same section is shown at higher power in C and D. (B) An adjacent section through the bud stained with 22/31, which in adult limbs stains some connective tissue cells and most blastemal cells at later stages of regeneration (Kintner and Brockes, 1985). All mesenchymal cells of early developing limb buds are stained with this reagent. (C) Section through the limb bud stained with 12/101 (a muscle marker) (Kintner and Brockes, 1985) and detected with fluorescein-labeled antibodies. A few differentiating myofibers of the lateral plate are stained, but no cells within the limb bud proper are 12/101-positive at this stage. The asterisk denotes a cluster of autofluorescent cells. (D) The same section as that shown in C but analyzed for rhodamine fluorescence that detects staining with 22/18. None of the limb mesenchymal cells are stained. Asterisk denotes the same autofluorescent cells as in C. Details of the double-labeling protocol can be found in Kintner and Brockes (1985). Bar for plates B—D = 50 μm.

FIGURE 6. Summary of the experiments that assess the ontogeny of 22/18 expression in response to amputation. Larvae of both *N. viridescens* and *P. waltl* were used in this study. The stages of limb development (Gallien and Durocher, 1957) are drawn schematically, along with an approximate indication of the state of innervation. The earliest appearance of nerves and myoblasts, detected by immunostaining, is indicated on the right. Each vertical arrow represents a single tadpole and begins at the stage of amputation. The length of the arrow indicates the stage to which the contralateral control limb develops during the course of the experiment. For each animal, the expression on the amputated side is compared to the control side, so that a minus sign indicates no induction of the 22/18 antigen, whereas plus signs (+ to +++) signify increasing induction and reflect the relative number of positive cells. Those cases where nerve fibers were detected in the blastema are indicated by an N. The single star denotes the case shown in Figure 7; the double star, the case in Figure 8.

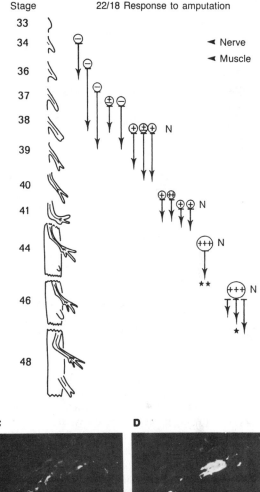

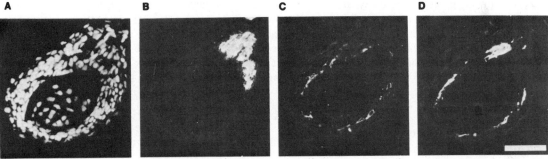

FIGURE 7. Association of 22/18 staining with nerve, but not muscle, in the regenerating limb of a stage-46 *P. waltl* larvae. The limb was amputated six days before, sectioned nearly transversely, and stained with a panel of antibodies. A, B, and C are the same section viewed with an epifluorescence microscope through different filters to detect nuclei with Hoescht dye (A), muscle with 12/101 (B), or 22/18 (C). The relatively cell-free zone in the center of the limb is the amputated humerus. It is flanked by blastemal cells near the distal tip of the limb (in the region devoid of muscle). Details of the staining protocol can be found in Kintner and Brockes, 1985. (D) A nearby section has been stained with a rabbit antiserum directed against the 70-kd neurofilament protein, followed by rhodamine-conjugated goat antirabbit Ig. The smaller, distalMost nerve bundles show the more complete overlap with 22/18 staining. Bar = 100 μm.

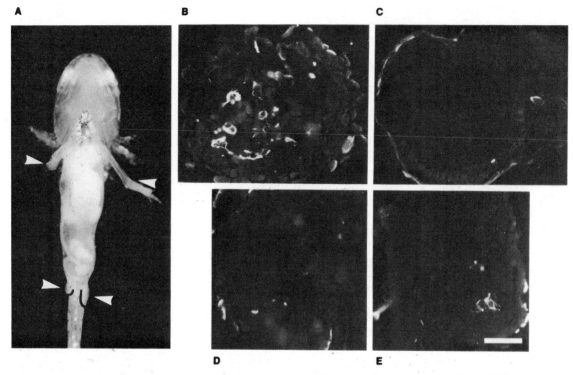

FIGURE 8. 22/18 staining is associated with the regenerating forelimb but not hindlimb of a N. *viridescens* larvae amputated at stage 44. The right limbs were amputated, leaving the left limbs as controls. (A) A low-power view of the larvae at the time of fixation. Arrows indicate the approximate plane of section for the sections in plates B–E. (B) Right forelimb eight days after amputation through the mid-humerus. Some of the blastemal cells at the distal tip of the limb are stained with 22/18. (C) Left forelimb showing staining only of the epidermis. (D) Right hindlimb five days after amputation of the distal half of the early bud. Staining with 22/18 is restricted to the epidermis and few fine filaments in the limb bud proper; it does not differ from the unamputated control limb shown in E. Forelimbs and hindlimbs were amputated at different times to allow for the faster regeneration of early buds. Bar for B–E = 50 μm. (For a more complete experimental series, see Fekete and Brockes, 1987.)

sponse, the results are very informative. They indicate that the 22/18 transition does not inevitably accompany amputation of the limb but is established during its development. If the specificity of the antibody in adult regeneration is also exhibited in the developing limb, then regeneration is initially nerve independent for proliferation and subsequently becomes nerve dependent. These observations clearly have relevance to discussions of the aneurogenic limb.

The Aneurogenic Limb

The method described by Yntema (1959) to derive aneurogenic forelimbs of axolotls can also be used successfully in newts, as illustrated in Figure 9. A pair of *P. waltl* embryos are fused in parabiosis at stage 21 or 22 (before the forelimbs emerge) by removing a patch of skin on each partner's side and pressing the wounds together. The following day an incision is made in one partner and most of the

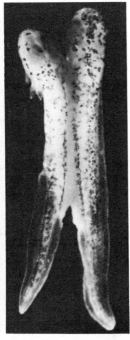

A   **B**

FIGURE 9. Parabiosis and neural tube removal of *P. waltl* larva. (A) A pair of larvae viewed shortly after the neural tube was removed from the left member. The notochord is clearly visible, flanked on either side by the somites. The wound will heal by migration of the epithelial cells over the exposed internal tissues. (B) Another pair of larvae viewed one week after the neural tube was removed from the left member. The dorsal crest fails to develop over the length of the extirpated spinal cord. The magnification is the same for A and B, and shows that the larvae continue to grow and differentiate after the surgery.

neural tube is removed, after which the operated embryo heals and is sustained by its neighbor. There is a definite tendency for nerves to sprout into the limb from remaining brain tissue, but in favorable cases the forelimb develops in the virtual absence of a nerve supply. When the aneurogenic limb is amputated, it regenerates in the absence of a nerve supply. In the experiments of Thornton and Thornton (1970) such limbs were substituted for the forelimbs of normal embryos, and the transplanted limb was maintained aneurogenic by cutting the host's brachial plexus. Under these conditions, the transplant regenerates in the absence of a nerve supply. If the brachial nerves are allowed to enter the aneurogenic limb, they take about 10 days to grow to the end, as judged by the appearance of sensory responses. The nerve dependence of limb regeneration is evaluated by denervation and amputation at the same time. When groups of animals are examined between 10 and 13 days after amputation, the limbs undergo a transition from nerve-independent to nerve-dependent regeneration during this period (Thornton and Thornton, 1970).

Most attempts to explain these results have centered on the neurotrophic factor and its regulation. One model suggests that before innervation all cells in the limb produce this factor, but after innervation the nerve suppresses its production and becomes the sole source (for review, see Chapter 2 in Wallace, 1982). Our proposal is quite different and is a simple extension of the observations on normal development summarized in Figure 6. The regeneration of the aneurogenic limb proceeds by the same system of growth control used by the early developing and regenerating limb. Before innervation, the cells that arise after amputation are able to respond to non-neural growth factor(s). After the arrival of the nerve (including axons and Schwann cells), amputation gives rise to cells that do not respond to the non-neural growth factors in the limb but do respond to the nerve-derived one. The generation of these cells after amputation involves a transition that is marked by the appearance of the 22/18 antigen. This hypothesis does not necessarily depend on any long-term modulation that the nerve may exert on the developing or aneurogenic limb. The control is exerted at the time of amputation, which in one case gives rise to progenitor cells depend-

TABLE 1. Summary of the Two Systems of Growth Control Proposed to Operate in the Limb

Nerve-Independent Proliferation (22/18-Negative Cells)	Nerve-Dependent Proliferation (22/18-Positive Cells)
Normal development	Normal regeneration
Regulation of the early limb bud	Aneurogenic/innervated regeneration[a]
Aneurogenic development	
Aneurogenic regeneration	

[a]Reactivity with 22/18 is not yet determined.

ent on the nerve and in the other case does not. Table 1 shows the circumstances under which the two systems are proposed to operate. Although it may be an oversimplification to group all the nerve-independent examples together, there is no clear justification at the moment for not doing so and it serves to emphasize the distinction that is being made. The clear experimental prediction, which we have recently verified, is that an aneurogenic blastema is 22/18-negative, while its innervated counterpart in the parabiotic pair is 22/18-positive (Fekete and Brockes, in press). This obviously involves leaving the pair to a stage (Figure 6) where the innervated limb gives an unequivocal 22/18-positive blastema.

An important task for the future is to determine the basis of the 22/18 transition. How does the presence of the nerve in conjunction with injury give rise to this process? One important insight is that the presence of axons at the amputation plane is not required—indeed, their transient absence after regeneration is probably important. Figure 10 shows a montage of a ten-day blastema arising from a limb that had been denervated four days before amputation. Despite the absence of axons, a feature that was checked by staining sections with the neurofilament antiserum, amputation has given rise to 22/18-positive cells in the blastema and the nerve. This experiment emphasizes the distinction between the role of the nerve in controlling proliferation and its role in the formation of the dependent population of blastemal cells. Although the findings rule out any requirement for the presence of axons at the time of amputation, they do not rule out some longer-term influence. We favor a different possibility, namely, that the denervated 22/18-positive Schwann cells at the amputation plane are the important element immediately contributed by the nerve, and that some product of the cells in conjunction with injury in turn triggers interstitial cells of the connective tissue to undergo the transition. One possible way of testing these hypotheses would be to derive larvae that are devoid of limb Schwann cells by virtue of neural crest removal early in development (Harrison, 1924).

In order to determine the molecular basis of this transition, in terms of both the molecules that induce it and the consequences for the cell, it will be essential to study the transition in dissociated culture. We have established culture conditions in which dissociated cells from the early blastema or cells that migrate from explants of the normal limb will stably express 22/18 (see Figure 2). We are currently culturing dissociated cells from the early limb in an attempt to establish the transition and study its regulation.

Summary

After immunizing mice with adult newt blastemas, a monoclonal antibody (termed 22/18) has been derived that reacts with the majority of early blastemal cells but not with the tissues of the normal limb. The antibody binds to an epitope present on a class of intermediate filaments or possibly on an intermediate filament-associated protein.

The 22/18 antigen is first detectable two days after amputation in Schwann cells of the nerve sheath and interstitial cells of the connective tissue. At early or midbud stages of regeneration, most of the blastemal cells react with the antigen. The antigen is subsequently lost from most cells in the blastema, although at later stages it is present on a population that differentiates into muscle. The mature regenerate, like the original limb, has little or no reactivity.

The 22/18 antigen is a good marker for cells whose division is dependent on the nerve supply. The labeling index (with tritiated thymidine) of 22/18-positive blastemal cells decreases sevenfold relative to control when assayed three days after denervation. The 22/18-negative population is less dependent, particularly at later stages of regeneration.

The appearance of 22/18 after amputation does not require that cells enter S phase and does not depend on the presence of a wound epidermis. The 22/18 antigen can be locally induced in Schwann cells and fibroblasts by injury or wounding, and appears in Schwann cells of the distal stump after cutting the brachial nerves. The appearance of the antigen is not a transient response to the events of injury, but appears to be a potentially stable transition both in vivo and in vitro.

Although 22/18 appears transiently in the embryonic epidermis, rather few internal cells (less than 1%) of the developing limb bud are reactive. There is thus a clear difference between the incidence of the antigen in development and the incidence in regeneration.

When the developing limb bud is amputated on one side, it shows morphological

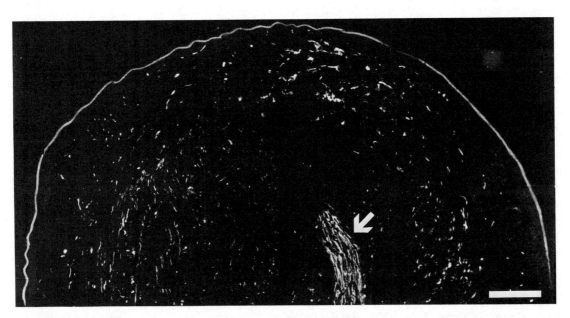

FIGURE 10. The 22/18 antigen appears after amputation in the absence of axons. Longitudinal section through the blastema of an adult *N. viridescens* forelimb stained with 22/18. The limb was denervated four days before amputation and allowed to regenerate for ten days. The Schwann cells of the denervated nerve (arrow) and many cells in the blastema are stained with 22/18. The boundary of the limb has been outlined in white for orientation purposes. The distal tip of the limb is towards the top. Bar = 200 μm.

regeneration at all stages tested. When the regenerating and the control buds are stained with 22/18, the antigen is not induced after amputation at the early stages of limb development. When 22/18 is first detected after amputation, it appears to be in cells associated with the nerve supply, as evidenced by double-label staining with an antiserum to a neurofilament protein. The stage of development at which 22/18 is induced after amputation is primarily a function of the development of the limb, as shown by comparison of the response of forelimbs and hindlimbs in the same larva. These latter results are consistent with a transition from a nerve-independent to a nerve-dependent mechanism of regeneration. We suggest that regeneration of the aneurogenic limb proceeds by a process equivalent to early regulation because the absence of innervation prevents the establishment of the epimorphic system. After innervation of the limb, amputation provokes a transition in the precursors of blastemal cells that leaves them refractory to the nerve-independent system and hence dependent on the nerve supply. The appearance of 22/18 after amputation occurs in the absence of axons and may depend on the conjunction of the injury response and the denervated Schwann cells.

Although it has been difficult to obtain a monoclonal antibody that distinguishes blastemal cells from cells of the normal limb, the properties of 22/18 seem sufficiently informative to justify the effort. The 22/18 antibody has provided new, distinctive information about aspects of epimorphic regeneration and in particular about the involvement of the nerve supply. Besides the nerve's classical neurotrophic role in controlling proliferation in the early blastema (Singer, 1952, 1974), it also contributes Schwann cells (albeit to an uncertain extent) to the population of blastemal cells that depends on it for proliferation and may play a role in the early events after amputation that give rise to this population. One of the major questions that cannot be answered at present is why the nervous system "intervenes" in regeneration, but we hope that further studies at the cellular and the molecular levels may shed light on this event.

The results discussed in this chapter offer an interesting perspective on more general issues of growth control. Although there is increasing information about the molecular diversity and mechanisms of action of mitogenic growth factors, there is rather little understanding of how they may act in vivo to regulate cell division. The nerve-dependent system of regeneration is a favorable model by which to investigate these issues. A central concern of this chapter has been to suggest how one system of growth control (regulation) is overlaid by another (nerve-dependent regeneration). This is accomplished not primarily by regulating the availability of growth factors but by changing the responsiveness of the progenitor population, a feature that is also shared with the normal time course of regeneration in the adult (see above). It may be possible to investigate these issues in a more detailed way by cloning the gene for 22/18, expressing it in cultured cells from the limb, and identifying the flanking sequences that control its expression. This might allow us not only to assess if the antigen is functionally important in regeneration, but also to use its control mechanism to express other genes of interest (for example, growth factor receptors) both in cells in culture and introduced into the blastema in vivo. Such an approach offers definite possibilities for tackling the complexities of epimorphic regeneration.

II

MOLECULAR BIOLOGY OF THE BRAIN

KATE F. BARALD

The application of new technology to a field often results in a sudden quantum leap in the amount of available information. However, an even more important test of the value of technology to any discipline is whether use of the new methods paves the way for the solution of questions that were previously unapproachable. Both criteria are satisfied when we consider the application of recombinant DNA technology to the study of the nervous system, but the marriage of the technique and the discipline has been longer in the arrangement than many expected.

There is no question that developmental neurobiology spells excitement to molecular biologists; some notable practitioners, including Gunther Stent, found neurobiology to be a new frontier even before the advent of recombinant DNA methods. Seymour Benzer (*Drosophila*), Sidney Brenner (*C. elegans*) and their colleagues and students (including Zipursky, Chalfie, Horvitz and Sulston) have demonstrated spectacularly how valuable the application of recombinant DNA methods in a genetic context is to the study of "simple" developing nervous systems. Others, like the late Ed Herbert and Jim Roberts (opiate peptides), Axel Ullrich (β-nerve growth factor), John P. Merlie, Jim Patrick, Steve Heinemann, and David Anderson (acetylcholine receptors) and their colleagues, as well as Thoenen, Numa, N. Davidson, Tobin, Rubin, and Axel, have cloned the genes for important nervous system components. The laboratories of

43

Richard Scheller, Corey Goodman, Lou Reichardt and Regis Kelly have also applied these techniques to advantage. However, neurobiologists, while not slow to recognize the value of recombinant DNA techniques, have generally hesitated to use them in their own work.

Two of the following chapters, those by Chikaraishi and by McKinnon et al., are the work of molecular biologists who have brought their considerable skills to studies of the brain. Chapter 5 is by a group of neurobiologists, Patrick et al., who a few years ago took their place in the avant garde by retooling their laboratory to apply recombinant DNA methods to studies of muscle and brain acetylcholine receptors. The results of these three studies have increased our knowledge of the brain's complexities and have raised some intriguing theories about brain-specific mechanisms of gene regulation. They have also contributed to our knowledge of the regulation of genes for specific molecules, including acetylcholine receptors and tyrosine hydroxylase, and their functions. They have also raised some controversial issues.

Brain Messenger mRNAs

Studies of both the level of expression and the location in the brain of messages for abundant, highly conserved transcripts such as actins may not tell us much about brain function. Ubiquitous transcripts for "housekeeping functions," common to a great many cells inside and outside the nervous system, have limited interest for researchers who would like to understand how different kinds of neurons fulfill their unique functions. Early studies (Bantle and Hahn, 1976; Chikaraishi, 1978) indicated that the mammalian brain expressed far more genes than other tissues, and that many of these genes were specific for neurons and glial cells in the brain. It is likely that a very rare class of transcripts (0.1 to 1 copy per cell) is responsible for the production of the equivalent of about 150,000 different average-sized messenger RNAs in the brain and reflects the diversity of the large number of cell types found in this highly complicated organ.

Chikaraishi's interest in high-complexity, low-abundance messenger RNAs in the brain led her to devise new methods for the detection and isolation of genes encoding these brain-specific, low-abundance RNAs. She has used a technique called transcription mapping to identify and isolate such RNAs and has found that the brain has the largest transcriptional diversity of any organ in the body: 33% of the RNA in brain is complex RNA, whereas that in other tissues such as liver (20%) and mammary gland (25%) is less complex. Unlike most messenger RNAs, at least half of these low-abundance, brain-specific messenger RNAs do not have poly A tails; the significance of this observation is unknown. These low-abundance poly A⁻ mRNAs are found *only* in brain, and at least one of them encodes RNA that appears to be on brain polysomes. In addition, these specific RNAs may be regulated by processing events that happen after transcription (post-transcriptional processing).

An example of such post-transcriptional processing is seen in the adrenergic to cholinergic switch of superior cervical ganglion (SCG) neurons in conditioned medium (Patterson and Chun, 1977b; Wolinsky and Patterson, 1983). For these studies, Chikaraishi has collaborated with Patterson's group. In the process of the adrenergic/cholinergic switch in tissue cultured SCG neurons, there is a three- to fourfold decrease in mRNA levels for tyrosine hydroxylase (TH), the enzyme responsible for production of catecholamines such as noradrenalin. This reflects a decrease in the number of TH mRNAs per cell from about 150–200 in adrenergic cultures to 50 in cholinergic cells. However, no such decrease is seen in the level of nuclear precursor RNA, heterogeneous nuclear RNA (hnRNA). This result is compatible with the idea that the adrenergic/cholinergic "switch" is probably not based on changes in rates of transcription but in post-transcriptional processing. However, other alternatives, among them stabilization of TH mRNA in noradrenergic cells, could also account for these results.

What Chikaraishi's studies tell us is that identification and isolation of the diverse, very low-abundance, brain-specific mRNAs should be pursued, despite their scarcity and the technical problems that this presents. Major goals include determining the protein products of these mRNAs and performing in situ hybridizations to localize the sites of expression of these rare transcripts among the cell types in the developing nervous system.

Neuronal Nicotinic Acetylcholine Receptors

The work presented by Patrick's group in Chapter 5 illustrates how gene cloning techniques, straightforwardly applied, can be used to bootstrap our knowledge of molecules that are important for nervous system function.

Having obtained cDNA clones for the subunits of the muscle nicotinic acetylcholine receptor (AChR), Patrick's group used low-stringency DNA/DNA hybridization to find homologous sequences in both cDNA and genomic sequences in the CNS. While the function of the protein encoded by the putative neuronal AChR receptor cDNA clone is unknown, its structure is very similar to the muscle α-subunit of the AChR. The differences in the structure again raise the question of the mechanisms by which tissue-specific gene expression might occur in this particular case. During the course of this work, the Salk group also found a number of α-subunit-like molecules, one or more of which may be related to α-bungarotoxin binding proteins in neurons (Ravdin and Berg, 1979) that are apparently related to, but do not serve the same function as, the nicotinic α-subunit of the receptor. They have also performed in situ hybridization localization studies to distinguish the areas of the brain that make transcripts for the neuronal receptor from those which bind α-bungarotoxin. This illustrates how molecular biological techniques both confirm and extend findings obtained by other methods such as antibody localization, protein isolation and analysis, and the physiological assays that have provided the main body of information about the complexities of the nervous system to date.

Neuronal Identifier Sequences

What makes a gene product brain-specific and what is its functional significance? Families of genes which produce one form of protein (i.e. one protein isoform) in the brain and another in a tissue such as muscle have been known for a very long time. Examples of such diversity are seen in the brain and muscle isozymes of the enzyme creatine kinase and in the muscle and nonmuscle forms of actin, both of which are the products of multigene families. Various members of a gene family, all of which are seen in genomic DNA, are differentially expressed in different tissues and at different stages of development. However, eukaryotes can also use single genes to produce more than one protein product from the same primary transcript in different cell types. These are examples of differential transcription and differential RNA processing, respectively (reviewed in Watson et al., 1987). Different, sometimes subtle, splicing patterns in the gene also can account for different forms of a protein.

Some genes contain two different promoter regions that are alternatively activated in a tissue-specific manner. Tissue-specific regulatory elements located in, for example, the promoter regions that precede a single gene can produce tissue specific transcription. Tissue-specific protein factors, RNA factors or RNA–protein complexes that bind to specific sequences of nucleotides or structures within the promoter region allow tissue-specific transcription. The binding of such factors, called *trans*-acting factors (proteins), to *cis*-acting regulatory elements called enhancers that are within the gene itself could facilitate or block transcription (Watson et al., 1987).

The experiments by McKinnon et al. reported in Chapter 6 indicate that, in the rat brain, sequences they call "identifier" sequences are present in large numbers. An ID sequence is a short sequence of base pairs that is present in, and specific for, RNAs of a given tissue, such as neurons. Identifier sequences may act like enhancers and may be transcribed by RNA polymerase III (Pol III), rather than RNA polymerase II (Pol II), to make two small poly A$^+$ RNAs, BC1 and BC2, which are found to be enriched in cytoplasmic extracts of neuronal tissues. The concentrations of BC1 and BC2 vary in different regions of the brain and appear to be developmentally regulated, increasing dramatically postnatally.

While the BC1 and BC2 RNAs are apparently without function, the identifier sequences which code for them are thought to be transcriptional enhancers that may be regulators of batteries of developmentally important genes whose expression is enhanced postnatally. Transcription of these ID elements by Pol III, which has been demonstrated in vitro, is thought to open regions of chromatin for transcription by Pol II. However, there is presently no supporting evidence for this function, and Sutcliffe's group as well as others have found that the ID elements are not restricted to the genes whose expression they are thought to regulate.

This work raises an enormous number of questions about the functions and mechanisms as well as the specificity of the putative brain-specific ID sequences, and, like all good speculative discussions which challenge as well as intrigue, provide researchers with a fertile experimental ground.

Glossary

The following definitions give an overview of the terminology used in molecular studies of the brain and nervous system.

A-box (A-block); B-box (B-block)   Internal control regions required for genes transcribed by RNA polymerase III. Numbering is from the 5′ end of the mature RNAs. In the consensus sequences for A block and B block, the nucleotides in subregions are essential for transcription.

Alu elements   In humans, these are highly repetitive sequences which are not found clustered together, but distributed evenly throughout the genome and interspersed with longer stretches of moderately repetitive or unique DNA. The Alu family is composed of sequences that are about 300 base pairs long and are not precisely conserved. They are so-named because most contain a single cleavage site for the restriction enzyme Alu I somewhere near the middle. Thus, cleavage of human genomic DNA by the restriction endonuclease Alu I generates an Alu I fragment present in high copy number that can be easily seen as a prominent band when such digested DNA is separated by size on agarose gels. In a human genome there are almost a million Alu sequences, constituting 3–6% of the total DNA. Such sequences in vitro are good templates for in vitro transcription by RNA polymerase III. Related sequences are found in other mammalian genomes. Alu sequences are in part homologous to a small and abundant mRNA called 7SL (300 nucleotides in length). 7SL is associated with six proteins to form a small cytoplasmic RNA-protein complex (SRP) that is essential for the moving newly synthesized proteins across rough endoplasmic reticulum membranes. 7SL RNA is also transcribed by pol III.

cDNA   A single strand of DNA complementary to mRNA (i.e., complementary to and capable of hybridization to mRNA) that lacks introns found in the genomic DNA (that contains both introns and exons). The poly A tail of mRNA allows the synthesis of cDNA; by hybridizing or annealing chains of oligo-dT to the poly A tails to provide a primer for "reverse transcriptase" (an RNA dependent DNA polymerase, obtained from RNA tumor viruses, called retroviruses). The reverse transcriptase uses the RNA as a template to synthesize the cDNA. The single-stranded cDNA can itself be used as a template for making double-stranded cDNA.

cis-acting regulatory element   A regulatory sequence in the flanking region of a gene or within a gene that acts in concert with *trans*-acting regulatory proteins to regulate the transcription of the gene.

consensus sequence (splice junction)   The sequences at the opposite ends of different introns are predominantly of two types: the first two bases at the 5' end of each intron are GT, and the last two at the 3' end are AG. They are important in binding protein factors that bring the opposite ends of introns together prior to splicing when introns are removed.

cRNA   An RNA copy of a cDNA, made (for example) by in vitro transcription of a cDNA cloned into a plasmid containing bacterial or bacteriophage promoters (riboprobe vectors). The cRNA either can have the same sequence as the mRNA [and is designated (+)] or can have the complementary sequence [designated (−)] which would anneal to mRNA.

end labeling   The addition of labeled nucleotides or phosphate moieties (either radioactive or biotinated) to the 5' or 3' ends created by restriction enzyme cuts in DNA or RNA or to the naturally-occuring ends of RNA.

exons   Functional gene sequences are known as exons because they "exit" from the nucleus to function in the cytoplasm.

heteroduplexes   Duplex molecules composed of complementary strands that contain non-homologous regions (i.e., single-stranded regions), which are seen as bubbles or loops along the duplex strand in the electron microscope.

heterogeneous nuclear RNA (hnRNA)   Also called pre-mRNA. Large unprocessed heterogeneous messenger RNA precursors, about 10 times the length of cytoplasmic mRNAs; less than 10% of their mass appears in the cytoplasm as spliced mRNA molecules. They have a short halflife ($t_{1/2}$ = 3 min).

identifier sequence   A short sequence of base pairs that is specific for RNAs of a given tissue such as neurons.

introns   gene sequences that are removed from mRNA but are present in genomic DNA. Introns exist in all mammalian and vertebrate genes and in eukaryotes such as yeast. The noncoding introns of a gene often contain many more nucleotides than the coding exons, which explains why so many primary RNA transcripts are large. There are specific base sequences found at exon-intron boundaries (see consensus sequence).

nuclear run "on"/run "off" transcription   A means of measuring transcription rates. The nuclei are isolated from the tissue or cells of interest. The isolated nuclei are then suspended and incubated with radioactively labeled UTP. The RNAs produced are those of sequences *already initiated* by RNA polymerase II, which under the conditions of the experiment is incapable of new initiation. The RNA is then isolated by phenol-chloroform/isoamyl alcohol extractions and alternating EtOH precipitations in the presence of yeast tRNA or some other carrier. The transcription rate is measured by dotting the cDNA onto nitrocellulose and hybridizing the radiolabeled RNA to the unlabeled DNA to determine the number of specific gene transcripts synthesized. This is done by quantitative autoradiography.

open reading frame   Long stretches of triplet codons beginning with ATG (which is, in DNA, the universal initiation codon for translation; AUG in mRNA), which are not interrupted by translational stop codons (UAA, UAG, UGA) in the same reading frame.

post-transcriptional modification   Mechanisms by which RNAs can be stabilized in certain cells, mechanisms presently unknown.

post-transcriptional processing   Includes the addition of poly A tails to some messenger RNAs in eukaryotes, which is accomplished by an enzyme called poly A polymerase. This enzyme does not add the tails directly to completed molecules, but instead adds them upstream from transcription termination sites onto newly created 3′ ends made by a specific endonuclease that cleaves the bound mRNA chains 11 to 30 bases downstream (3′) to the AAUAAA recognition sequence. Some mRNAs lack poly A tails (see Chapter 4). Splicing (removal of) introns is also an example of post-transcriptional processing.

primer extension assays   A method for locating the 5′ end of a particular transcript in a mixture of RNA molecules. A short sequence is chosen which is located entirely downstream from the promoter. This is labeled at the 5′ ends by an enzyme, polynucleotide kinase. The short piece is then hybridized to the RNA. Reverse transcriptase is then used to extend the 3′ end of the DNA primer until it reaches the 5′ end of the RNA used as a template. The extended product is analyzed on DNA sequencing gels with appropriate size markers. The size of the labeled product can then be determined to measure the distance from the known restriction site to the start of transcription.

quantitative liquid hybridization   Described in detail in Chapter 4. Hybridization without filters, in which the DNA/RNA to be hybridized and the probe are suspended in buffer; method for obtaining high complexity–low copy number RNA (see RNA transcription mapping).

restriction site mapping   Restriction enzymes are used to cut DNA into fragments. The fragments are easily separated and revealed on agarose or polyacrylamide gels since the rate at which DNA pieces move is dependent on their lengths. The order of DNA restriction fragments along a chromosome or chromosomal segment gives rise to restriction maps; the order is determined by studying the appearance of incompletely cut fragments that get progressively smaller as digestion proceeds to completion.

retroposons (retrotransposons)   Like the transposon, these elements move to new chromosomal locations by transfer of information, not from DNA to DNA but from reverse transcription of the RNA to yield a complementary DNA chain. After conversion of the single stranded cDNA into a double stranded DNA helix, it is inserted into a new chromosomal location, presumably through a transposition-like event.

RNA transcription mapping   A technique in which an excess of labeled probe (either cRNA or cDNA) is annealed to an unlabeled RNA preparation. The non-annealed nucleic acid is degraded with nucleases that are specific for single stranded nucleic acid; the annealed probe, which is protected from nuclease digestion because it is in a double-stranded duplex, can be detected by gel electrophoresis. Transcription mapping is 10- to 100-fold more sensitive than Northern blots and therefore is well-suited for detecting non-abundant RNAs. This technique has been especially useful for isolation and identification of unique DNA sequences that are found in large numbers in the brain. For example, total rat brain DNA is digested to completion with a selected restriction endonuclease. Unique sequence DNA is then isolated from the total digest by allowing the repetitive sequences to reassociate. The *unique* sequences are isolated by hydroxylapatite chromatography. These unique sequences are then ligated into a specific restriction endonuclease site in a bacteriophage, the phage is grown, and recombinant plaques are selected. The rat DNA inserts are then excised with the appropriate restriction enzymes and the resulting fragments are end-labeled. The labeled DNA is hybridized to RNA, and S1 nuclease is used to digest single-stranded nucleic acid not hybridized to RNA. S1-resistant fragments are then separated by electrophoresis on polyacrylamide gels. If no bands are seen, it means that the S1 nuclease completely digested the single-stranded nucleic acid, since no RNA hybridization protected the labeled 5' end. Nonhomologous RNA is used as the control.

RNPs   Ribonucleoprotein granules, attached by "stalks" to the thin chromatin thread. Specific proteins attach to the nascent RNA chains leading to the formation of ribonucleoprotein granules.

Rot value   Kinetics of hybridization involving reannealing of RNA and DNA to measure the concentration of the hybridizing RNA species. This is done in an excess of RNA; The Rot value itself is the product of the initial RNA concentration (R_O) $\times$ time.

saturation hybridization   DNA/DNA, DNA/RNA, or RNA/RNA hybridization on filters or under liquid conditions in which a saturating concentration of nonradioactive DNA or RNA is used, such that all of the probe capable of hybridizing is driven into the duplex form by the end of the annealing period.

S1 nuclease digestion   S1 nuclease specifically degrades single-stranded unpaired DNA or RNA. It is used to clip the hairpin that forms when double-stranded cDNA is synthesized, and can also be used to analyze mRNA-DNA hybrids (S1 analysis).

trans-activator protein   A diffusable protein that acts on another gene or chromosome than that which is responsible for its production.

viral transcriptional regulatory elements; positive transcriptional regulatory elements; enhancers   Cis-acting DNA sequences important for increasing a promoter's transcriptional activity in the cell; these are outside elements that alter the efficiency of transcription by RNA polymerase II. Enhancer sequences function in both a position- and an orientation-dependent manner. That is, they can act over long distances—e.g., several thousand base pairs—and they can be located within or downstream from the region transcribed. Enhancer sequences were first found in DNA tumor viruses, such as polyoma and SV40 in the region that contains both the origin of DNA replication and the sites of initiation of RNA synthesis (promoters for both early and late genes (those expressed at early and late times after infection). Such enhancer regions can augment transcription, usually in a tissue-specific manner, from almost all other promoters to which they are linked. Mammalian and avian genes contain these regulatory elements as well. A particular enhancer element is functional exclusively or preferentially in certain cell types. Such tissue-specific enhancers may prove to be the basis for differential expression of genes during development or cell-specific gene expression.

CHAPTER FOUR

Characteristics of Brain Messenger RNAs

DONA M. CHIKARAISHI

Introduction

Models of gene regulation have focused on genes whose transcripts are relatively abundant, at least in certain cells. However, the vast majority of genes expressed in a cell or tissue belong to a "high complexity–low abundance" class composed of transcripts that are present at 1 to 10 RNA copies per cell and that represent almost all of the *different* kinds of proteins in a cell. This chapter reviews data indicating that the mammalian brain expresses an even rarer class of transcript (present at 0.1 to 1 copy per cell) that encodes three- to five-fold more information than is found in other tissues. This level of diversity is equivalent to 150,000 different average-sized RNAs and presumably reflects a mixture of cell types within the brain such that a given neural cell may only express a subset of genes appropriate for its function. Because of their scarcity, these transcripts cannot be detected by many standard techniques such as Northern blot or plaque/colony hybridizations. This chapter focuses on the cloning of genes encoding brain-specific, low-concentration RNAs and two controversial aspects of their nature, namely, that they may be largely regulated by post-transcriptional processing and that half may be not be polyadenylated.

RNA Complexity: High mRNA Diversity in the Brain

Although a number of methods can be used to estimate the number of active genes, the most common approach used to determine the expression of genes transcribed at low frequency is saturation hybridization. In this technique, genomic DNA is radiolabeled, denatured, and annealed to RNA. If a region of DNA is transcribed, it will anneal to RNA and be converted to an RNA:DNA hybrid that can be distinguished from unhybridized (single-stranded) DNA on a hydroxylapatite column or by resistance to single-strand specific nuclease digestion. The percentage of the total genomic DNA (discounting repetitive DNA elements) that can be annealed to RNA is then measured. By multiplying by the number of nucleotides in the genome, this value can be converted to the total length of nucleotides transcribed, which is the complexity of RNA expressed in this tissue. The total number of active genes can then be estimated by dividing the total RNA complexity by the average length of a gene. Estimates of the number of active genes vary, depending upon what value is used for the average size of a gene.

The rate at which a given RNA anneals to

its DNA complement reflects its concentration in the hybridization reaction. Abundant species will find their complement more rapidly than scarce RNAs. Therefore, the rate at which DNA is driven into hybrid can be used to determine the abundance of those RNAs responsible for the hybridization. By knowing the amount of RNA per cell, one can then estimate the average number of RNA copies per cell for those RNAs that anneal at a given point during the reaction.

Complexity studies, mainly by Davidson and colleagues in the sea urchin and by Chikaraishi and coworkers in the rat, suggest that the majority of rare transcripts may be regulated post-transcriptionally. In the sea urchin, RNA complexity measurements (Kleene and Humphreys, 1977; Wold et al., 1978; Ernst et al., 1979) as well as hybridization using tissue-specific clones (Lev et al., 1980) have shown that almost all tissues in the adult animal, as well as tissues at each stage of development, share a common set of nuclear RNAs, the primary products of transcription. From this ubiquitous pool, different sets of mRNAs are processed in a cell-specific manner.

In the rat, RNA complexity studies using saturation hybridization showed that different adult tissues expressed 10 to 33% of the single-copy DNA as heterogeneous nuclear RNA (hnRNA). Of the eight tissues examined (Chikaraishi et al., 1978; Beckmann et al., 1981; Brilliant et al., 1984), the brain exhibited the highest complexity (33%), equivalent to 5.9×10^8 nucleotides (Figure 1). The nuclear RNAs fell into a nested set, such that the majority of nuclear sequences expressed in a given tissue were also found in other tissues with a higher absolute complexity (Chikaraishi et al., 1978; unpublished data). For example, the kidney sequences (10% complexity) were found in liver hnRNA (20% complexity), the liver sequences were expressed in mammary gland hnRNA (25% complexity), and the

mammary sequences were expressed in brain hnRNA (33% complexity). This suggests a hierarchy of expression, in which brain has the largest transcriptional diversity. Based on the kinetics of annealing, these complex hnRNAs were present at an average of 0.1 to 1 copy per cell (Chikaraishi et al., 1978).

Laboratory studies have shown that the rodent brain compared to other tissues expresses more genetic information as *messenger* RNA (mRNA) as well as hnRNA (Chikaraishi, 1979; Bantle and Hahn, 1976; Van Ness et al., 1979). The level of expression is equivalent to 100,000 to 150,000 different mRNAs (assuming mRNA lengths of 1.5 to 2 KB, or 50,000 transcripts if the average length is 4 KB), which is equal to 16 to 20% of the single-copy coding capacity of the genome as assayed by saturation hybridization and kinetics of cDNA annealing. In contrast, other tissues (liver, kidney, or cultured cells) express three- to fivefold fewer sequences as mRNA (Beckmann et al., 1981; Chaudhari and Hahn, 1983) (see Figure 1). Recently, Kaplan and colleagues have shown high RNA complexity in neural tissue as compared to non-neural tissue in nonmammalian species such as the goldfish and the squid (Kaplan and Gioio, 1986; Capano et al., 1986), which suggests that even in invertebrates, diverse neural gene products may be evolutionarily related to the increased complexity of neural function.

Although we do not know why neural tissue expresses more genetic information than other cell types, Grouse et al. (1978) have suggested that high RNA complexity results from sensory stimulation gained by interaction with the environment. In their study, environmentally deprived rats (no toys, no littermates) had one-third *less* complex brain RNA than did experientially enriched rats. In a similar study, kittens whose eyelids had been bilaterally sutured at birth expressed 25% fewer different RNAs in their visual cortex than did normal-sighted kittens, while show-

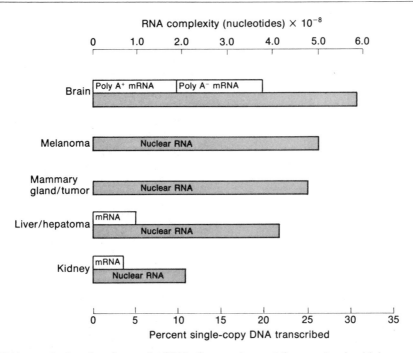

FIGURE 1. RNA complexity of nuclear and mRNAs from various rat tissues, showing high complexity in brain as compared to other tissues. The data are expressed as either the percentage of the single-copy DNA or the absolute number of nucleotides transcribed. RNAs from various adult tissues were hybridized to completion with iodinated single-copy rat DNA, and the percentage of the DNA hybridized was determined by digestion of unannealed DNA by S1 nuclease or by hydroxylapatite chromatography. (Data are adapted from Chikaraishi et al., 1978; Chikaraishi, 1979; Brilliant et al., 1984.)

ing no difference in RNA in the nonvisual cortices (Grouse et al., 1979). Although the data showing decreased RNA were compromised by technical difficulties, there is strong evidence that upon environmental deprivation, areas of the brain such as the upper occipital cortex that are involved in higher functions showed a decreased weight, dendritic field size, and synaptic density (Greenough, 1984 and Chapter 19).

The entire spectrum of 150,000 different mRNAs is probably not found in every neural cell, since the mRNA complexity of clonal neural cell lines has been shown to be one-third that found in the total brain (Beckmann et al., 1981). This suggests that the high diver-

sity in the brain may be the sum of different mRNAs from many heterogeneous cells. Figure 2 summarizes data showing the nuclear and mRNA complexities of clonal cell lines of neural origin. In the same study (Beckmann et al., 1981), the mRNA from individual brain sections was shown to be less complex than that of the whole brain, especially in regions like the cerebellum that are composed of a restricted number of cell types. However, the *nuclear* complexities of cell lines and all brain sections were equal to that of the whole brain. Furthermore, when additivity experiments were performed, the nuclear RNAs from the clonal cell lines and brain sections were essentially the same set as expressed in the whole brain. This implies that, as in the sea urchin, cell-specific post-transcriptional processing may be important in selecting appropriate mRNAs (Beckmann et al., 1981). Similar data have also been obtained for nuclear and mRNAs from various sections of sheep brain (Deeb, 1983; Hitti and Deeb, 1984).

Tyrosine Hydroxylase: A Marker for Post-transcriptional Regulation

To test the possibility that post-transcriptional regulation modulates neuronal function by RNA processing, we have determined the level of RNA for tyrosine hydroxylase, the rate-limiting enzyme in catecholamine biosynthesis, during conversion of noradrenergic sympathetic ganglion cells to a cholinergic phenotype induced by conditioned media.

Under standard culture conditions, pri-

mary superior cervical ganglion (SCG) cultures retain noradrenergic characteristics. However, co-culture with cells that normally receive cholinergic innervation (fetal heart cells) induces acetylcholine synthesis, as does media conditioned by such cells (Patterson and Chun, 1977a). A 45-KD glycoprotein has recently been isolated and shown to be responsible for the switch (Fukuda, 1985), although other serum factors may also influence transmitter choice. Under strongly "cholinergic" conditions there is a 100- to 1000-fold shift in the ratio of acetylcholine to norepinephrine synthesized. This is mainly due to a dramatic 100- to 1000-fold increase in the activity of choline acetyltransferase, the enzyme responsible for acetylcholine synthesis, and a modest (three- to fivefold) decrease in the tyrosine hydroxylase enzyme activity (Wolinsky and Patterson, 1983). However, even under strong-

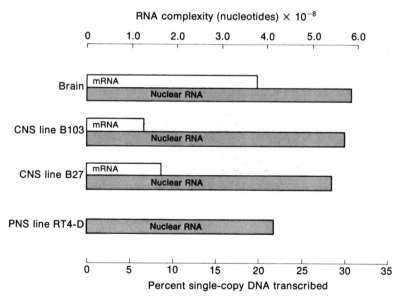

FIGURE 2. RNA complexity of clonal cell lines of neural origin. B27 and B103 are glial and neuronal-like rat central nervous system (CNS) lines described by Schubert et al., 1974. RT4 is a peripheral neural cell line described by Imada and Sueoka (1978). (Data are adapted from Beckmann et al., 1981, and unpublished data.)

ly cholinergic conditions, the high-affinity up-take mechanism for catecholamines is re-tained (Johnson et al., 1976).

Using "weakly cholinergic" conditions, cul-tures retain dual expression for as long as the cultures are maintained (Iacovitti et al., 1981). Under such conditions, individual cells syn-thesize, store, and release both acetyl-choline and catecholamines.

The transmitter switch in mass culture is not due to growth selection of a small sub-population of cholinergic cells or to an activa-tion of previously silent cells (Patterson and Chun, 1977a; O'Lague et al., 1974). Indeed, the SCG cells are postmitotic when initially plated. In a remarkable set of experiments, Potter et al. (1983, 1986) followed individual neurons as they underwent the switch from ad-renergic to cholinergic. By recording from mi-crocultures containing a single neuron and co-cultured myocytes, they showed that an in-dividual neuron that was initially adrenergic could pass through an interim period of dual expression and finally become purely cholin-ergic. However, the majority of neurons cul-tured in a conditioned medium changed toward a more cholinergic state, but retained dual transmitter expression even after three to four months (Potter et al., 1986).

The ability to switch transmitters is largely limited to cultures prepared from young ani-mals (less than two weeks old), and probably reflects an in vivo developmental window. In newborns, the SCG is weakly adrenergic. Dur-ing the following month, both in vivo and in culture, adrenergic properties become fully dif-ferentiated, after which time these cultures can no longer switch transmitters (Patterson and Chun, 1977b; Ross et al., 1977; Johnson et al., 1980). Depolarization of young SCG cultures also renders them refractory to switching (Walicke et al., 1977).

Landis and others have shown that this plasticity may occur in vivo (see Landis, 1983, for review). The sympathetic innervation to the rat eccrine sweat gland is catecholaminergic in the newborn, but becomes cholinergic during the first postnatal month. Again, this is not due to a selection of previously cholinergic fibers, and these mature cholinergic terminals still re-tain the re-uptake mechanism for catechol-amines.

In collaboration with Patterson's group, we have assayed the level of tyrosine hydroxylase RNA in noradrenergic and cholinergic cul-tures. Primary SCG cultures from newborn rats were grown for five weeks under one or the other culture condition. Figure 3 shows an RNA dot blot in which RNA from both cul-tures was hybridized to a cDNA probe for rat tyrosine hydroxylase (Lewis et al., 1983). Based on densitometric measurements, the choliner-gic cultures contained three- to fourfold less TH mRNA than the noradrenergic cultures. This decrease paralleled the reduction in in vivo catecholamine biosynthesis assayed by [^{3}H] tyrosine conversion to dopamine and norepinephrine. Since the majority of hyb-ridizing sequences represent mRNA mole-cules, the reduction in catecholamine syn-thesis can be accounted for by a proportional decrease in the TH mRNA. By comparing the hybridization signal of the primary cultures with that of PC12 cells where the level of TH RNA had been previously measured by quan-titative liquid hybridization, and by counting the radioactive dots, the number of tyrosine hydroxylase RNA molecules can be estimated to be about 150 to 200 per cell in noradrenergic cultures and about 50 per cell in the choliner-gic cells. TH RNA from whole ganglia taken from newborns is about 80 copies of RNA per neuron, assuming 40,000 principal neurons per ganglion. These data imply that from the initial level in newborn ganglia, TH mRNA increases in culture under noradrenergic con-ditions and decreases under cholinergic in-fluences.

In preliminary experiments, we have at-tempted to assay the level of nuclear precursor RNA for TH in both cultures. Using a radioac-tive cRNA probe that includes both intron and

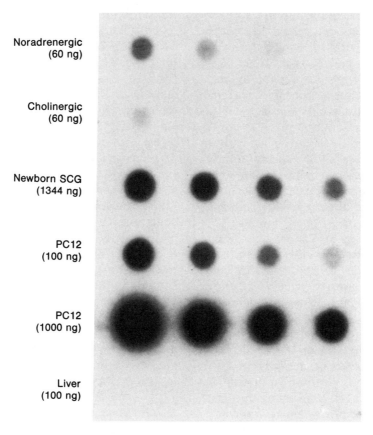

Noradrenergic
(60 ng)

Cholinergic
(60 ng)

Newborn SCG
(1344 ng)

PC12
(100 ng)

PC12
(1000 ng)

Liver
(100 ng)

FIGURE 3. Hybridization of RNA from primary superior cervical ganglion (SCG) cultures to tyrosine hydroxylase cDNA. The first row contains RNA from noradrenergic cultures; the second row, RNA from cholinergic cultures; the third row, RNA from newborn SCG; the fourth and fifth rows, RNA from PC12 cells; and the sixth row, RNA from liver. Primary cultures were prepared as described (Walicke and Patterson, 1981) and grown for 5 weeks in either noradrenergic conditions (20 mM K^+) or cholinergic conditions (50% conditioned media + 1 mM butyrate). Total RNA was extracted in LiCl-urea as described (Auffray and Rougeon, 1980) and applied in twofold dilutions to a Nytran membrane after formaldehyde denaturation (White and Bancroft, 1982). The first dot in the first row contained 60 ng of total RNA from noradrenergic cultures; first dot, second row, contained 60 ng of total RNA from cholinergic cultures; first dot, third row, contained 1344 ng of total RNA from whole SCG from newborn rats; first dot, fourth row, contained 100 ng of total RNA from PC12 cells; first dot, fifth row, contained 1000 ng of PC12 RNA; first dot, sixth row, contained 100 ng of total RNA from liver. The filter was hybridized to 2.3×10^6 cpm/ml of TH cRNA (2×10^9 cpm/µg) for 40 hr and washed to $0.2 \times$ SSC at 60°C.

exon TH sequences, we assayed the level of mRNA and unprocessed nuclear precursor RNA by RNA transcription mapping (Melton et al., 1984). A length of probe corresponding to exon sequences was protected from RNase digestion by mRNA, while the entire probe (intron plus exon) was protected by unspliced hnRNA. While the level of mRNA was three- to fourfold less in cholinergic cultures, in agreement with the dot blot analysis, the level of precursor RNA was not significantly different in the two cultures (data not shown).

These preliminary data suggest that this well-characterized phenotypic switch may not involve changes in transcription, but may be primarily due to post-transcriptional RNA processing. However, since the rate of transcription has not been directly measured, these data are also consistent with an increased rate of transcription offset by an increased rate of hnRNA processing or by a stabilization (increase in half-life) of TH mRNA in noradrenergic cultures. Further experiments designed to measure the rate of TH RNA transcription directly would distinguish between these alternative explanations.

In contrast to the results in the SCG cultures, when TH mRNA levels in PC12 cells are increased by agents such as glucocorticoids and cAMP, the level of nuclear precursor RNA is coordinately increased, matching or exceeding the increased mRNA levels. In the case of glucocorticoids and cAMP, the increase in nuclear precursor RNA reflects increased rates of transcription as assayed by in vitro RNA synthesis ("run-on" transcription) in isolated nuclei (E.J. Lewis et al., in press). Unlike the phenotypic switch in transmitter synthesis, which requires weeks, an increase in TH transcription with glucocorticoid or cAMP is detectable within 10 minutes and the maximal increase occurs in 30 to 60 minutes. These data imply that transcriptional activation may be a rapid process, while post-transcriptional modulation may be relatively slow.

Cloning and Characterization of Brain-Specific Genes Encoding Rare mRNAs

Detection of clones whose transcripts are present at 0.1 to 1 copies per cell requires a method of screening different from that used for selecting abundantly transcribed clones. Since colony, plaque, Northern, or dot blots are not sensitive enough to identify clones whose RNAs are present at less than 10 copies per cell (at best), we developed a modified transcription-mapping technique that can identify clones whose transcripts are present at 0.05 copies/cell (approximately 1×10^{-8} of the total RNA in a typical mammalian cell) (Brilliant et al., 1984). Figure 4 diagrams this cloning scheme. A genomic library of unique-sequence DNA was prepared in a single-stranded DNA vector, M13 mp7, such that digestion of the single-stranded DNA by Sau 3A preferentially cleaves out the rat insert. The mp7 vector has a series of palindromic cloning sites, such that insertion of Sau 3A fragments into the Bam site creates a 14-base pair duplex containing the Sau 3A site. Therefore, Sau 3A preferentially cleaves this duplex sequence over other single-stranded Sau 3A sites in the remainder of the vector, generating a fragment containing the rat DNA insert. The digested DNA is 5' end labeled and is used as the excess component in hybridizations with total brain RNA. If an RNA transcript anneals to the rat insert, it protects the labeled 5' end of the DNA from S1 nuclease digestion, and the resulting DNA fragment can be detected on a gel.

The sensitivity of the assay was estimated by preparing a β casein cDNA fragment in the analogous position in M13 and annealing it to known amounts of β casein mRNA. This technique is superior to Northern or dot hybridizations, because in order to detect a positive signal, at least 30 nucleotides must be engaged in a resistant hybrid that includes the 5' end of the DNA fragment. Adventitious annealing between short or GC-rich regions, which could give a background signal in Northern or dot hybridizations, is not scored in this assay.

To improve the sensitivity of the assay, the initial isolates have been recently recloned into SP6 vectors (Melton et al., 1984). The insert can be specifically transcribed into radioactive "riboprobe" RNA in an in vitro reaction. Using "riboprobe" RNA for transcription mapping, detection can be increased to 0.005 copies of RNA per cell.

From the initial library of unique-

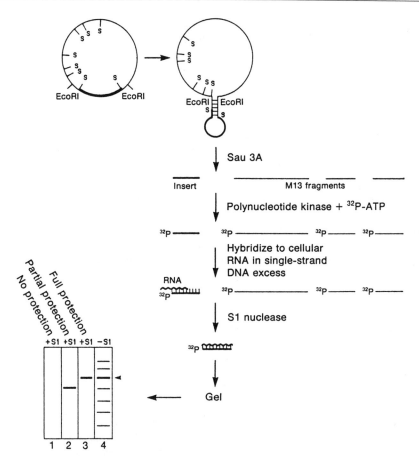

FIGURE 4. Transcription mapping of unique-sequence DNA cloned into M13. Lane 1, no bands are seen, illustrating complete digestion by S1 nuclease (no RNA hybridization that protects the labeled 5′ end); lane 2, partial protection (the protected fragment is less than the full length of the insert); lane 3, full protection from S1 nuclease digestion; lane 4, pattern obtained without S1 nuclease treatment. The absence of labeled phage fragments after S1 nuclease digestion is an important internal control. An additional control in which a non-homologous RNA (*E. coli* tRNA) gives no bands was also performed for each assay. Total rat DNA was digested to completion with the restriction endonuclease MboI or Sau 3A. Unique-sequence DNA was separated from repetitive DNA by three cycles of reassociation to cot (concentration of DNA × time) 200, followed by hydroxylapatite chromatography. The unique-sequence DNA (200 to 400 bases in length) was ligated into the BamHI site of the replicative form of the bacteriophage M13mp7, and recombinant (clear) plaques were selected from indicator plates. DNA from an individual clone of single-stranded phage M13, strain mp7, with a rat DNA insert (indicated by the heavy line) is drawn schematically (top left). The inverted repeat (14 nucleotides) between the EcoRI and the Sau 3A sites (S) is shown as double stranded. The rat insert is preferentially excised by Sau 3A, and the resulting fragments are end-labeled with [γ^{32}P]ATP. After hybridization to RNA, S1 nuclease digestion eliminates single-stranded nucleic acid not hybridized to RNA. The S1 nuclease–resistant fragments are separated by electrophoresis on polyacrylamide gels and visualized by autoradiography. Schematic examples of gel results are shown. The rat insert fragment indicated by the arrowhead is easily distinguished from the known Sau 3A fragments generated by M13, by both size and specific activity.

sequence DNA, individual clones were randomly selected and tested for their ability to anneal to brain RNA. Of 50 random clones selected for the initial screening, 7 clones were expressed in brain RNA and all belonged to the very low concentration class (0.05 to 5 copies/cell) (Brilliant et al., 1984). Complexity estimates predict that one out of three (33%) random clones should be positive for brain expression using a double-strand vector and one out of six (16.6%), using a single-strand vector. Our value of 7/50 (14%) is therefore very close to that prediction.

The overall pattern of these clones fits that predicted by the earlier complexity measurements with regard to tissue expression. We isolated four clones that are specifically expressed in brain mRNA. One clone (rg34) is interesting in that its sequence is expressed in the nuclear RNA of all tissues examined, but is found expressed in polysomal mRNA only in brain tissue, suggesting that it is regulated post-transcriptionally. Another clone (rg40) is restricted to brain and melanoma cells (which are of neural crest origin), and in the brain it is found in the cortex but not the cerebellum. Three other clones were processed to polysomal mRNAs in all tissues and presumably represent low-abundance "housekeeping" functions. One clone was expressed in the nuclear RNA of all tissues, but was not processed to mRNA in any tissue examined. The nested hierarchy seen with rat tissue was not violated by any of the clones, that is, (1) all those expressed in kidney are also transcribed in all other tissues, (2) two clones (rg13 and rg40) that are not in all tissues are restricted to those having the highest complexities, and (3) rg40 found in melanoma (complexity 25%) is also expressed in brain (complexity 33%).

We also explored whether abundant transcripts fell into a nested hierarchy. If so, one would expect to find albumin sequences in the nuclear RNAs of mammary gland and brain tissues, since both tissues have higher absolute complexities than that of liver. However, using three tissue-specific probes (albumin, casein, and prostate glycoprotein), expression was not detected (at the level of 0.1 copies/cell) in any tissue but that which was appropriate for the gene. Therefore, while the rare clones fall into the expected hierarchy, the abundant transcripts do not, suggesting that the two classes may be regulated differently.

Because only 7 out of 50 randomly selected clones were transcribed in brain RNA and only 3 of the 50 showed neural-specific messenger expression, we are developing a "subtraction chromatography" technique to isolate rare transcripts differentially expressed in one tissue and not another. In brief, *total* genomic DNA was digested with Sau 3A and cloned into M13 mp7. Single-stranded, M13-recombinant DNA from about 5,000 different clear plaques was prepared and annealed to excess liver RNA. Clones containing liver-expressed sequences and repetitive DNA elements, which should have become double stranded, were removed by hydroxylapatite chromatography. The remaining single-stranded DNA was then hybridized to excess brain polysomal RNA, the double-stranded fraction collected, and this fraction transfected into *Escherichia coli*. In our preliminary screen, several hundred recombinant clones were recovered. Fifteen of these were selected at random and characterized for their pattern of tissue-specific expression. All had inserts and 10 were transcribed in brain: 4 of the 10 were brain specific or greatly enriched in brain, while the remaining 6 were common to both brain and liver. Most importantly, the 10 transcribed clones encoded RNAs that ranged in abundance from about 0.1 to 30 copies/cell, suggesting that we are still dealing with rare transcripts. Libraries of clones expressed in brain and not in liver or expressed in the adult but not in the newborn rat are currently being prepared and tested for their transcriptional pattern. These libraries may be useful for selecting cell-specific, region-specific, or developmentally regulated genes.

Characterization of Clones Encoding Poly A⁻ RNAs

A surprising finding is that half the brain mRNAs measured by saturation hybridization appear to be poly A⁻ in both the rat (Chikaraishi, 1979) and the mouse (Van Ness et al., 1979). Since almost all eukaryotic mRNAs (with histone messages being a notable exception) contain poly A on their 3' ends, it is crucial to verify that these poly A⁻ RNAs are true messages. At present it is not known which proteins are encoded by these poly A⁻ RNAs in brain, and therefore their messenger function can only be inferred. The RNAs are found on active polyribosomes because they can be released from high-molecular-weight polysomes by EDTA or puromycin (Chikaraishi, 1979; Van Ness et al., 1979). In particular, puromycin is thought to be quite specific in terminating nascent protein chains by disaggregating only polysomes in the act of protein synthesis. In addition, purified poly A⁻ RNA can be translated in vitro using reticulocyte lysates with efficiencies equal to that of poly A⁺ mRNA (Chikaraishi, 1979; Hahn et al., 1983) and it has been shown that poly A⁻ RNAs fractionate with mRNPs rather than nuclear RNPs on cesium chloride gradients after fixation with formaldehyde (Van Ness et al., 1979).

Hahn and colleagues have shown that poly A⁻ mRNAs are brain specific and appear after birth in the mouse (Chaudhari and Hahn, 1983; Hahn et al., 1983). At birth, the poly A⁺ mRNAs are already present and correspond to the same population as those in the adult, while the poly A⁻ mRNAs are largely absent and accumulate slowly over the subsequent 45 days of development. Interestingly, the appearance of the nuclear poly A⁻ RNA precedes that of the cytoplasmic poly A⁻ mRNAs by about 25 days, suggesting temporal post-transcriptional regulation. In addition, the postnatal appearance of poly A⁻ species (both nuclear and polysomal) argues that they are *not* precursors or remnant introns from the poly A⁺ RNAs, which are already fully expressed at birth. Because the studies of environmental deprivation were performed with postnatal rats (Grouse et al., 1978), the observed reduction in nuclear complexity might have been due to a selective lack of induction of the poly A⁻ sequences.

For clarity, it is important to distinguish between abundant and rare poly A⁻ RNAs. Abundant poly A⁻ mRNAs are found in all cells—HeLa (Milcarek et al., 1974), L cells (Greenberg, 1976), sea urchins (Nemer et al., 1974)—and comprise two classes: those that are both A⁺ and A⁻, e.g., β actin (Hunter and Garrels, 1977), and those that are largely poly A⁻, e.g., mRNA for the histones. As would be expected, brain has both classes of poly A⁻ messengers. Bimorphic mRNAs can be easily seen in the translation products of A⁺ and A⁻ mRNAs (Chikaraishi, 1979; Morrison et al., 1979). Abundant poly A⁻ mRNAs, which are not found in the poly A⁺ fraction, represent at least 60% of the mass of poly A⁻ mRNA, as estimated by cDNA annealing (Van Ness et al., 1979). However, these reflect abundant species that are not brain specific, as demonstrated by Hahn and colleagues who cloned these RNAs and found most to be expressed in other tissues (Hahn et al., 1983).

In contrast, the experiments demonstrating that brain tissue expresses enough genetic information to code for 50,000 to 75,000 *different* average-length poly A⁻ mRNAs measure only rare RNAs. In fact, the kinetics of hybridization indicate that the average RNA copy number for any individual brain mRNA is about 0.1 copy per cell (Chikaraishi, 1979; Bantle and Hahn, 1976; Van Ness et al., 1979) when total brain RNA is assayed. Since this class appears to be brain specific, we have characterized two clones—rg13 and rg100—that appear to be typical of this class. The rg13 clone has been most fully characterized and has been shown by transcription mapping to be expressed only in the brain (Figure 5), where it appears to be enriched in neuronal nuclei as opposed to nonastrocytic glial nuclei (M.H. Brilliant et al., in preparation). Clone rg13 encodes RNA that appears to be on brain polysomes (Brilliant et al., 1984), though more of the sequences seem to be restricted to the nucleus. This is in contrast to most mRNAs, in which most of the homologous RNA is messenger, with only a few percent of the sequences residing in the nucleus. When polysomes are fractionated into membrane-bound or free forms, rg13 sequences are associated with membrane-bound and free polysomes, suggesting that the putative rg13 product may be a membrane or secreted protein (M.H. Brilliant et al., in preparation). The sequence of the clone contains one open reading frame that translates into a peptide that is not found in the current protein sequence banks. However, the original genomic clone is only 75 nucleotides, so that an open reading frame over such a short sequence could be fortuitous. Unfortunately, we do not know the size of the rg13

RNA because of its low abundance. We are in the process of selecting longer rg13 clones to evaluate putative protein products. In addition, we are beginning to perform in situ hybridization to localize the site of expression of these rare transcripts in the central nervous system.

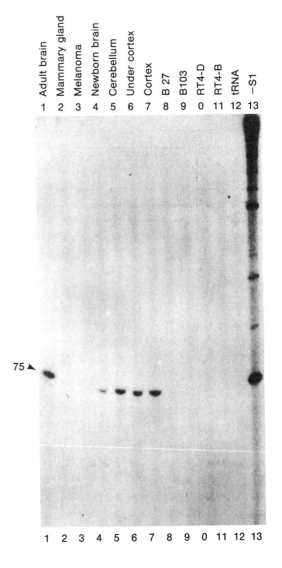

FIGURE 5. Transcription mapping of clone rg13. The 75-nucleotide rat insert fragment of clone rg13 (indicated by arrow) is protected in this assay by RNA from whole adult brain (lane 1), newborn brain (lane 4), cerebellum (lane 5), regions underlying the cortex (lane 6), and cortex (lane 7) samples. Lane 13, pattern obtained without S1 digestion. The protected bands are all 75 nucleotides (the gel has "smiled"). Exposure was for 5 days. Labeled Sau 3A fragments of clone rg13 were annealed to 50 or 100 µg of total cellular RNA from the indicated sources. The hybrids were digested with S1 nuclease, separated by electrophoresis on denaturing polyacrylamide-urea gels and visualized by autoradiography.

Two Poly A⁻ Clones Expressed at Low Levels as Poly A⁺ RNAs

We originally showed that rg13 transcripts were nonpolyadenylated by preparing poly A$^+$ and poly A$^-$ RNA from equivalent amounts of tissue and hybridizing them to rg13 probes. The only detectable annealing occurred with the poly A$^-$ fraction (Brilliant et al., 1984). By this criteria two clones, rg13 and rg100, were classified as encoding brain-specific poly A$^-$ RNAs. However, when 20-fold more poly A$^+$ RNA is annealed to these probes, a clear signal is seen in the poly A$^+$ fraction. Figure 6 shows the results of annealing rg13 sequences to 2 μg and 20 μg of poly A$^+$ of adult brain RNA and 50 μg of poly A$^-$ RNA from the same preparation. Since poly A$^+$ RNA is about 2% of the total RNA, 20 μg of poly A$^+$ RNA is equivalent to 1 mg of poly A$^-$ RNA, which should give 20 times the signal seen in lane 3. The other well-characterized clone, rg100, gives the same results, namely, that about 90% of the homologous sequences fractionate as poly A$^-$ and about 10% as poly A$^+$. The initial RNA complexity measurements would not have detected these sequences in the poly A$^+$ fraction, since tenfold more RNA (or a 10 times higher rot value) would have had to have been used.

The finding that a small proportion of transcripts originally found in the poly A$^-$

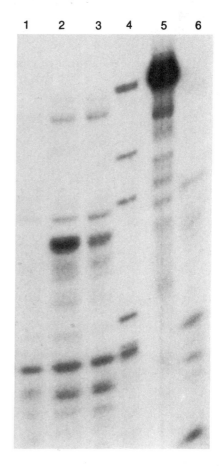

FIGURE 6. Distribution of rg13 transcripts in poly A$^+$ or poly A$^-$ RNA from adult rat brain. Lane 1, rg13 probe hybridized with 2 μg poly A$^+$ RNA; lane 2, rg13 hybridized with 20 μg poly A$^+$ RNA; lane 3, rg13 hybridized with 50 μg poly A$^-$ RNA; lane 4, markers (SV40 DNA cut with EcoRII); lane 5, sample with no nuclease digestion, showing the original size of the run-off transcript; lane 6, rg13 hybridized with 10 μg tRNA. Total RNA from adult rat brain was prepared using 3M LiCl/6M urea/8 mM vanadyl ribonucleoside as described by Auffray and Rougeon (1980) and separated into poly A$^+$ and poly A$^-$ fractions by two passages over oligo dT cellulose. The antisense strand of rg13 was transcribed to high-specific activity with ^{32}P-UTP (2×10^9 cpm/μg) by run-off from rg13 cloned into "riboprobe" vector pSP65 (Melton et al., 1984). The labeled probe was gel purified and used in nuclease protection assays to determine the amount of rg13 message present in various RNA preparations. Liquid hybridization of RNA and probe (35,000 cpm) was performed in 20 μl of 72% formamide/40 mM PIPES pH 6.8/0.4 M NaCl/1 mM EDTA at 50°C overnight after boiling for 3 minutes. The unhybridized sequences were removed by pancreatic RNaseA digestion (40 μg/ml), and the protected fragments were denatured and electrophoresed on 8% polyacrylamide gels containing 8 M urea.

fraction are also in poly A$^+$ RNA opens a number of interesting possibilities. Conceivably, the proportion of poly A$^-$ to poly A$^+$ molecules is regulated by a difference in splicing pattern among the two populations. In addition, different areas in the brain may preferentially express the sequence as poly A$^+$ or poly A$^-$. Alternatively, the subcellular location may be regulated such that the poly A$^-$ species are primarily nuclear, while the poly A$^+$ species are mainly cytoplasmic.

One possibility is that the genomic clones that we have described as poly A$^-$ are actually sequences very near the 5$'$ end of very long poly A$^+$ mRNAs. Under our standard isolation conditions, brain poly A$^+$ mRNAs 4 to 5 KB in length remain intact. However, if the brain contains unusually long messenger transcripts (perhaps on the order of 10 to 20 KB), it is possible that these RNAs are always nicked during isolation and the 5$'$ most sequences are separated from the 3$'$ poly A tract. They therefore lack poly A$^+$ tails and fractionate as poly A$^-$ on oligo dT cellulose affinity columns. The small fraction of poly A$^+$ species would then reflect the proportion of intact molecules that survived isolation. In this regard, Milner and Sutcliffe (1983) have suggested that less abundant brain transcripts tend to be longer than those that are more abundant and common to other tissues. However, because of their scarcity, RNAs constituting the rare transcript class in brain tissue have not yet been measured. Since expression of complex poly A$^-$ sequences occurs only in brain tissue, these long transcripts would be a unique characteristic of postnatal neural transcription. The individual clones designated as poly A$^-$ are not expressed in either the poly A$^+$ or the poly A$^-$ fraction of liver, kidney, melanoma, or mammary gland RNA. Because many neural genes have been shown to use alternative splicing to generate different polypeptides (Amara et al., 1984 [calcitonin/cGRP]; Nawa et al., 1983 [substance P/K]; de Ferra et al., 1985 [myelin];

Tsou et al., 1986; Spindel et al., 1986 [gastrin]), perhaps these long brain-specific transcripts represent an extreme example by which alternative processing generates different forms of 3$'$ ends.

Summary

A number of different approaches have been used to estimate the number of genes expressed in neural tissues. The early population studies using saturation hybridization (Bantle and Hahn, 1976; Chikaraishi et al., 1978; Chikaraishi, 1979; Van Ness et al., 1979; Coleman et al., 1980; Grouse et al., 1978) or cDNA hybridization (Ryffel and McCarthy, 1975; Young et al., 1976) clearly indicated that the mammalian brain expresses more genes than other tissues. These predictions have been borne out by selection of brain-specific clones for the more abundantly expressed genes (Milner and Sutcliffe, 1983; Hahn et al., 1983; see Chapter 6) as well as for the rare class genes (Brilliant et al., 1984). Because of the vast number of different cell types within the brain, it is not surprising that neural tissues might require more gene products than tissues composed of few cell types and that there are cell-specific gene products that distinguish one neuron from another, as assayed both with antibodies and with cloned sequences.

Perhaps even more intriguing is that there may be mechanisms of gene regulation, such as post-transcriptional processing via alternative splicing or 3$'$ end selection, that are highly specialized in neurons. Some hypotheses postulate that a DNA replication is required to reshuffle chromatin in order to transcriptionally activate or repress genes. However, since many important events in neural development occur after neurons have ceased mitosis, the possibility for a terminal differentiating division is lost. Therefore, neural cells may express in hnRNA all the genes that they may ever require and sort out by RNA processing those mRNAs necessary for their final phenotype. In addition, multiple proteins can be generated from a single primary RNA transcript, greatly increasing the repertoire of potential gene products encoded by a fixed amount of DNA.

How these mechanisms are regulated is not known. Whether the 45-KD protein involved in transmitter switching (Fukada, 1985) acts to alter gene expression or RNA processing is an issue that may be approached using molecular biological techniques. However, elucidation of the function of the large number of putative neural gene products is what is ultimately important and will rely on a multidisciplinary approach from many aspects of neurobiology.

CHAPTER FIVE

Neuronal Nicotinic Acetylcholine Receptors

JIM PATRICK, JIM BOULTER,
DANIEL GOLDMAN, PAUL GARDNER,
AND STEVE HEINEMANN

Introduction

It seems very likely that the formation and maturation of proper synaptic connections is crucial to the functioning of the nervous system. This is clearly the case in the peripheral nervous system, where the lack of synaptogenesis obviously leaves endorgans such as muscle without a source of activation. The case is clear but less obvious in the central nervous system, where the failure to form the proper synaptic connections during development may forever preclude certain afferent sensory pathways (for reviews, see Wiesel and Hubel, 1983; LeVay et al., 1980; Changeux and Danchin, 1976; Purves and Lichtman, 1980). It has also recently become apparent that a functional maturation of existing synapses may account for the ability of the central nervous system to recruit neurons to new functions (Merzenich et al., 1984). Likewise, the potentiation of certain pathways through repeated use may be a consequence of a use-dependent modification or maturation of ex-

isting synapses (Levy, 1985). If we are to understand the nervous system, we must understand synaptic transmission and how the developmental process and sensory experience control the formation and maturation of synapses.

The majority of the efforts to understand synaptic function have centered on the neuromuscular junction. Several decades of elegant physiological studies (for reviews, see Katz, 1966; Purves and Lichtman, 1985; Hille, 1984) have provided a good description of how an action potential in a motorneuron is converted to an action potential in a muscle fiber. The release of acetylcholine by the motorneuron and its reception by the acetylcholine receptor in the muscle membrane has been described in detail. We know that acetylcholine is packaged in vesicles and released as quanta following the voltage-dependent influx of calcium into the nerve terminal. Although the biochemical mechanisms are not yet known, the various channels and vesicle release and uptake elements are being purified and may one day be reconstituted in vivo. The acetylcholine receptor from muscle has been purified and its primary structure is known (for reviews, see Stroud and Finer-Moore, 1985; McCarthy et al., 1986; Popot and Changeux, 1984; Conti-Tronconi and Raftery, 1982; Karlin, 1980). It is still not clear, however, exactly what portions of the receptor form the channel, which portions bind ligand, or how binding of acetylcholine leads to the formation of an ion channel. However, we can now see ways to answer the important outstanding questions about synaptic transmission at the neuromuscular junction. Part of the success of these studies of synaptic transmission has been due to the development of techniques, theories, and instruments for the isolation and dissection of the various components of the neuromuscular junction. It seems quite likely that much of the progress made at the neuromuscular junction will be applicable to other synapses, including those in the central nervous system. It is the case, however, that our current appreciation of synaptic mechanisms in the central nervous system lags behind that of the neuromuscular junction. This difference is probably a consequence of the relative ease of studying the neuromuscular junction as compared to the central nervous system.

It is also the case that our understanding of synaptogenesis at the neuromuscular junction is more detailed than its counterpart in the central nervous system. Many studies have documented the ability of the innervating motorneuron to determine the biochemical properties of the muscle fiber, including the amount, distribution, and properties of the acetylcholine receptor (for review, see Cold Spring Harbor Symposium on Quantitative Biology, Vol. 48). These studies have made it clear that innervation is much more than the simple establishment of a point of conjunction of nerve and muscle. Both the biochemical interaction of the nerve and muscle and the electrical activity induced in the muscle by the innervating neuron play major roles in the establishment of a mature functional synapse. We suspect it is the case but we do not know to what extent either of these events control the properties of the postsynaptic cell in the central nervous system. We have even less information about the molecular mechanisms that might be responsible for central nervous system events such as long-term potentiation or synaptic plasticity. Although insights gained at the neuromuscular junction may well be applicable to problems in the central nervous system, there may also be important additional mechanisms of which we are not now aware.

We began by directing our attention to the nicotinic acetylcholine receptor found at the neuromuscular junction. The goals of this work were to determine the structure of a neurotransmitter receptor and to learn how binding of a neurotransmitter to this structure led to an ionic current. We also wanted to understand how innervation of muscle regulates the distribution, synthesis, and properties of the receptor protein. Research which began by studying the biochemical properties of the purified receptor is now conducted on recombinant DNA clones containing sequences coding for the four polypeptides that comprise the mature receptor oligomer. Early programs that studied the effect of denervation on the properties and distribution of receptor relied heavily upon the use of elapid neurotoxins to detect the receptor. The availability of appropriate recombinant DNA clones now allows detection of gene transcripts and will probably soon lead to the identification and purification of the genomic sequences that are involved in regulation of expression of the receptor coding genes. These powerful new approaches should provide us with a very sophisticated understanding of the receptor and its regulation by innervation.

Attention was focused on the nicotinic acetylcholine receptor at the neuromuscular junction in part because a good ligand, α-bungarotoxin, was available for its study, in part because the *Torpedo* electric organ provided quantities of the protein suitable for biochemical studies, and in part because the neuromuscular junction is well suited for developmental studies. We now know a great deal about the structure, function, and regulation of this receptor at the neuromuscular junction, and the new studies using recombinant DNA technology promise to teach us much more. We have less information about the structure, function, and regulation of the receptor at other types of synapses. We know that nicotinic acetylcholine receptors are present in the postsynaptic membrane of synapses between neurons, but we do not know how many types of neuronal nicotinic receptors there are, whether they differ dramatically in their pharmacology, or whether they gate channels of different ion selectivities. There may also be nicotinic acetylcholine receptors in the presynaptic membrane of neuron–neuron synapses. Without insight into these basic questions, it will be difficult to study neural regulation of receptor properties during neuron–neuron synaptogenesis or maturation. Initial attempts to move from the neuromuscular junction to neuron–neuron synapses depended upon α-bungarotoxin to identify nicotinic acetylcholine receptors in ganglia and the central nervous system. This approach was pursued less avidly when it was found that the toxin did not block function of these receptors and in some cases could be shown to bind to molecules other than the receptor molecule (Patrick and Stallcup, 1977a, b; for review, see Oswald and Freeman, 1980). Study of nicotinic receptors on neural tissues are made more complex by the clear presence on neurons of α-bungarotoxin binding components with a pharmacology similar to nicotinic receptors, but with different properties and distributions. There clearly exists a need to understand the diversity, properties, and regulation of the nicotinic acetylcholine receptor and related proteins on neurons.

Muscle Nicotinic Acetylcholine Receptors

Our approach to the general problems of synaptogenesis and synaptic transmission has been through the acetylcholine receptor at the vertebrate neuromuscular junction. This neurotransmitter receptor is an oligomeric glycoprotein composed of four different subunits in the stoichiometry of $\alpha_2\beta\gamma\delta$. The mature receptor is an integral membrane protein in which each of the four subunits spans the postsynaptic membrane. The binding sites for the agonist acetylcholine and the antagonist α-bungarotoxin are found on the α-subunit. Binding of acetylcholine to the α-subunit leads to the formation of a cation-selective ion channel presumably through a conformational change in the receptor oligomer.

The primary structure of the nicotinic acetylcholine receptor made in the electric organ of the ray *Torpedo* has been deduced from the nucleotide sequence of cDNA clones encoding each of the receptor subunits (Claudio et al., 1983; Noda et al., 1982, 1983a, b; Sumikawa et al., 1982; Devillers-Thiery et al., 1983). Analysis of the deduced amino acid sequences led to highly predictive models of the distribution of receptor across the membrane (Claudio et al., 1983; Noda et al., 1982; Devillers-Thiery et al., 1983), of a proposed ion channel (Finer-Moore and Stroud, 1984; Guy, 1984) and of potential agonist binding sites (Noda et al., 1982). The earliest model (Claudio et al., 1983) predicted four hydrophobic membrane-spanning regions per subunit, based upon the presence of four stretches of hydrophobic amino acids in the *Torpedo* γ-subunit. These sequences are found in all four subunits and are highly conserved both between subunits and across species. If these were the only transmembrane sequences and if the amino terminus were on the outside, then the receptor would have 20 membrane-spanning regions, four per subunit, and the five carboxy terminals would be on the outside of the cell. Both Finer-Moore and Stroud (1984) and Guy (1984) identified a sequence, present in each of the four subunits, that is amphipathic. This sequence, located between transmembrane sequences three and four, could form an α-helix in which one side of the helix has charged residues and one side has uncharged residues. They proposed that the uncharged face might interact with the lipid bilayer or with other hydrophobic sequences and that the charged face might, by combining with the charged face of amphipathic helices from other subunits, form an ion channel. In this model the mature receptor oligomer would have 25 transmembrane segments, 20 of which were amphipathic. This model places the five carboxy terminals on the cytoplasmic face of the membrane. The location of the carboxy terminus has been addressed by immunological techniques that show that antibodies raised against a carboxy-terminal sequence bind to the cytoplasmic face of the membrane (Ratnam and Lindstrom, 1984; Young et al., 1985). Likewise, antibodies that bind to a peptide corresponding to a sequence in the amphipathic helix seem to bind to the cytoplasmic side of membrane vesicles (Ratnam et al., 1986). A more recent model has added two membrane-spanning regions between the proposed transmembrane segment one and the amino terminus, and disposed of the original proposed fourth transmembrane sequence (Ratnam et al., 1986). Given the current uncertainty about which sequences actually span the membrane, it is not surprising that the sequences lining the ion channel remain to be identified.

The α-subunit was first identified as the ligand-binding subunit on the basis of labeling a reactive cysteine with an affinity-labeling reagent (Karlin and Cowburn, 1973). This suggested that there was a cysteine at or near the acetylcholine binding site. The first published sequence of an α-subunit revealed four cysteines in the proposed extracellular domain of the protein: two at positions 128 and 142 and two contiguous cysteines at positions 192 and 193. It was proposed that the cysteines at positions 128 and 142 participate in the disulfide bond near the acetylcholine binding site and that the sequence between 128 and 142 could be folded to accommodate a molecule of acetylcholine (Noda et al., 1982; Luyten, 1986). It was later proposed that the contiguous cysteines at positions 192 and 193 form disulfides with cysteines 128 and 142 and in so doing form two loops in the extracellular domain of the protein (Boulter et al., 1985). However, recent evidence shows that cysteines 192 and 193 form a disulfide between themselves (Karlin et al., 1986) and that the affinity-labeling reagent forms covalent bonds with cysteines 192 and 193 (Kao et al., 1984), which suggests that the contiguous cysteines at 192 and 193 are near the acetylcholine binding site and not obligatorily linked to the sequences between cysteines 128 and 142.

We have focused our attention not on the *Torpedo* acetylcholine receptor but on the receptor from mouse muscle. The nicotinic acetylcholine receptor from the BC_3H-1 cell line has been well studied physiologically (Sine and Taylor, 1980; Sine and Steinbach, 1985), its synthesis and degradation have been well characterized (Patrick et al., 1977; Merlie and Sebbane, 1981), and the protein has been purified (Boulter and Patrick, 1977). The cells are also a comparatively rich source of the receptor, and since the receptor is turned over rapidly ($\sim t_{1/2} = 8$ hours), we expected the cells would also be a good source for receptor coding mRNA. We have isolated and sequenced cDNA clones coding for the α-, β- and γ-subunits of the BC_3H-1 nicotinic acetylcholine receptor (Boulter et al., 1985, 1986a, b, and in preparation). These clones all code for the entire precursor to the mature protein and have a

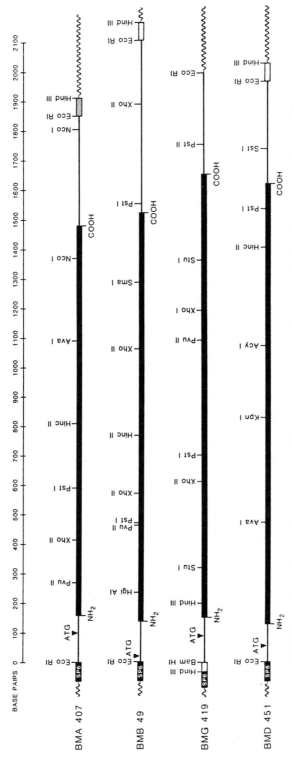

FIGURE 1. Partial restriction maps of cDNA clones coding for the four subunits of mouse muscle nicotinic acetylcholine receptor. The clones are in the vector pSP65 (indicated by wavy lines) downstream of the phage SP6 promoter. The ATG indicates the start of translation at the amino terminus of the leader sequence, and the heavy dark line indicates the portion of the clone coding for the mature protein.

portion of the 5′ untranslated sequences. We also obtained from Dr. Norman Davidson a clone coding for the precursor to the δ subunit (LaPolla et al., 1984), which we used to isolate an additional clone with 5′ untranslated sequences suitable for expression studies.

The clones encoding the mouse muscle nicotinic acetylcholine receptor subunits have been placed behind the SP6 promoter, as shown in Figure 1, and used to transcribe the four mRNA species. These have been injected into *Xenopus* oocytes, where they direct the synthesis of an acetylcholine receptor that can be detected both by measurement of the α-bungarotoxin binding activity of the oocytes and by electrophysiological measurement of the acetylcholine-induced change in conductance.

Neuronal Nicotinic Acetylcholine Receptors

A major goal in obtaining clones coding for the mouse muscle nicotinic receptor was to use these clones to detect, by low-stringency DNA/DNA hybridization, sequences coding for possible neuronal nicotinic acetylcholine receptors. We began this program using the pheochromocytoma (PC12) cell line as a source of neural-type nicotinic acetylcholine receptors (Greene and Tischler, 1976). This cell line has on its surface both a neural-type nicotinic receptor and an α-bungarotoxin binding molecule. Both pharmacological (Patrick and Stallcup, 1977a) and immunological (Patrick and Stallcup, 1977b) studies have shown that these are distinct molecular species. The physiological role of the α-bungarotoxin binding component on these cells is not known. We prepared poly A$^+$ RNA from the PC12 cell line and found that a probe prepared from the mouse muscle α-subunit clone hybridized to two size classes of RNA species, as seen in Figure 2 (Boulter et al., 1986a). The hybrids were not stable at high stringency, suggesting that the bands we detec-

ted represented homologous but not identical sequences.

cDNA was prepared from the PC12 poly A$^+$ RNA and cloned into the vector λgt10, and the resulting library was screened with a probe corresponding to a conserved region of the mouse α-subunit. This screening, which was done at low stringency, should have detected molecules that differed by as much as 30% in their sequences. Three clones were found and purified. Subsequent restriction site mapping suggested that the clones were similar if not identical and the longest, PCA48, was chosen for sequencing.

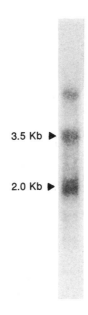

3.5 Kb ▶

2.0 Kb ▶

FIGURE 2. Autoradiograph of Northern blot hybridization using radiolabeled cDNA encoding the mouse muscle α-subunit and poly A$^+$ RNA isolated from the PC12 cell line. Eighteen micrograms of poly A$^+$ RNA were loaded and electrophoresed on a 1% agarose gel. Hybridization was carried out in 5X SSPE (1X SSPE is 180mM NaCl, 9mM Na$_2$HPO$_4$, 0.9mM NaH$_2$PO$_4$, 1mM EDTA pH 7.4), 50% formamide, 1% sodium dodecylsulfate (SDS), and 1X Denhardt's at 42°C, and the blot was washed in 5X SSPE and 0.05% SDS at 65°C.

Clone PCA48 was sequenced and found to encode the precursor to a protein, the sequence of which is shown in Figure 3. We propose, for the following reasons, that the protein is an α-subunit of a neuronal nicotinic acetylcholine receptor. The encoded protein has the four hydrophobic sequences thought to form transmembrane segments in the muscle α-subunit and in this respect is structurally similar to the muscle receptor α-subunit. The

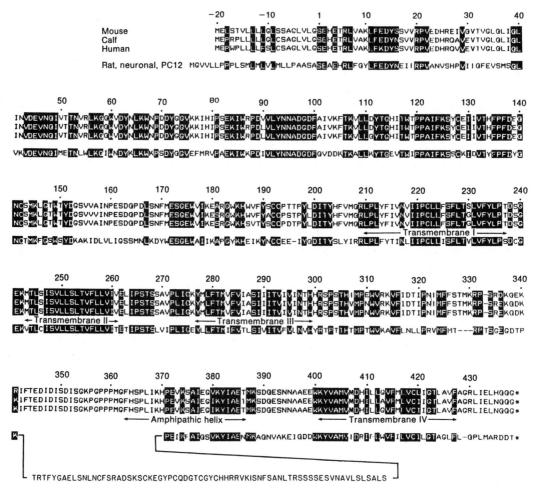

FIGURE 3. Interspecies comparison of amino acid sequences for muscle and nerve acetylcholine receptor α-subunits. Positions in which the residue found in the neuronal α-subunit is found in any of the muscle α-subunits are shown as white letters on a black background. Proposed hydrophobic and amphipathic membrane-spanning regions are indicated. The additional sequences found in the neuronal α-subunit are looped out in the bottom of the figure. The residues between position 359 and 379 in this region are different than those published (Boulter et al., 1986a), because a frameshift was introduced into the sequence due to a sequencing artifact. (The mouse muscle sequence is taken from Boulter et al., 1985, the calf and human sequences from Noda et al., 1983b, and the rat neuronal sequence from Boulter et al., 1986a.)

protein encoded by PCA48 also contains a sequence homologous to the amphipathic helix found in the α-subunits of muscle receptor. Although the sequence homology is not as great as is seen in the hydrophobic membrane-spanning regions, the distribution of charged and uncharged residues about the helix is similar. The glycosylation site found in the proposed extracellular domain of the muscle α-subunit is also found in the PCA48 protein. There is an additional potential glycosylation site in the PCA48 protein at asparagine 24. Finally, the four cysteine residues in the proposed extracellular domain that are conserved in all muscle α-subunits sequenced to date are also found in the PCA48 protein.

While these observations are all consistent with the idea that the PCA48 protein is a neural nicotinic acetylcholine receptor α-subunit, they do not rule out other alternatives. For example, the protein encoded by PCA48 might be a rat muscle α-subunit. However, this possibility was ruled out by the observation that a probe prepared from PCA48 does not hybridize at high stringency to RNA prepared from innervated or denervated rat muscle. Furthermore, rat muscle RNA does not protect PCA48 DNA from digestion by endonuclease S1. In contrast, a probe prepared from PCA48 recognizes RNA species in rat brain and in rat adrenal medulla but not in rat adrenal cortex.

Clone PCA48 was obtained from a library created from poly A$^+$ RNA prepared from the PC12 cell line that has both a nicotinic acetylcholine receptor and an α-bungarotoxin binding component. Although the protein encoded by PCA48 is homologous to the mouse muscle nicotinic receptor α-subunit, it might in fact represent the α-bungarotoxin binding component of this cell line. We tried to test this possibility using in situ hybridization to determine the distribution in mouse and rat brain of the RNA homologous to PCA48 (Goldman et al., 1986). Sections of rat and mouse brain were probed with radiolabeled single-strand RNA

derived from the rat neuronal clone PCA48 and the mouse muscle clone BMA407. Figure 4 shows autoradiograms of mouse brain hybridized with the rat neural probe (A), a mouse muscle probe (B), or a control probe consisting of the message sense strand of the rat neuronal clone (D), and rat brain hybridized with the rat neuronal probe (C).

The results from Figure 4 show that probes prepared from muscle and neuronal clones hybridize to different regions of the brain. The neuronal probe hybridizes most strongly to the medial habenula and to a clear but lesser extent to the ventral tegmental area, substantia nigra pars compacta, anteroventral nucleus of the thalamus, the medial geniculate nucleus, and the neocortex. In contrast, the muscle α-subunit probe shows little hybridization to these areas. The dentate gyrus and hippocampus show a positive signal with both the neural and the muscle probes but also show hybridization with the control probe, corresponding to a sense strand that should not hybridize to the mRNA. The hyridization seen in these areas might therefore represent nonspecific binding.

The hybridization results show that RNA species homologous to the neuronal clone are found in the brain but do not show that the RNA is identical. This question was addressed by S1 analysis of poly A$^+$ RNA isolated from the habenula. This RNA, along with poly A$^+$ RNA isolated from the PC12 cell line, was size fractionated on denaturing formaldehyde agarose gels and transfered to Gene Screen Plus. Transcripts hybridizing to the neuronal probe were found in both RNA preparations. Heteroduplexes formed between the neural α-subunit and RNA isolated from either the PC12 cell line or the habenula afforded complete protection of the cDNA clone to S1 nuclease digestion, as seen in Figure 5. This shows that cells in the medial habenula express the same gene as is expressed in the PC12 cell line and suggests that the hybridiza-

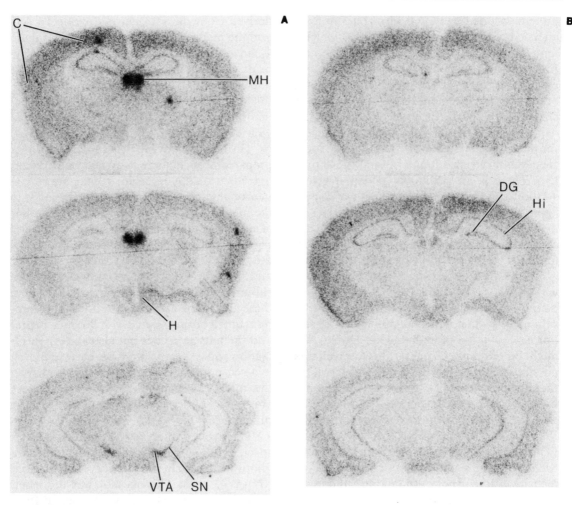

FIGURE 4. In situ hybridization of mouse (A, B, and D) and rat (C) brain sections with a rat neural α-subunit probe, τPCA48 (A and C) and a mouse muscle α-subunit probe, pMARα15 (B). (D) represents in situ hybridization of mouse brain sections with a "sense" strand probe prepared from the rat neural α-subunit clone to detect nonspecific binding. AV, anteroventral nucleus of the thalamus; C, neocortex; DG, dentate gyrus; H, hypothalamus; Hi, hippocampus; MG, medial geniculate nucleus; MH, medial habenula; VTA, ventral tegmental area; SN, substantia nigra pars compacta. Mouse and rat brain sections were exposed to x-ray film for 4 days at room temperature.

tion to this area of the brain is a consequence of this specific RNA species.

The fact that the medial habenula expresses the gene from which the PCA48 neuronal clone was derived suggests that this clone encodes an α-subunit of a nicotinic acetylcholine receptor rather than the α-bungarotoxin binding component present on the PC12 cell line. Pharmacological studies show that the medial habenula binds cholinergic ligands but does not bind α-bungarotoxin (Hunt and Schmidt, 1978; Clarke et al.,

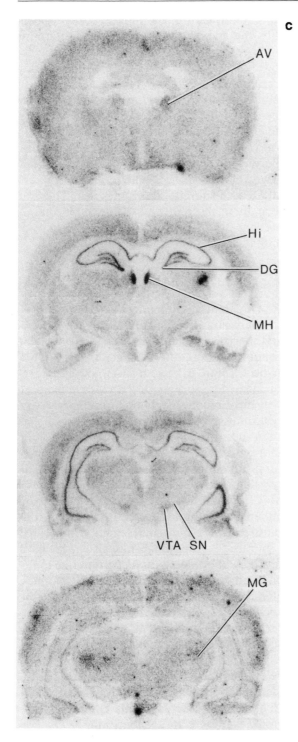

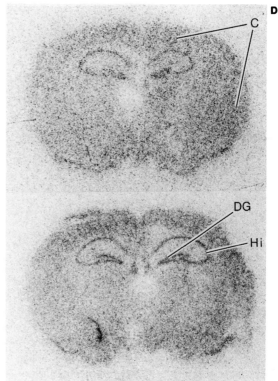

1985). The argument, however, is slightly more complex because cholinergic neurons in the medial habenula project via the fasciculus retroflexus to the interpenduncular nucleus (Herkenham and Nauta, 1977; Gottesfeld and Jacobowitz, 1979), where the amount of α-bungarotoxin binding is controversial (Hunt and Schmidt, 1978; Clarke et al., 1985). Hence the hybridizing RNA species seen in the medial habenula could code for the nicotinic receptor in the habenula or for the α-bungarotoxin binding component thought to be in the interpeduncular nucleus. Antibodies directed against the protein encoded by the PCA48 clone should help distinguish between these possibilities.

The in situ hybridizations, Northern blots, and S1 protection experiments suggest that the PCA48 sequence is expressed in the medial

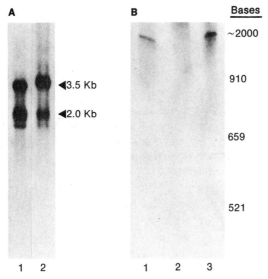

**FIGURE 5. (A) Northern blot analysis. Poly A⁺
RNA from either the PC12 cell line (lane 1) or a
region of the brain containing the habenula (lane 2)
was size fractionated on denaturing agarose gels,
transferred to Gene Screen Plus and hybridized with
[32]P-labeled τPCA48 insert. (B) S1 nuclease analysis.
τPCA48 was subcloned into M13mp18 and
M13mp19. Single-strand DNA was prepared and
used to form heteroduplexes with poly A⁺ RNA
isolated from the PC12 cell line (lane 1) and a region
of the brain containing the habenula (lane 3). Lane 2
represents a control where RNA was omitted. Hybridization reactions were incubated with nuclease S1
and those molecules surviving digestion were fractionated on denaturing acrylamide gels and elec-
troblotted to Gene Screen Plus. Heteroduplexes surviving digestion were visualized by hybridizing with
[32]P-labeled τPCA48. PBR322 restriction fragments
run in a parallel lane served as size markers. Blots
were exposed to x-ray film for 18 hours with intensify-
ing screens at −70°C.**

habenula and that homologous or identical
sequences are expressed in other brain regions. In addition, Southern blots of genomic
DNA suggest that there are several different
homologous sequences in the genome (Boulter et al., 1986a). Since these sorts of analyses

can be misleading, the only sure way to identify the hybridizing species and determine
their relatedness is by cloning. Therefore, we
purified poly A⁺ RNA from specific brain
regions, made cDNA libraries, and isolated
clones hybridizing to either the muscle or the
neuronal α-subunit clones. We have now
found hybridizing homologous but nonidenti-
cal clones in libraries prepared from the hippocampus, the thalamus, and the cerebellum.
These clones define two and possibly four additional α-subunit–like molecules. The role
that these molecules play in synaptic transmission in the central nervous system is
unknown.

Summary

Our approach to understanding synaptic
transmission and synaptogenesis in the central nervous system has been through the
genome. We used low-stringency DNA/DNA
hybridization to detect cDNA and genomic sequences homologous to the clones encoding
the nicotinic acetylcholine receptor of the
neuromuscular junction. This approach has
been successful, in that it has produced a clone
that encodes a protein with considerable structural homology to the muscle α-subunit. The
physiological function of this protein, however, is unknown and awaits its expression
either with the muscle type β-, γ- and δ-sub-
units or with a new complement of neuronal
subunits. The approach has also revealed a
diversity of homologous α-subunit–like molecules that await analysis. Analysis of these
clones will provide the primary structure of the
encoded protein and provide a means of expressing the protein in an environment more
suitable for experimentation than is the brain.

It will be interesting to compare the molecules identified by our approach with those
identified through alternative approaches.
Analysis of chicken genomic clones has suggested that there are sequences in the chicken

genome that encode proteins similar to the chicken muscle α-subunit (Mauron et al., 1985). A protein that binds α-bungarotoxin has been purified from chicken brain (Norman et al., 1982), and the sequence of its 12 terminal amino acids (Conti-Tronconi et al., 1985) suggests that it is structurally related to the mouse muscle α-subunit. A neurotoxin that blocks activation of nicotinic acetylcholine receptors on chicken ciliary ganglia (Ravdin and Berg, 1979) and rat sympathetic ganglia (Loring et al., 1983) has been purified, but the purification of the binding component has not yet been accomplished. Antibodies raised against the muscle-type nicotinic receptor bind to a protein in chick ciliary ganglia (Jacob et al., 1984) and have been used to purify from chicken brain a molecule that binds nicotine but not α-bungarotoxin (Whiting and Lindstrom, 1986). Antibodies against this protein block activation of nicotinic receptors on ciliary ganglia (Stollberg et al., in press). Although it is not clear that the purified protein is a receptor, it is clear that the antibodies recognize a receptor activity of neurons. It seems likely that the several approaches to receptors on neurons will result in the identification of new molecules homologous to the muscle acetylcholine receptor. Whether these molecules function as neurotransmitter receptors in the postsynaptic membrane, as presynaptic receptors, or in some as yet unknown function remains to be seen.

The Neuronal Identifier Sequence as a Positive Regulatory Element for Neuronal Gene Expression

RANDALL D. MCKINNON,
PATRIA DANIELSON, MARY ANN BROW,
MARTIN GODBOUT, JOSEPH B. WATSON,
AND J. GREGOR SUTCLIFFE

Introduction

In 1982, we described experiments demonstrating that a small RNA molecule, subsequently termed BC1, was abundant in extracts of rat brain but undetectable in extracts of liver and kidney (Sutcliffe et al., 1982). We had constructed cDNA libraries of rat brain poly A$^+$ RNA and then screened, by RNA blot analysis, clones randomly selected from the libraries (Milner and Sutcliffe, 1983). We expected that such an analysis would provide an overview of gene expression in the mammalian brain, identify sets of genes whose expression was under coordinate molecular control, and produce clones of many brain-specific mRNAs whose structures could be studied to provide information about brain mRNA generation and the amino acid sequences of brain-specific proteins. From more than 190 cDNA clones screened initially, five had the peculiar property that they shared a common sequence, although each was a copy of an RNA transcribed from a gene distinct from any of the others (Sutcliffe et al., 1982; Milner et al., 1984). We realized that this probably meant that the common sequence was repeated in the genome. Because the common sequence hybridized to a small brain-specific RNA, this genomic DNA element was termed an "identifier sequence" (Sutcliffe et al., 1982).

At that time we posed three questions: were identifier (ID) sequences related either to cis-acting viral transcriptional regulatory elements called enhancers or to highly repeated genomic DNA sequences such as human Alu elements; was the brain-specific BC1 RNA a primary transcription product or a processing vestige; and did the BC1 RNA itself have a brain-specific molecular function (Sutcliffe et al., 1982)? A large number of subsequent experiments in this and other laboratories have greatly increased our knowledge about the structure and expression of BC1 RNA, and the structure, distribution, and possible function of genomic ID elements. These experiments indicate that ID elements apparently have spread by transposition through the genome and are found preferentially, but not exclusively, in the introns of a large set of genes whose expression is highly enriched postnatally in cortical neurons. The ID elements are transcribed by RNA polymerase III postnatally in cells enriched in cerebral cortex to produce BC1 and a related RNA, BC2. ID elements exhibit the properties of positive transcriptional regulatory elements. These observations have led to a model that proposes that ID elements regulate a battery of neuronal genes during postnatal brain development. They have also led to some controversy as to whether their distribution within the genome is consistent with such a model. In this chapter we assemble the data that have emerged about ID elements and BC1 RNA, evaluate the evidence for and against the hypothesis that ID elements regulate gene expression, and raise questions whose answers would provide considerable insight into this problem.

Postnatal Developmental Appearance of Small Brain-Specific RNAs

Initially, cDNA clones containing ID elements were found to hybridize in Northern blot experiments to an alkali-sensitive molecule with slightly disperse 160-nucleotide gel mobility that was present in brain extracts but not detectable in extracts of liver or kidney (Sutcliffe et al., 1982). Subsequent analyses on high-resolution gels revealed a second small disperse brain RNA of about 110 nucleotides (Sutcliffe et al., 1984a). These two species were termed BC1 and BC2, respectively, because they were enriched in brain cytoplasmic RNA extracts compared to nuclear RNA extracts. Both BC1 and BC2 RNA species were retained on oligo dT cellulose columns, indicating that each contained runs of oligo A. A blot hybridization survey of RNAs extracted from a large number of organs revealed that BC1 RNAs were detectable in extracts of brain, pituitary, and a peripheral nervous ganglion (solar plexus), but not in several other tissues (adrenal, spleen, testis, lung, heart, muscle, gut, and thymus), which suggests that these small RNAs are neuronal specific. Subsequent analyses utilizing more sensitive primer extension assays indicated that BC2 was present in some non-neuronal peripheral tissues at very low levels relative to brain concentrations (McKinnon et al., in press). BC1 has not been detected in these non-neuronal organs. In more limited studies, several groups have confirmed the brain specificity of the BC RNAs (Owens et al., 1985; Sapienza and St. Jacques, 1986; Lone et al., 1986).

Within the brain, the concentration of BC1 varies from region to region. For example, its concentration in the cerebral cortex is several-fold higher than its concentration in the cerebellum (Sutcliffe et al., 1982). During development, BC1 and BC2 RNAs first become detectable in the brains of 18- to 20-day-old rat embryos. Their concentration dramatically increases during the first week of postnatal development, achieving adult levels by postnatal day 10 (Sutcliffe et al., 1984b). Despite this apparently tight developmental and anatomical regulation observed in differentiated cells in vivo, both BC1 and BC2 RNAs are readily detected in established rat cell culture lines of both neuronal and non-neuronal origins (McKinnon et al., 1986a and in press). When rat kidney cells are explanted, BC RNAs appear after a week in culture. Thus, cell culture results in aberrant expression of these RNAs. Because their expression in vivo is highly specific, at least to the extent we have checked, their expression in cultured cells should not be taken as lack of tissue specificity but rather as evidence that regulation differs between cells in vivo and cells in culture.

Small brain-specific cytoplasmic polyadenylated RNA molecules are found in other mammalian species. There are two mouse analogues of sizes comparable to BC1 and BC2 (Owens et al., 1985; Sapienza and St. Jacques, 1986; McKinnon et al., in press), a single hamster analogue with apparent size of 140 nucleotides (Sapienza and St. Jacques, 1986; McKinnon et al., in press), and a monkey analogue of 200 nucleotides (McKinnon et al., in press; Watson and Sutcliffe, in preparation). Detection of these RNAs using the rat ID sequence as a probe is improved when the stringency of RNA hybridization is reduced, indicating that their sequences are related but not identical to the rat RNA sequences. The mouse sequence is closest to that of the rat, whereas the monkey sequence is the most distant. The mouse, hamster, and monkey RNAs are expressed in established cultured cell lines of non-neuronal origins from these three species (McKinnon et al., 1986a and in preparation; Watson et al., in preparation). Thus, the generation of such BC RNAs is a phylogenetically conserved process, although neither the sequence nor the size of these RNAs has been highly conserved.

Structure of BC RNAs

The primary sequences of rat BC1 and BC2 RNAs have been established by DNA sequence analyses performed on several types of molecules. In the initial experiments (Sutcliffe et al., 1982; Milner et al., 1984), we used cDNA clones of large brain RNA molecules. These clones were cDNA copies of different RNAs but shared a common region—the ID sequence. This sequence can be defined by a "core" region consisting of 75 nucleotides and

an oligo A-rich tract that can exceed 20 nucleotides. Rat genomic clones from several origins have been found to contain ID elements (Barta et al., 1981; Lemischka and Sharp, 1982; Dhar et al., 1982; Watanabe-Nagasu et al., 1983; Schuler et al., 1983; Lone et al., 1986), and a comparison of these has led to a "consensus" ID sequence (Milner et al., 1984). The structures of the small RNAs themselves have been deduced from 20 cDNA clones of these small RNAs and sequence analysis of primary reverse transcripts of BC1 and BC2 (Brow et al., in preparation).

Because the BC RNAs were retained on oligo dT cellulose columns, we suspected that synthetic DNA oligonucleotides of the sequence $d(T_{12}VX)$—where V is A, G, or C and X is A, G, C, or T—would serve as primers that could be extended by the enzyme reverse transcriptase to form cDNAs complementary to BC1 and BC2 RNAs. Several oligonucleotides from this collection were found to prime the synthesis of small brain-specific cDNAs (Brow et al., in preparation). Size fractionation of the template brain RNA revealed which cDNA molecules were copies of BC1 and which of BC2. Both BC1- and BC2-derived cDNAs migrated on high-resolution gels as families of several distinct nucleotide lengths. Several of these cDNA molecules were eluted and their sequences determined. The sequences of the BC2 cDNAs using the primer $d(T_{12}GG)$ indicate that some individual BC2 species contain the ID "core" sequence at its 5′ end (beginning ± 1 nucleotide from the beginning of the consensus core), followed by an A tract beginning with the sequence $A_7 GAACCA_{12}$. The sequence of the extreme 3′ end of BC2 is not yet known. The sequences of the BC1 cDNAs show that individual BC1 species contain the ID "core" sequence, followed by a tract of A residues that varies in length from 20 to 30 nucleotides, followed by the sequence $GACA_7CA_5GACCA_{12}$ (Brow et al., in preparation). The sequence of the extreme 3′ end of BC1 is also not yet known. Thus, BC1 is probably encoded in the genome rather than being simply a polyadenylated version of BC2, and each BC1 gene can vary with respect to the length of its internal oligo A region. Since several different oligonucleotides of the $d(T_{12}VX)$ set function as primers, the BC RNAs exhibit at least some sequence heterogeneity. Many cDNA clones of these molecules have been isolated and their sequences determined. Most of these sequences differ from one another, and from the consensus sequence arrived at earlier, by one or a few nucleotides. The set of sequences has not yet begun to converge; that is, each new sequence determined is more often a unique sequence than a duplicate of a previously determined sequence. Thus the picture emerges that the BC1 and BC2 RNAs are products of many separate genes. For BC2 the number of genes is very large, and for BC1 there are at least several.

BC RNAs Are RNA Polymerase III Transcripts

Examination of the ID core sequence revealed regions with similarities to the so-called A- and B-box consensus regions of RNA polymerase III (Pol III) promoters (Lemischka and Sharp, 1982; Milner et al., 1984). Evidence that the small BC RNAs are produced by brain-specific transcription of genomic ID elements by Pol III comes from three experiments. First, cDNA and genomic clones of ID elements are potent templates (about 20-fold stronger than tRNA genes) for Pol III transcription in vitro in HeLa S100 extracts (Sutcliffe et al., 1984a; Gutierrez-Hartmann et al., 1984; Sakamoto and Okada, 1985a). Second, in nuclear run-on experiments, these small RNAs are synthesized in brain nuclei by Pol III, based on their resistance to low levels of the Pol II inhibitor α-amanitin (Sutcliffe et al., 1984a). Third, both the primer extension experiments and the nucleic acid sequence

analysis of cDNA clones of BC RNAs discussed above indicate that the 5′ ends of the brain-specific molecules observed in vivo are the same as the mapped 5′ ends of the products of transcription of ID clones by Pol III in vitro. Thus these small brain RNAs are direct transcriptional products of the genome rather than products of RNA processing of Pol II products.

The concentrations of BC1 and BC2 in brain extracts have been estimated from both the strength of hybridization reactions and the frequency of cDNA copies in libraries to be about 2% of the poly A^+ RNA (which itself is 1.5 to 2% of the total brain cytoplasmic RNA). In nuclear run-on experiments with exogenous Pol III provided by HeLa extracts, BC2 is a major transcription product of brain nuclei, whereas BC1 is only faintly visible. Because of its prevalence in run-on experiments but lower concentration in steady-state blotting experiments, it is likely that BC2 is fairly unstable in vivo, relative to BC1. Also, because BC1 and BC2 are transcribed from different genes, there are probably many fewer BC1 than BC2 genes.

Primate BC RNAs

Poly A^+ RNA isolated from monkey brain extracts, but not RNA isolated from liver extracts, hybridizes at low (but not at high) stringency to the rat ID element. Primer extensions utilizing the $d(T_{12}VX)$ primer set demonstrated that several primers gave rise to small cDNAs. The primer that worked with highest efficiency was used to examine RNAs from brain and liver and was found to prime synthesis from brain poly A^+ RNA but not from brain poly A^- RNA or from any liver RNA sample. The primer extended on brain poly A^+ RNA was isolated and subjected to DNA sequence analysis. The small monkey brain-specific poly A^+ RNA shares slightly less than 60% sequence identity with the rat BC RNAs, and much of the sequence correspondence is in the putative A-box and B-box structures required for Pol III transcriptional initiation (Watson and Sutcliffe, in preparation). The monkey sequence resembles, with 77% sequence identity, the left module of the highly repeated human *Alu* element. Several non-neuronal monkey tissues and human brain have been examined, and it appears that in primates a brain-specific analogue of the rat BC RNA is expressed from a repetitive genomic sequence by Pol III. However, when compared to the rat sequence, its primary sequence has greatly diverged.

A natural conclusion from the data discussed thus far is that the brain-specific Pol III transcripts perform some brain-specific molecular function, perhaps much as the ubiquitous 7SL RNA serves as a component of the signal recognition particle in the translation of mRNAs for secretory and membrane proteins (Walter and Blobel, 1982). The rat BC RNAs show distant sequence similarities to some tRNA molecules, suggesting that they might have evolved from a tRNA, although it is not clear which one (Gutierrez-Hartmann et al., 1984; Lawrence et al., 1985; Daniels and Deininger, 1985; Sakamoto and Okada, 1985b). However, unlike tRNA molecules, the BC RNAs have long stretches of oligo A and probably do not have CCA amino acid receptors. In addition, these RNAs are not found associated with polysomes (J. Brosius, personal communication). Thus we believe it unlikely that the BC RNAs have any tRNA function. To date we have been unable to find the rat BC RNAs associated with any large ribonuclear protein particles, and their lack of sequence conservation between mammalian species, as well as their high degree of sequence microheterogeneity within the rat, suggests that the BC RNAs might be nonfunctional. Nonetheless, the tissue specificity of their transcription is striking and occurs in different mammalian species. Thus we entertained the hypothesis that Pol III transcription of ID elements might be of functional importance, but that the transcription products themselves as stable cytoplasmic RNAs might be functionless. In other words, the ID elements might have *cis* functions mediated by Pol III.

ID Elements Function as Enhancers

Because it was suspected that ID elements might have *cis*-acting genetic regulatory functions, they were tested for their influence on Pol II transcription in both stable and transient gene expression assays by constructing recombinant plasmids carrying ID elements or control sequences linked at various positions and in both orientations relative to the reporter genes *neo* (neomycin resistance) or *CAT* (chloramphenicol acetyltransferase). The recombinant and control plasmids were used to transfect a variety of cultured rodent cell lines (McKinnon et al., 1986). Relative to controls, the ID element increased the expression of the adjacent reporter gene in both long-term and transient assays. It functioned in all three tested positions and in both alternative orientations. Its effect on gene expression did not multiply the effect of a known transcriptional enchancer, that of virus SV40. Thus the ID element appears to be a *cis*-acting positive regulator of transcriptional initiation.

The ID element functioned as a transcriptional enhancer in all established rodent cell lines tested, and each of these lines constitutively expressed high levels of the BC Pol III RNAs. ID has also been observed to function as a transcriptional enhancer in transformed primate cell lines (D. Vacante and E. Major, personal communication). However, primary rat kidney cultures do not express high levels of these BC RNAs, and in primary kidney cells the ID element does not have enhancer properties (McKinnon et al., 1986). Similar results were obtained with primary human glial cultures (D. Vacante and E.

Major, personal communication). Thus, Pol III transcription of ID elements may be required for them to influence Pol II gene expression. Bak and Jørgensen (1984) have shown that several enhancer elements share primary sequence similarities with Pol III transcription units.

ID Sequences Are Dispersed, Repetitive Retroposons

Given that the BC RNAs are primary transcription products exhibiting a high degree of sequence microheterogeneity and that ID elements have been found within many cDNA and genomic clones, it was clear that IDs must be repeated many times in the rat genome. A rat genomic library was screened with an ID probe and 43% of the clones with inserts that averaged 15 Kb contained at least one ID element (Milner et al., 1984; Sapienza and St. Jacques, 1986). This indicated that there are between 1 to 1.5×10^5 ID elements per haploid genome. Furthermore, because such a large percentage of genomic clones contain an ID element, this repetitive species is highly dispersed through the genome. While it is impossible with the present data to discern if the element is uniformly spread through the genome, it is clear that it is not restricted to a small subset of genes.

The sequences of several genomic regions containing ID elements have been determined (Barta et al., 1981; Lemischka and Sharp, 1982; Dhar et al., 1982; Watanabe-Nagasu et al., 1983; Schuler et al., 1983; Lone et al., 1986), including the sequence of the gene for neuronal membrane glycoprotein 1B236 (Sutcliffe et al., 1983; Milner et al., 1984). In each case the ID elements are flanked by direct DNA repeats of 7 to 15 nucleotides. Such structures suggest that at least most of the ID elements in the rat genome have spread through the process of retroposition—reverse transcription of Pol III transcripts of ID elements and integration into

the germline DNA (Rogers, 1983, 1985a,b; Weiner et al., 1986). The microheterogeneity of the sequences of the BC RNAs suggests that a large portion of these retroposed DNA copies are transcribed. The monkey analogue of the rat BC RNAs shows homology to human *Alu* elements. These elements are known to be highly repeated and dispersed in primate genomes and are flanked by direct repeats, indicating that these elements have also spread by retroposition (Rogers, 1983, 1985a,b; Weiner et al., 1986).

ID Elements Are Enriched in Cortical Neuronal Genes Expressed Postnatally

If ID elements influence Pol II gene expression in vivo and their effect is mediated by Pol III, then both the temporal and the spatial expression of BC RNAs should correlate with the expression of genes controlled by ID elements. In our original characterization of ID elements (Sutcliffe et al., 1982) we noted that they occupied positions within introns and thus that they are contained as passengers within unspliced hnRNAs found in the nucleus. Subsequent analysis of nuclear run-on transcripts from brain, liver, and kidney (Sutcliffe et al., 1984a) showed that ID elements were enriched in brain nuclear RNAs compared to nuclear RNAs from other tissues. However, quantitation of these differences was difficult because of the mixture of both Pol II and Pol III transcription in the assay. We concluded that ID elements were transcribed specifically in brain by both Pol II and Pol III and, because of the probable neuronal specificity of the Pol III transcription, suggested that ID elements were preferentially located on postnatal-onset neuron-specific genes (Sutcliffe et al., 1984a,b).

Several authors (Milner et al., 1984; Whitney and Fuarano, 1984; Owens et al., 1985; Sapienza and St. Jacques, 1986) have observed that ID elements are located within introns of some genes that are not neuron specific. This might not be surprising, given that there are more than 10^5 elements that hybridize to ID probes in the genome. Thus, there is not a one-to-one correspondence between ID elements and brain-specific genes, as was discussed at length by Milner and colleagues (1984). However, for ID elements to merit serious consideration as possible regulatory elements for postnatal-onset neuron-specific genes, the locations of many ID elements should correlate with such genes. A few examples of postnatal-onset neuronal genes containing ID elements are known, such as neuron-specific enolase and 1B236.

To extend the results of the nuclear run-on experiments discussed above, we prepared nuclear RNA from brain, liver, and kidney of 16- to 25-day-old rats. We also performed dissections and purified nuclear RNA from cortex and cerebellum. To determine whether ID expression in Pol II hnRNAs was neuronal or glial, we separated nuclei of cortical neurons and cortical glia by sedimentation through dense sucrose and isolated their RNA. We then displayed the nuclear RNA samples from each preparation on gels and determined by blot hybridization what the relative content of ID elements was in the various samples (Brown et al., in press). Equal gel loading was controlled for by hybridization with probes for ribosomal RNA and oligo dT. Liver, kidney, whole-brain, cerebellum, and cortical glial samples had similar contents of ID elements in their nuclear hnRNA. These were similar results to those found by Owens et al. (1985) and Sapienza and St. Jacques (1986). However, cortex nuclear RNA was enriched about 7-fold over the other samples, and cortical neuronal hnRNA was enriched 14-fold. Because equal masses of RNA were analyzed in each sample, the greater complexity of brain RNA was not a contributing factor. Thus, the nuclear run-on experiments somewhat underestimate the relative concentration of ID elements in steady-state hnRNA isolated from peripheral

tissues, possibly because neuronal nuclei are much more active transcriptionally than glial nuclei (Thomas and Thompson, 1977). Nevertheless, cortical neuronal hnRNA is significantly enriched in these elements.

Nuclear RNA was isolated from cortical neurons of rats at various stages of early postnatal development and assayed for its content of ID elements. Cortical neuronal nuclear RNA isolated during the first postnatal week contains ID levels comparable to glial or nonbrain samples. However, during the second week, the ID content in hnRNA from cortical neurons increases 11-fold, while the content in nuclear RNA from glia or nonbrain tissues remains constant. Thus, ID elements are preferentially, but not exclusively, located in the introns of genes whose expression in cortical neurons increases dramatically over other cell types, including glia, during the second week of postnatal development (Brown et al., in press). Hence, the Pol II expression of genes that contain ID elements follows developmentally and spatially the transcription of these elements by Pol III.

Nuclear RNA was prepared from mice, and a similar (10-fold) developmental increase in the ID content of hnRNA from mouse cortical neurons was measured. The hybridization had to be performed at decreased stringency (65 to 80%) in order to detect the phenomenon in mice (Brown et al., in press). Thus there is a disperse repetitive genomic element enriched in, but not exclusive to, the introns of mouse postnatal-onset cortical neuronal genes. The element apparently is not highly homologous to the rat sequence, hence the relevance of measurements on murine ID analogues (Owens et al., 1985; Sapienza and St. Jacques, 1986), which were performed at high hybridization stringency, must be questioned.

Structure of ID Elements in Chromatin

Within the nuclei of cells in vivo, DNA is complexed with nucleosomal proteins to form chromatin. The arrangement of nucleosomes on a gene and the state of chromatin in that region varies, depending upon whether a gene is active within the tissue studied. Brain tissue, and in particular cortical neurons, is different from peripheral tissues in that a greater portion (42%) of the genome is involved in active transcription (Bantle and Hahn, 1976). In most tissues, bulk DNA (which is mostly non-transcribed) is found associated with nucleosomes spaced, on average, about every 190 to 200 nucleotides. Such DNA is coiled into supernucleosomal solenoid structures and, when isolated, is relatively resistant to experimental nuclease digestion. The bulk DNA isolated from cortical neurons differs from that isolated from cortical glia and peripheral tissues in that the nucleosomes are spaced, on average, every 160 to 165 nucleotides (Thomas and Thompson, 1977). This DNA does not form highly ordered solenoid structures, and it is relatively sensitive to experimental nuclease digestion.

We examined the nucleosomal-association state of ID elements within rat chromatin isolated from cortical neurons and peripheral organs (Brown and Sutcliffe, 1987). Isolated chromatin was mildly digested with micrococcal nuclease, nucleosome ladders were displayed by gel electrophoresis, and repeat lengths were calculated by measuring the gel mobilities of the 3 to 8 nucleosome peaks. Bulk chromatin, as assayed by either ethidium bromide staining or hybridization with another known repetitive genomic element, was found to have the previously described pattern of spacing, namely, 160 nucleotide repeats in cortical neurons and 200 nucleotide repeats in peripheral tissues. However, a considerable fraction of the ID elements were found in the 160-nucleotide chromatin conformation in peripheral tissues as well as in brain tissue, indicating that ID elements within chromatin are peculiar compared to other DNA sequences.

If ID elements function in vivo as transcriptional enhancer elements, then they must be accessible, when called upon, to interact with specific *trans*-activator proteins. Bulk chromatin is relatively inaccessible for protein-DNA interactions; however, chromatin in the 160-nucleotide repeat form is more accessible and hence reactive. The finding that a significant subset of ID elements in peripheral tissues is packaged in the reactive form (Brown and Sutcliffe, 1987) suggests that these elements might be predisposed to interact with *trans*-activator proteins.

Regulators of Postnatal-Onset Cortical Neuronal Genes

ID elements appear to have the critical properties that a *cis*-acting regulator for a gene battery must possess. They are abundant enough to be positioned near or within a large number of genes. They are preferentially, although not exclusively, associated with a particular class of genes whose expression increases dramatically in early postnatal development in neurons, especially a class of neurons that is enriched in cerebral cortex compared to cerebellum. The elements have a recognizable functional property: they are templates for Pol III transcription, and the transcription begins during early postnatal development and is enriched in cerebral cortex compared to cerebellum. In transfection experiments, the ID elements have the properties of transcriptional enhancers in cells that are permissive for their Pol III transcription but not in cells that do not transcribe the ID elements. Other mammalian species contain elements sharing many if not most of these critical properties, but the sequences of the analogues between species have not been highly conserved. Thus we suggested (Sutcliffe et al., 1984b) that Pol III transcription of ID elements during postnatal development might potentiate the expression of a class of postnatal-onset neuronal genes. We can further refine this hypothesis to specify genes whose expression is enriched in cortical neurons compared to cerebellum.

Studies on brain chromatin structure (Thomas and Thompson, 1977; Brown, 1978; Ermini and Kuenzle, 1978) have shown that cortical neuronal, but not glial, chromatin undergoes a general transition at the end of the first postnatal week. The packaging of nucleosomes changes from a rigid structure in which their spacing averages from 190 to 200 nucleotides to a more flexible form in which their repeat length is 160 to 165 nucleotides. After the transition, the neuronal DNA is more accessible than previously to soluble proteins (assayed by nuclease cleavage). This transition is not evident in cerebellum nor is the shorter repeat length characteristic of the general chromatin structure in non-neuronal tissues. Thus it seems that the transition in nucleosome structure might signify a general transcriptional activation in cortical neurons. Indeed, the complexity of brain transcripts increases shortly thereafter (Chaudhari and Hahn, 1983), indicating that new gene expression is occurring. These newly expressed genes may be those genes that are enriched in ID elements, as detected in Northern blot experiments of nuclear RNA (Brown et al., in press).

Apparently the chromatin transition precedes the onset of Pol II transcription of ID elements. We propose a causal relationship between Pol III transcription of ID elements and the chromatin transition that leads to gene activation. Pol III transcription of ID elements beginning shortly after birth in rodents, presumably because of the appearance of a specific transcriptional factor (Sutcliffe et al., 1984a) or possibly because of an increase of the concentration of a general factor, might induce a reorganization of the chromatin surrounding each ID element. This in turn could lead to the observed transition in nucleosome arrangement. The open state of the DNA after the transition would render it suddenly readily accessible to the transcriptional initiation factors that control the expression of individual genes, thus facilitating a wave of new gene expression. Thus we envision that gene activation occurs in two stages. First, each member of the gene battery goes through a transition from an inactive to an active state. In cortical neurons, this step is mediated by Pol III transcription of ID elements. Second, the newly activated genes are now suitable templates for Pol II initiation when the appropriate factors become available in particular neurons.

ID Promiscuity and Evolution

This model of gene activation has considerable appeal; it seems consistent with the known data and specifies an element with Britten–Davidsonian properties (Britten and Davidson, 1969) active in developmental gene regulation and suggests a mechanism for its operation. However, it may seem unappealing because of the apparent sloppiness of the hypothetical system. There are seemingly many more ID elements in the rat genome than there are neuronal genes whose expression is higher in postnatal cerebral cortex than in cerebellum. (Actually, this is an empirical, rather than a formal, definition of the gene battery. There may be members of this gene battery expressed in cerebellar neurons, as well as neurons in other regions, but they are less abundant than cortical neuronal genes in terms of their hnRNA representation.) If all 10^5 ID elements are activated as enhancers, then a substantial portion of the genome would become involved, since 43% of genomic clones carrying inserts that average 15 Kb contain ID elements (Milner et al., 1984). In practice, cortical neuronal chromatin becomes generally euchromatic in adults, as compared to its generally heterochromatic form in most cell types; thus most of the genome is involved in a transition (Thomas and Thompson, 1977). Furthermore, it has been estimated from RNA complexity studies (Bantle and Hahn, 1976) that as much as 42% of the rodent genome is expressed as nuclear transcripts in the brain, and much of the brain RNA complexity has postnatal onset (Chaudhari and Hahn, 1983). Thus, what might appear at first glance to be a broad and perhaps random distribution throughout the genome of putative enhancer elements for a class of postnatal-onset neuronal genes might instead reflect the large number of genes in a gene battery that occupy a major fraction of the genome. The finding that ID elements occupy intrinsic positions in genes that are not neuron specific (Milner et al., 1984; Whitney and Fuarano, 1984; Owens et al., 1985; Sapienza and St. Jacques, 1986) at a frequency of about 10% of their frequency within cortical postnatal neuronal genes (Brown et al., in press) might signify that the mechanism of dispersing this putative control element through evolution to its regulatory destination has been a sloppy one, perhaps utilizing a retroposon intermediate (Lemischka and Sharp, 1982; Rogers, 1983). Thus ID elements located in "wrong" positions might be considered as a nonlethal tolerated load (Milner et al., 1984). To activate almost half of a genome for transcription might require a somewhat brute force—and thus a necessarily sloppy mechanism. In this general sense it may be more appropriate to consider ID elements as general activators of almost the entire genome rather than as specific activators of a particular defined battery of genes.

The issue of the inconstancy of the sequence of the ID putative control element through phylogeny must also be considered. Initially, this fact suggested that brain-specific Pol III transcripts might not themselves be functional. Is it possible that an element that putatively controls the expression of a large battery of genes could drift so dramatically in sequence? Because the ID element correlates with postnatal-onset genes whose expression is especially enriched in cortical neurons, one need not look far down the phylogenetic tree, since this class of genes is likely to represent the latest of evolutionary additions. If our model is correct, then some regulatory element must have evolved coincident with the debut of these cortical genes as a battery. It seems highly unlikely that such elements could be altered en masse at each step throughout mammalian speciation and even more unlikely that the rat ID element was altered step wise to become the *Alu* left module. However, both rat ID and human *Alu* elements are clearly transposable (Weiner et al., 1986), and thus these elements could have (and clearly have) spread throughout the genome. Since the mechanism of retroposition requires that these elements already be Pol III transcription units, it would only be necessary that the specificity of their transcription evolve to become neuron specific. It is possible they had this property before they were dispersed. Alternatively, this could be accomplished by altering the specificity of a single Pol III transcriptional regulatory factor rather than changing 10^5 elements. At that stage the new invading elements would have an excellent chance of supplanting the domestic regulatory element for this extensive gene battery.

It is difficult to imagine how cumulative mass retroposition could occur repeatedly during evolution if the specific set of targets were only a few hundred genes. That would require some targeting to neuron-specific genes to occur in the germline. However, if essentially half of the genome is the target, and in the other half it is only required that exons cannot be transposition targets and that a few genes (which theoretically could not be allowed high expression in brain) be avoided, then it is quite conceivable that events could occur repeatedly through the ascent of mammals. In fact, such events would constantly (in the evolutionary sense) recruit new genetic regions into service in neurons of the cortex, a rapidly evolving mammalian organ. Such recruitment would occur one gene at a time when the ID element remained constant, but would be a wholesale phenomenon if a new element were to supplant the old one.

Issues and Questions

The main issue concerning ID elements is whether they or their BC RNA transcription products or both have a biological function in the central nervous system. We have already discussed our reasons for believing that the small BC RNAs might not have a function: their sequences are not conserved among species, in rat there is microheterogeneity, and no association with RNPs has been detected. While these observations make a function for BC RNAs seem unlikely, they do not rule one out. A more thorough search to rule out associated proteins would strengthen the position. If the RNAs do not have function, then do the genomic ID elements themselves have a function? The ID element has been shown in two laboratories to have the properties of a transcriptional enhancer in in vitro transfection assays. Are ID elements truly regulators of gene expression in vivo, and if so, what would constitute a critical test of this function?

There are more than 10^5 ID elements in the rat genome. Are all these transcribed by Pol III or is only a subset active? Might there be sequences outside the core ID sequence, possibly upstream, that determine whether a particular element is transcriptionally active and, possibly, whether it has enhancer activity? It has been shown that sequences upstream from a 5S gene (Morton and Sprague, 1984) and a 7SL gene (Ullu and Weiner, 1985) are required for their efficient transcription. The ID element within the second intron of the rat growth-hormone gene is not an active Pol III template in vitro, even though it is identical in sequence with an ID element that is transcribed in the same extracts (Gutierrez-Hartmann et al., 1984). This suggests that a sequence outside the ID core inhibits its template activity. Is Pol III required for enhancer function? Are other known enzymes, such as a topoisomerase, required to alter chromatin structure? Do ID elements cluster within genes they putatively control, increasing their potency? How many genes for BC1 are there, and how does their structure differ from BC2 genes (which are presumably the same as known ID elements)?

At the gross level, Pol III transcription of ID elements to produce BC RNAs is brain specific, yet low levels of BC2-like RNAs are detected in peripheral tissues by primer extension assay. What is the origin of this background? Could it be peripheral neurons or is "brain-specific" Pol III transcription occurring in these tissues at low levels? Within the brain, are BC RNAs truly neuronal, or are they expressed in glia at an intermediate level? Are they present in all neuronal cell types? Does their appearance in different brain regions follow the general development of the CNS in the early postnatal period? Why does transfer to cell culture activate expression of the BC RNAs in non-neuronal cells? What accounts for the difference measured between steady-state blotting experiments and nuclear run-on transcription of the ID concentration in hnRNA molecules? Recent studies have shown that ID elements might also be elevated in muscle RNA molecules (Lone et al., 1986) and in RNA of myoblast cell lines that have differentiated in culture to form myotubes (Herget et al., 1986). These reports should be viewed with caution because of the known difficulty of preparing DNA-free RNA from muscle tissue and determining its concentration. In our own studies in which we have controlled for possible variation in RNA concentration, we have not detected a higher level of ID elements in muscle nuclear RNA than in nuclear RNA from other peripheral tissues. Nonetheless, it is possible that neurons and muscle cells share some properties in ID expression in Pol II molecules.

Is it possible, based on evolutionary considerations, that our provisional model of ID elements regulating a battery of postnatal-onset cortical neuronal genes is valid? The mouse genome contains fewer elements with high homology to the rat ID element than the rat genome, but it still contains about 20,000 (Sapienza and St. Jacques, 1986), which could be an adequate number. There may be more elements that could be detected at lower hybridization stringencies (Brown et al., in press). Primates clearly have enough copies of *Alu* elements, although it is not yet known how many classes of *Alu* repeats are transcribed or whether the genomic *Alu* Pol III template consists of both the left and right *Alu* monomers. Does the *Alu* element have enhancer properties? What constitutes a formal definition of the gene battery putatively regulated by ID elements, and when did it arise during evolution?

III

AXONAL PATHFINDING

KATHRYN TOSNEY

During development a neuron is confronted with a considerable challenge: it must extend a neurite through a complex and changing environment to reach a specific target that may be tens or hundreds of cell diameters distant. This specific outgrowth of neurites is called pathfinding, and the chapters in this section focus on the identification of mechanisms that mediate pathfinding in different neuronal populations. Pathfinding has been most thoroughly characterized in the accessible and relatively simple nervous systems of invertebrates (see Goodman et al., 1982, and Palka, 1986, for reviews); however, enough progress has been made in more complex systems (see Landmesser, 1984, and 1986 for reviews) that researchers are now able to ask to what extent mechanisms are common throughout the animal phyla. Possible differences and similarities in pathfinding between invertebrates and vertebrates is a coherent theme throughout this section.

Pathfinding is predicated on the navigation of the tip of the outgrowing neurite—the "growth cone." Growth cones have an unusual ability, shared with a few migratory cell populations like the neural crest, to invade and advance through tissues that are themselves coherent and immobile. The growth cone is the only motilely active portion of the neuron; it advances and extends cellular processes that probe the embryonic environment as the neurite elongates and is consolidated behind it (see Bunge, 1986, for a review). Growth associated proteins (GAPs) are required for the elongation of many vertebrate neurites; their presence permits but may not guide outgrowth (Kalil, Chapter 10).

Studies in the simplified environment of tissue culture have elucidated some of the pathfinding capabilities of growth cones. For instance, when neurons are seeded on a culture substratum of different substances that are arrayed in a geometrical pattern, their growth cones will trace out the pattern of the substance to which they adhere best (Letourneau, 1975), suggesting that growth cones can be guided by interactions with molecules on the surfaces of cells or within the extracellular matrix. Cellular interactions like those that mediate other morphogenetic processes thus lie at the heart of pathfinding as well. Growth cones may also be guided by diffusible gradients of specific substances. For instance, growth cones of sensory axons in culture will move toward a source of nerve growth factor (Gunderson and Barrett, 1980).

Although tissue culture studies have been informative, we ultimately want to know how growth cones navigate in developing embryos. Judicious use of specific labeling techniques in the last decade has allowed the growth cones of specific neurons or neuronal populations to be studied relatively directly in embryonic systems.

Specific molecular cues must be postulated to explain a number of growth cone behaviors in the embryo. In organisms like the grasshopper and zebrafish, in which uniquely identifiable neurons can be examined, different growth cones make divergent and cell-specific pathway choices when confronted with an identical environment (Harrelson et al., Chapter 7; Westerfield and Eisen, Chapter 8). In vertebrate embryos in which neurons are more numerous, populations of neurons also display specific identities and behaviors. For instance, populations of motoneuron growth cones in the chick are uniquely specified, project with a high degree of discrimination down divergent pathways toward different targets in the limb, and can seek out the correct pathway when experimentally displaced from their normal starting point (Landmesser, Chapter 9). Furthermore, in the vertebrates as in the invertebrates, growth cones individually alter their trajectories in regions where possible pathways diverge (Kalil, Chapter 10; see also Landmesser, Chapter 9), suggesting that each growth cone responds independently to specific cues in the local environment, even when it is a member of a much larger population.

The most promising systems for the elucidation of specific molecular cues have been organisms with simple, well-characterized, and easily manipulated nervous systems. In Chapter 7, Harrelson et al. describe insect embryos in which individual growth cones display selective and specific affinities for particular axonal surfaces. Growth cones confronted with an orthogonal scaffold of axon bundles in the embryonic neuropil grow selectively down subsets of axon fascicles. The axons have been postulated to be differentially labeled by surface molecules that growth cones recognize and use for specific pathfinding (the "labeled pathways" hypothesis). Harrelson et al. describe two molecules, Fasciclin I and II, that are dynamically expressed on certain axon fascicles during the periods in which specific interactions are required for proper pathfinding. This observation raises hope that some specific molecular cues have been identified.

Zebrafish also have relatively simple, albeit vertebrate, nervous systems, and the results described by Westerfield and Eisen (Chapter 8) suggest that mechanisms similar to those that mediate pathfinding in insects are indeed important in some vertebrate nervous systems. In the zebrafish, individual primary motoneurons can be observed in the living embryo. Their growth cones make specific pathway choices and innervate their correct targets without error, displaying features similar to those of growth cones that pioneer pathways in invertebrate embryos.

While researchers generally agree that specific cues guide growth cones in all systems, the identity and disposition of cues is likely to vary among neuronal populations and among species. For instance, in vertebrates, specific cues have been phenomenologically localized to certain regions along pathways; however, in contrast to the invertebrate CNS, the pathways that are "labeled" by specific cues in most cases do not appear to consist of axons (Chapters 8, 9, and 10). Specific fasciculation among vertebrate axons has been demonstrated in culture (Kapfhammer et al., 1986) but may play a less central role in pathfinding in vertebrates. In addition, in all systems the relative importance of different mechanisms and the nature of specific cues may vary as any particular growth cone makes sequential pathway decisions on the way to its target. For instance, the first growth cones to grow out in the peripheral nervous system of the grasshopper appear to sequentially use a number of different cues (Caudy and Bentley, 1986) as do the growth cones in the vertebrate systems described in this section.

Pathfinding in invertebrates appears, however, to require a small number of mechanisms, each of which is often indispensible. For instance, in the grasshopper CNS, preferences for axonal surfaces are absolute rather than hierarchical: when the preferred axons are ablated or delayed, growth cones remain stranded and do not find an alternative way to their target (reviewed by Harrelson et al., Chapter 7). In contrast, a multiplicity of mechanisms appear to be used during pathfinding in many vertebrate systems, so that if some mechanisms are experimentally curtailed, correct connections can still develop. For instance, during regeneration (Kalil, Chapter 10) or when experimentally displaced (reviewed by Landmesser,

Chapter 9), growth cones in vertebrate systems can often take novel pathways to the correct target, evidently relying on subsidiary mechanisms or on a wider spatial availability of specific cues. While some mechanisms may be common to pathfinding in invertebrates and vertebrates, these may not be of equal importance during the development of vertebrate neuronal specificity.

In many vertebrates there appears to be a system of general pathways—regions delineated by developing tissues that provide paths of least resistance or of greater adhesion for growth cone outgrowth. For instance, growth cones may elongate down channels through a neuroepithelial matrix (the "blueprint hypothesis;" Singer et al., 1979) or follow "substrate pathways" (Katz and Lasek, 1979) composed of aligned glial cells or other elements. Such pathways may be relatively nonspecific in that growth cones from many populations are able to follow them; for instance, embryonic retinal axons advance through stereotyped regions if forced to grow into the spinal cord (Katz and Lasek, 1978), and wing axons experimentally channeled into the chick hindlimb trace out an anatomically correct hindlimb pattern (Straznicky, 1983). General pathways that are distinguishable from specific pathway cues have not been demonstrated in invertebrate systems. Even in vertebrate systems in which large numbers of growth cones grow out virtually simultaneously, general pathways are insufficient, by themselves, to explain the precision of outgrowth, but they may be important in channeling axons into regions where specific cues are available (Landmesser, Chapter 9). Indeed, the general pathway that is normally used during development can sometimes be dispensed with under experimental conditions, as described above.

An accretion, during evolution, of additional mechanisms that subserve the development of neuronal specificity is illustrated by the importance of target interactions in the development of specific connections in vertebrates (see the next section, Intercellular Contacts). Once pathfinding has led vertebrate neurites into the target region, they often interact with the target to adjust the number of neurons, hone the initial pattern of projection, and form patterned connections within the target (Landmesser, Chapter 9). Furthermore, mechanisms in addition to pathfinding and target interactions must subserve some of the remodeling of neural projections, such as the elimination or addition of side branches that is described in the motor axons of the zebrafish by Westerfield and Eisen (Chapter 8).

Future work will doubtless define the actual cellular mechanisms and molecular bases of pathfinding, perhaps, as forecast by one researcher (Landmesser, 1986), within the next decade. This optimism is not necessarily overstated, given the immense progress made in the previous decade and given the advent of a number of useful preparations and techniques. Preparations in which many problems may be resolved are provided by simple and accessible systems, such as the grasshopper, fruitfly, and zebrafish embryos described here, and by complex systems that are simplified by the use of slice preparations (Landmesser, Chapter 9) or tissue culture. Molecules important to pathfinding may be characterized using monoclonal and polyclonal antibodies raised to rare and transiently expressed cell surface molecules, following the example of Harrelson et al. in Chapter 7. Advances in molecular biology coupled with a variety of genetic, biochemical and immunological approaches are likely to rapidly elucidate how growth cones find their way through complex embryonic environments.

CHAPTER SEVEN

From Cell Ablation Experiments to Surface Glycoproteins: Selective Fasciculation and the Search for Axonal Recognition Molecules

ALLAN L. HARRELSON,
MICHAEL J. BASTIANI, PETER M. SNOW,
AND COREY S. GOODMAN

Introduction

The studies reviewed here are aimed at understanding how neurons recognize other neurons, that is, how during development neurons (and in particular their growth cones) are able to distinguish particular neuronal surfaces with which to selectively interact. Of all of the forms of neuronal recognition, one of the most amenable to cellular and molecular studies is the process of selective fasciculation, in which neuronal growth cones display selective affinities for the surfaces of specific bundles (or fascicles) of axons as they navigate towards their targets (Goodman et al., 1984; Bastiani et al., 1985; Kuwada, 1986).

Of course, in addition to axonal surfaces, many other surfaces (e.g., basement membranes, epithelial cells, glial cells, and mesodermal cells) are used as substrates for growth cone extension and guidance at different times and places throughout development. But, although many mechanisms are involved in guiding growth cones to their targets, our focus here is the ability of growth cones to recognize specific axonal surfaces, and our model system for these studies is the relatively simple and highly accessible developing central nervous system of the grasshopper embryo.

In the grasshopper embryo, many growth cones are confronted with an orthogonal scaffold of longitudinal and commissural axon fascicles in each segment. Although their filo-podial extensions contact many fascicles, these growth cones invariably choose to extend along particular bundles of axons, giving rise to the stereotyped patterns of selective fasciculation (Raper et al., 1983b; Bastiani et al., 1984, 1986a).

Previous studies proposed the labeled pathways hypothesis, which predicts that axon fascicles in the embryonic neuropil are differentially labeled by surface recognition molecules that are used as guidance cues by neuronal growth cones (Ghysen and Janson, 1980; Goodman et al., 1982; Raper et al., 1983a,b; Bastiani et al., 1984). This model was supported by specific cell ablation experiments in the grasshopper embryo (Raper et al., 1983c, 1984; Bastiani et al., 1986a; du Lac et al., 1986; Doe et al., 1986). Similar types of descriptive and experimental studies have provided support for this hypothesis in a simple vertebrate embryo as well: the developing spinal cord of the fish embryo (Kuwada, 1986).

We begin this chapter by briefly describing a set of cell ablation experiments in the grasshopper embryo that are aimed at testing the selective affinity of the aCC and pCC growth cones for the axons in the U and MP1 fascicles, respectively.[1] We then review our recent studies using monoclonal antibodies to identify and purify two surface glycoproteins, called fasciclin I and II, that are expressed on different subsets of axon pathways in the grasshopper embryo (Harrelson et al., 1986; Bastiani et al., 1986b). Because of their patterns of expression (e.g., fasciclin I is specifically expressed on the axons in the U fascicle), both of these proteins are good candidates for axonal recognition molecules, as predicted by the cellular studies.

[1]All of these initials are abbreviations for particular attributes of the identified neurons. None of these attributes are essential to the description of development, so we do not provide them here, and designate the neurons by letters alone.

Selective Affinities of the aCC and pCC Growth Cones

The first three longitudinal axon fascicles in the grasshopper embryo initially contain the axons of seven identified neurons, as shown in Figure 1A (Bastiani et al., 1986a). From medial to lateral, these pathways are the vMP2 fascicle (containing the vMP2 axon), the MP1/dMP2 fascicle (containing the MP1, dMP2, and pCC axons), and the U fascicle (containing the U1, U2, and aCC axons). These growth cones are also involved in pioneering the intersegmental nerve (ISN), one of the two major peripheral nerve roots exiting the central nervous system (CNS). The growth cones of the sibling aCC and pCC neurons make divergent choices: the pCC growth cone turns anterior and fasciculates with the MP1/dMP2 axons, whereas the aCC growth cone turns posteriorly and fasciculates with the U1/U2 axons (Figures 2 and 3). The role of glial surfaces in these early interactions is discussed elsewhere (Bastiani and Goodman, 1986; Doe et al., 1986).

To provide a cellular test of the labeled pathways hypothesis, cell ablation experiments were performed, as summarized below.

Experiment 1. When the U1 and U2 neurons are ablated, the aCC does not extend along any other axon pathway (Figure 1B) (du Lac et al., 1986).

Experiment 2. When the MP1 and dMP2 neurons from the same and next anterior segments are ablated, the pCC does not extend along any other axon pathway (Figure 1C) (Bastiani et al., 1986a).

Experiment 3. When the MP1 and dMP2 neurons from the same segment only are ablated, once the MP1 and dMP2 axons from the next anterior segment extend posteriorly to within reach of the pCC growth cone, the pCC

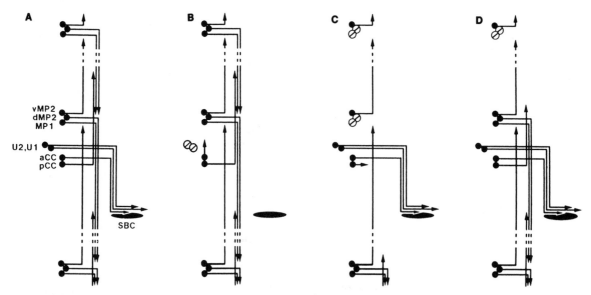

FIGURE 1. Schematic diagram summarizing the axons in the first three longitudinal axon fascicles and inter-segmental nerve in the grasshopper embryo (A), and the results of three different cell ablation experiments (B-D; in text, experiments 1, 2, and 4, respectively). (B-D) In vitro ablation experiments in which the U neurons (B), MP1 and dMP2 neurons in the test segment and next anterior segment (C), and MP1 and dMP2 neurons in the next anterior segment only (D) had been ablated. The results demonstrate the selective affinity of the aCC growth cone for the U fascicles and of the pCC growth cone for the MP1/dMP2 fascicle. SBC, segment boundary cell. (Adapted from du Lac et al., 1986.)

then extends anteriorly along them (not shown) (Bastiani et al., 1986a).

Experiment 4. When the MP1 and dMP2 neurons from the next anterior segment only are ablated, the pCC extends anteriorly and then stops where it reaches the anterior extent of the MP1/dMP2 axons from its own segment. The pCC does not extend along any other axon pathway in the middle of the developing neuropil (Figure 1D) (Bastiani et al., 1986a).

Experiment 5. When only one of these two neurons (either the MP1 or dMP2) is ablated, the pCC extends anteriorly along the remaining cell (not shown) (Bastiani et al., 1986a).

The results from these studies led to several conclusions. First, in the developing grasshop-

per embryo, neuronal growth cones do not indiscriminately fasciculate with any axons they encounter. Rather, growth cones display selective affinities for specific axonal surfaces, giving rise to the stereotyped patterns of selective fasciculation.

Second, these results support the labeled pathways hypothesis (Goodman et al., 1982; Raper et al., 1983a–c, 1984; Bastiani et al., 1984) by showing that individual growth cones can distinguish among different axon fascicles within filopodial grasp; growth cones invariably choose a specific axonal pathway upon which to extend.

Third, growth cones require the specific neuronal surfaces of axon pathways as continuous influences rather than as simple passive substrates. Such labeled pathways provide more than just a specific signal to initiate

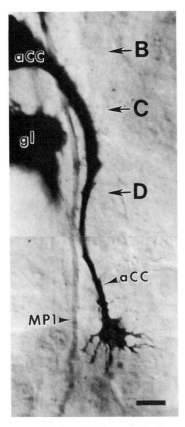

FIGURE 2. Morphology of the aCC growth cone and axon at 35% of development, as shown in this whole-mount embryo in which the aCC neuron—and additionally the MP1 neuron and a glial cell (gl)—had been filled with horseradish peroxidase (HRP). The aCC growth cone is about to turn laterally along the intersegmental nerve at the segment border. This embryo had been prepared for electron microscopy. The letters B–D indicate the levels of the sections shown in panels B–D in Figure 3. Bar = 10 μm.

an otherwise nonspecific axon outgrowth; instead, the specificity of interaction is required all along the length of the pathway. The requirement of the pCC growth cone for contact with the MP1 and dMP2 neurons goes beyond the initiation of axon extension by the pCC. Rather, these neuronal surfaces are required

as a specific axonal pathway upon which the pCC growth cone extends anteriorly.

Fourth, more than one axon appears to share the same recognition label. Although the pCC growth cone displays a selective affinity for the MP1/dMP2 fascicle, it does not distinguish between the MP1 and dMP2 axons in its own segment, nor between these axons from its own segment and those from other segments.

Fifth, individual growth cones appear to demonstrate more of an absolute preference for a specific axonal pathway than a hierarchical preference among a set of permissible axonal pathways.

Sixth, the precise timing of neuronal contact is not absolutely critical for proper pathway choice. In the absence of the MP1 and dMP2 axons from its own segment, the pCC growth cone, after a temporal delay, is ultimately contacted by the MP1 and dMP2 neurons from more anterior segments. The pCC growth cone then extends anteriorly along these axons.

In summary, these results argue against (1) the simple location of axons, (2) the simple timing of axon outgrowth, or (3) simple quantitative differences in the expression of a common surface label as the major determinant of the selective affinities of the aCC and pCC growth cones.

The pCC growth cone appears to display a high preference for the MP1 and dMP2 axons, and the aCC growth cone displays a similarly high preference for the U1 and U2 axons. These experiments support the notion that the surfaces of the MP1/dMP2 axons and the U1/U2 axons have special recognition labels that allow the aCC and pCC growth cones to distinguish between them and the many other axons that develop within the neuropil.

The experiments summarized above, together with other studies on these (Doe et al., 1986) and other neurons (Raper et al., 1983a–c, 1984; Bastiani et al., 1984), support the hy-

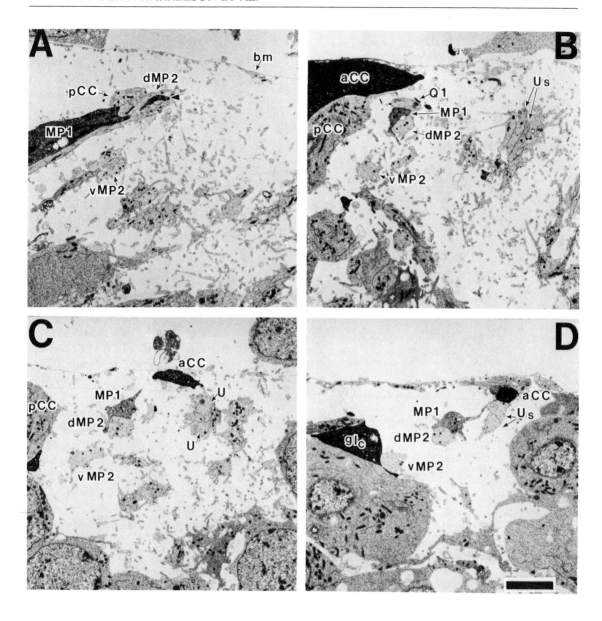

pothesis that recognition molecules are differentially expressed on the surfaces of different embryonic axon fascicles or subsets of axons within them. These postulated recognition molecules are likely to guide neuronal growth cones as they make specific choices to follow particular pathways and tracts to their appropriate targets.

How many recognition molecules exist remains an open question. Certainly every neuron does not have its own unique chemical label; the model predicts unique axon path-

◀ **FIGURE 3. Selective affinity of the pCC growth cone for the MP1/dMP2 axons and of the aCC growth cone for the U axons, as revealed by HRP injections and serial section electron microscopy. The MP1 and aCC neurons, and an identified glial cell (gl), were filled with HRP, as shown in Figure 2. (A) Section at the level of the MP1 cell body shows the axon of the filled MP1 extending dorsolaterally from its soma to join the MP1/dMP2 fascicle that at this level also contains the pCC axon. The vMP2 fascicle is just ventral and medial to the MP1/dMP2 fascicle and at this stage and level contains the vMP2 from the next posterior segment (labeled) and the vMP2 from the same segment (unlabeled). (B) Section through the aCC cell body shows its axon extending laterally just under the basement membrane (bm) over the MP1/dMP2 fascicle and toward the U fascicle. The pCC axon extends around the aCC to join the MP1/dMP2 fascicle, as shown in A. The axon of the Q1 neuron contacts MP1 as it crosses medially to form one of the first fascicles in the posterior commissure. The vMP2 fascicle contains only the vMP2 axon from the next posterior segment at this level. (C) At the level of the pCC cell body, the aCC axon is still extending along the dorsal basement membrane toward the U fascicle. The ventromedial vMP2 fascicle contains only the vMP2 axon. The centrally located MP1/dMP2 fascicle contains only the filled axon of MP1 and the unfilled axon of dMP2. Note the densely filled profile of the glial cell just ventral to the pCC soma. (D) Section through the longitudinal connective shows the three longitudinal bundles (the vMP2, MP1/dMP2, and U fascicles). The aCC fasciculates with the two U axons. The vMP2 axon is in close apposition to the filled profile of the glial cell. Bar = 2 μm.**

way labels, not unique neuronal labels. Passive and spatiotemporal constraints no doubt help reduce the number of molecules needed by channeling growth cones into certain pathways. But in addition, active mechanisms are required; growth cones make specific and divergent choices that implicate different molecules in pathway and target recognition.

The Expression of the Fasciclin I and II Glycoproteins

To identify candidates for such axonal recognition molecules, we generated monoclonal antibodies (MAbs) that recognize surface antigens expressed on subsets of axon fascicles in the grasshopper embryo (Harrelson et al., 1986; Bastiani et al., 1986b). The 3B11 and 8C6 MAbs were used to identify and study the expression of two different surface glycoproteins, called fasciclin I and II. These two glycoproteins are expressed on different subsets of axon fascicles during development in a spatiotemporal pattern consistent with the predictions of the hypothesis. Thus, these cell surface glycoproteins represent good candidates for molecules involved in the events of selective fasciculation.

The antigens recognized by the 3B11 and 8C6 MAbs are localized on different subsets of axon bundles in the grasshopper embryo (Figure 4). Within the developing segmental ganglia, the 3B11 MAb stains a small subset of commissural pathways in an embryo having completed 43% of development (a 43% embryo) (Figure 4B,E). In addition, at this stage, the 3B11 MAb stains one of the longitudinal axon bundles between the developing segmental ganglia (Figure 4E), all of the axons in the intersegmental nerve (Figure 4E), and two small bundles in the segmental nerve root. In contrast, in 45% and older embryos, the 8C6 MAb primarily stains longitudinal pathways (Figure 4C,F). At early stages, the 8C6 MAb transiently stains several commissural bundles, but not the same subset as is stained by the 3B11 MAb.

The 3B11 and 8C6 MAbs immunoprecipitate single proteins of 70 kD and 95 kD, respectively (Figure 5). Both membrane proteins are glycosylated, as indicated by their binding to the lectin Concanavalin A. Because these two glycoproteins are expressed on different

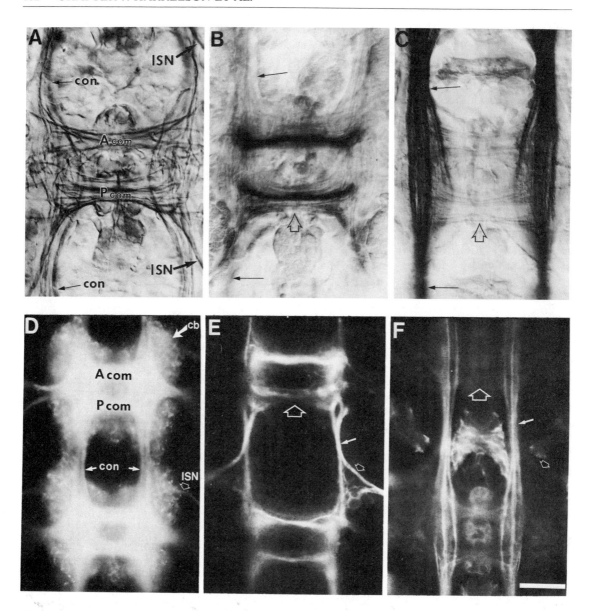

subsets of axons, we call them fasciclin I and fasciclin II, respectively.

We wondered whether the restricted spatio-temporal distribution of the two antigens that are recognized by the 3B11 and 8C6 MAbs actually represents the restricted expression of the two core proteins or alternatively whether it simply reflects the restricted expression of particular epitopes on otherwise more ubiquitously expressed proteins. Antisera against the purified proteins might help resolve this dilemma, because they would recognize multiple epitopes on the core protein. To this end, large quantities of solubilized embryos were

◄ FIGURE 4. 3B11 and 8C6 Monoclonal antibodies recognize specific subsets of axon pathways in the grasshopper embryo. (A–C) Dorsal views with Nomarski optics of single focal planes of the whole-mount neuroepithelium of single segments of 45% grasshopper embryos stained with particular monoclonal antibodies (MAbs) and visualized with an HRP-conjugated second antibody and HRP immunocytochemistry. (A) The embryonic axon scaffold in a single segment of the grasshopper embryo, showing the entire orthogonal array of longitudinal, commissural, and lateral axon fascicles in the neuropil of the second thoracic segment, as revealed with the I-5 MAb (Chang et al., 1983). (B) The 3B11 MAb stains a specific subset of axon fascicles (open arrow) in the anterior and posterior commissures (the 3B11 MAb also stains a longitudinal fascicle in the connectives and the ISN, both out of the plane of focus in this photomicrograph; see E). (C) In contrast, the 8C6 MAb stains most of the longitudinal axon fascicles in the neuropil and in the connective (solid arrow). (D–F) Dorsal views with epifluorescence of single focal planes of the whole-mount neuroepithelium of pairs of segments of 40–45% grasshopper embryos stained with particular antibodies and visualized with an FITC-conjugated second antibody. (D) Staining of all axon fascicles and cell bodies in two segments of the CNS using anti-HRP serum antibody (Jan and Jan, 1982). (E) By comparison, the 3B11 MAb stains only a small subset of commissural (large open arrow) and longitudinal (solid arrow) axon fascicles, and the ISN (small open arrow). (F) The 8C6 MAb stains all the major longitudinal axon fascicles in the connective (solid arrow) and the ISN (small open arrow), but few of the commissural axons (large open arrow). The 8C6 MAb also binds to the neuroepithelial cells at the segment border and some along the midline. A com, anterior commissure; P com, posterior commissure; ISN, intersegmental nerve; con, connective; cb, cell bodies. Bar = 30 μm (A, B, C); 50 μm (D, E, F).

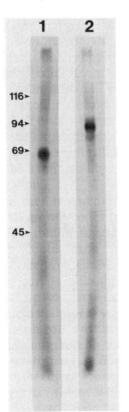

FIGURE 5. Analysis by SDS-PAGE of the proteins immunoprecipitated by the 3B11 and 8C6 MAbs and visualized autoradiographically. Membranes were prepared from adult CNS tissue and labeled with ^{125}I. Membrane proteins were solubilized in NP-40–containing buffer and subjected to immunoprecipitation using preformed antibody complexes made with either the 3B11 or the 8C6 MAbs. Immunoprecipitates were analyzed by SDS-PAGE on an 8.5% gel under reducing conditions. (Lane 1) Target antigen immunoprecipitated by the 3B11 MAb, called fasciclin I. (Lane 2) Target antigen immunoprecipitated by the 8C6 MAb, called fasciclin II. Molecular weight markers are indicated on the left in kilodaltons (kD).

used to purify microgram quantities of each glycoprotein using affinity chromatography, followed by preparative gel electrophoresis. Antisera were generated in rats against each of the gel-purified glycoproteins.

Immunocytochemical studies revealed that the two antisera recognize the same subsets of axon pathways, respectively, as do the two MAbs (Figure 6). These results indicate that fasciclin I and II are indeed expressed on restricted subsets of axon pathways.

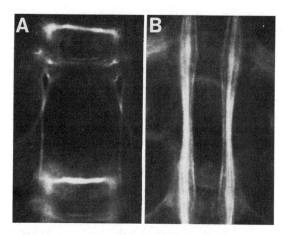

FIGURE 6. (A) Anti-fasciclin I antiserum; (B) anti-fasciclin II antiserum. The two antisera recognize the same subsets of axons pathways as do the two MAbs (see 3B11 MAb staining in Figure 4E and 8C6 MAb staining in Figure 4F for comparison). These photomicrographs show dorsal views of single focal planes of the whole-mount neuroepithelium of pairs of segments stained with the specific antisera and visualized with an FITC-conjugated second antibody. The fasciclin I and II proteins were purified using immuno-affinity chromatography and antisera generated in male rats. Immunohistology with these antisera was performed on 40–45% grasshopper embryos. Bar = 50 μm.

Regional Expression of Fasciclin I on Specific Axonal Processes Correlates with Axon Fasciculation

During the early stages of axonal outgrowth (at about 32% of development), a small subset of neurons, including two well-characterized pairs of neurons, begin to express fasciclin I on their cell bodies and growth cones (Figure 7A). Of the many axon pathways forming at this stage, these first neurons to express fasciclin I pioneer two particular axon pathways. The more anterior pair of neurons pioneers a single commissural pathway in the anterior commissure (Figure 7A,B). The more posterior pair of neurons (U1 and U2) pioneers a single longitudinal pathway (the U fascicle), which turns laterally to establish the intersegmental nerve (Figure 7C).

The anterior group of fasciclin I–positive neurons appears to express the glycoprotein regionally, in that it is localized only on portions of their surfaces. While pioneering their commissural pathway by looping up to the dorsal basement membrane and then extending across the midline, the neurons express the protein on their growth cones and all along their axons. However, when they turn onto longitudinal axon pathways at the lateral

FIGURE 7. HRP immunocytochemistry using the 3B11 MAb to reveal the expression of fasciclin I in a single ▶ segment (A, B) or between two segments (C). (A) Fasciclin I is expressed on a small subset of neurons at about 32% of development. The arrows point to the growth cones of two neurons that turn medially and pioneer an axon fascicle in the anterior commissure. The arrowheads point to the growth cones of the U1 and U2 neurons that turn posteriorly and pioneer the longitudinal U fascicle. (B) At 38% of development the major axon pathways expressing fasciclin I have formed in the anterior commissure (A com) and posterior commissure (P com), the longitudinal connective (con), and the segmental nerve (SN) and intersegmental nerve. This focal plane shows that the aCC neuron, which grows along the U fascicle (Uf), expresses fasciclin I on its surface, while its sibling, the pCC neuron, which grows along the MP1 fascicle (MP1f in C), does not express fasciclin I. The arrowheads show the position of the electron micrograph section in Figure 8A, C. (C) Photomicrograph of the longitudinal axon fascicles in a single connective (con) between two embryonic ganglia at 38% of development. Of the first three longitudinal axon fascicles (the vMP2, MP1, and U fascicles), only the U fascicle (Uf) expresses fasciclin I. The stained axon (thin arrow) from the aCC cell body can be seen crossing the unstained axon pathways to reach the U fascicle. The intersegmental nerve (ISN) turns laterally from the longitudinal pathways at the level of the segment boundary. MNB, median neuroblast. Bar = 50 μm (A, B); 25 μm (C).

edges of the commissure, they stop expressing the protein on their growth cones, and expression only persists where their axons fasciculate together in the commissure (see Figure 4B).

An interesting correlation between axon fasciculation and fasciclin I expression is observed for the aCC and pCC neurons. The pCC growth cone extends anteriorly along the MP1/dMP2 fascicle, which does not express fasciclin I; the aCC growth cone extends posteriorly along the U fascicle, which does express fasciclin I (Figures 7C and 8D). After the aCC growth cone begins to extend posteriorly along the U fascicle, the aCC also begins to express fasciclin I on its cell body and axonal surface (Figure 7B). In contrast, its sibling, the pCC, never expresses fasciclin I on its surface.

The initial evidence for the surface expression of fasciclin I came from the binding of the MAb to the surface of neurons in living embryos. The ultrastructural localization of fasciclin I confirms that it indeed occurs on the surface of axons that fasciculate together (Figure 8). Moreover, at very early ages, before axons first begin to express the protein, some filopodia already express fasciclin I (Figure 8B).

At the light microscope level, the fasciclin I protein is seen on the surface of axons in only one of the 10 or so major pathways in the anterior commissure at 43% of development (see Figure 4B,E). Electron microscope analysis revealed that the fasciclin I-positive pathway in the anterior commissure consists of three tightly associated bundles of axons (Figure

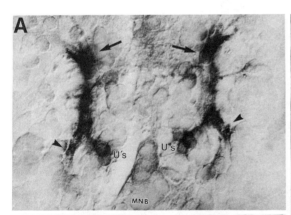

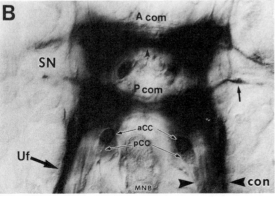

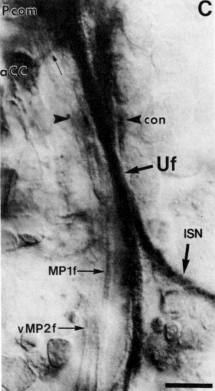

8A,C). The protein appears to be more abundant on axons in one of these associated bundles (filled arrow in Figure 8A) and less abundant, although present, on axons in the other two associated bundles (open arrows). In these other two bundles, the protein appears to be more abundant on the surfaces of the axons that are in closest proximity to the bundle that expresses higher amounts of the protein (Figure 8A). The different levels of expression of fasciclin I on axons within these three associated bundles has been consistently observed in analysis of several different embryos and several different segments within each embryo.

The differential expression of fasciclin I on these three associated bundles may be due to differences in intrinsic expression and/or subsequent stabilization of the protein on particular axon surfaces. Alternatively, the protein might be intrinsically expressed on the surface of axons in only one of the three bundles and extrinsically acquired by the asso-

ciated axons through contact. Further studies will be required to distinguish between these two alternatives. Whatever the mechanism, this differential expression may suggest that fasciclin I is partly responsible for the pattern of selective affinities both within and between associated bundles.

Expression of Fasciclin I in the Adult

In the peripheral nervous system (PNS), many sensory neurons express fasciclin I. In the longitudinal connectives of the CNS at 45% of embryonic development, in addition to the dorsolateral U fascicle, a small ventro-

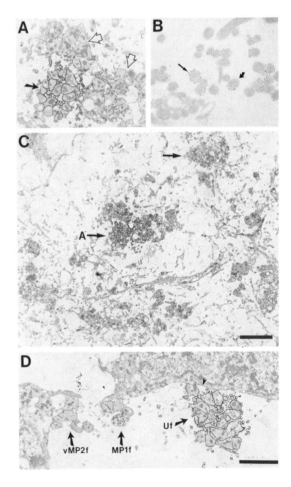

FIGURE 8. Immunoelectron microscopy showing localization of fasciclin I on specific axon bundles. HRP (A, C, D) and gold (B) immuno-EM labeling using the 3B11 MAb to reveal the expression of fasciclin I. (A, C) A is an enlargement of the axon pathway labeled A in part C; C is a lower-magnification view showing most of the axon pathways in the anterior commissure of a 42% embryo. The fasciclin I–positive pathway in the anterior commissure consists of three tightly associated bundles of axons; other pathways in the anterior commissure do not express the protein (e.g., top right arrow in C). (B) At a very early age (31%) before axons first begin to express fasciclin I, some filopodia have the protein on their surface, as shown with 5-nm gold bound to goat antimouse IgG. This suggests that these filopodia directly express the protein rather than acquiring it from other surfaces. (D) In the connective of a 38% embryo, all of the axons in the U fascicle (Uf) express fasciclin I, while the axons in the more medial fascicles (vMP2f and MP1f) are unlabeled. Bar = 5 μm (A, B, D); 10 μm (C).

medial fascicle begins to express fasciclin I. This fascicle continues to increase in size as it is joined by the axons of many fasciclin I-positive sensory neurons that grow into the CNS along the segmental nerve (SN) and turn anteriorly along this pathway.

In the adult segmental nervous system (T1 through A5), only the ventromedial longitudinal pathway expresses fasciclin I (Figure 9); this is the same sensory pathway that earlier expressed the protein. Apparently all the interneurons and motoneurons in the CNS and some of the sensory neurons in the PNS (particularly those entering the ISN) have stopped expressing fasciclin I by this time.

In the adult CNS, no commissural, dorsolateral longitudinal (Figure 9), or ISN pathways express fasciclin I, as they did in the embryo. Moreover, much of the fasciclin I expression disappears by the end of embryogenesis.

Summary

In the first part of this chapter we briefly reviewed some of the cell ablation experiments that support both the labeled pathways hypothesis and the prediction of axonal recognition molecules. In the next part, we reviewed our recent studies aimed at identifying molecular candidates for these axonal recognition molecules. We generated MAbs that recognize surface antigens expressed on subsets of axon fascicles in the grasshopper embryo. We used two MAbs, 3B11 and 8C6, to identify and study the expression of two different membrane glycoproteins (70 kD and 95 kD respectively) and to purify these proteins for the generation of specific antisera. These two glycoproteins, called fasciclin I and II, are localized on different subsets of axon fascicles during development in a spatiotemporal pattern that suggests that they play a role in the events of selective fasciculation. Parallel ex-

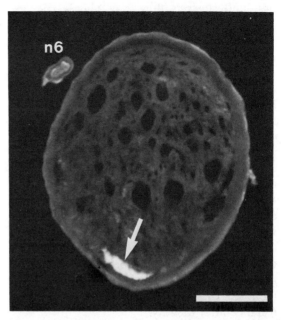

FIGURE 9. Restricted expression of fasciclin I in the adult CNS. Fluorescent labeling using the 3B11 MAb and FITC-conjugated second antibody to reveal the expression of fasciclin I in the adult CNS. In the connective between the first and second thoracic ganglia, only the ventromedial sensory axon pathway expresses fasciclin I. There is no labeling of a dorsolateral fascicle or of the nerve 6 (n6) branch of the intersegmental nerve. Bar = 100 μm.

periments in our laboratory have used MAbs to reveal a different surface glycoprotein, called fasciclin III, that is expressed on a different subset of axons in the *Drosophila* embryo (Patel et al., 1987).

The best evidence supporting the notion that the fasciclin I and II glycoproteins function as axonal recognition molecules during selective fasciculation is their regional expression. Both glycoproteins are expressed on particular axon pathways. While they are not necessarily expressed over the entirety of an individual neuron, their regional expression correlates with the patterns of axon fasciculation and suggests they may play a role as pathway labels.

For example, fasciclin I is expressed on the surface of axons in a particular commissural pathway. At the point where the distal axons of these neurons leave this commissural pathway to extend onto other longitudinal pathways, they stop expressing fasciclin I. Similarly, fasciclin II is localized on most longitudinal pathways but not on most commissural pathways, even though most of the neurons whose axons run in these longitudinal pathways are interneurons whose axons first extend across one of the commissural pathways before joining a longitudinal axon bundle. Thus, the way that these glycoproteins are regionally distributed on particular segments of axons suggests that their expression correlates with the identity of axon pathways and not the neuronal cellular identity.

The expression of fasciclin I on groups of neurons appears to be correlated with their patterns of fasciculation rather than their lineage. For example, the aCC and U neurons arise from different neuroblast lineages, share a common axon pathway (the U fascicle), and both express fasciclin I. In contrast, the aCC and pCC neurons arise as siblings from the same neuroblast lineage, choose different axon pathways (e.g., the U fascicle and MP1 fascicle, respectively), and differ in their expression of fasciclin I: the aCC does and the pCC does not.

From the onset of axonal outgrowth, fasciclin I is expressed on the surface of specific commissural, longitudinal, and sensory axon pathways. In contrast, by the end of embryogenesis, the commissural and longitudinal pathways containing the axons of CNS interneurons and motoneurons, and some sensory axon pathways, no longer express the protein. This correlates with the observation that all interneurons and motoneurons are born and differentiate during grasshopper embryogenesis. Thus, fasciclin I

is not expressed on the surface of mature CNS neurons, but rather is only expressed on their axonal surfaces during the period of axonal outgrowth. Moreover, many sensory pathways (including one ventromedial longitudinal pathway in the CNS) continue expressing fasciclin I throughout life, in correlation with the continued addition of new sensory neurons. Thus, fasciclin I expression persists when such a putative pathway label is still needed as a guidance cue and disappears when it is no longer needed.

Fasciclin I and II are both also expressed on the surface of a variety of neuronal and non-neuronal cells at different times and places throughout embryonic development. For example, in addition to its occurrence on a subset of axon pathways, fasciclin I is transiently expressed in a segmentally repeated pattern during neurogenesis and is also expressed in a segmentally repeated pattern on the ectoderm of the body wall and limb buds (Bastiani et al., 1987). Thus, fasciclin I appears to be used at several different times and places during embryonic development, suggesting that it may serve different but perhaps related functions in different tissues. In each case, the expression of fasciclin I defines a particular subset of cells and/or regions of cells within a tissue.

If our conclusions were based solely on staining with the MAb, one might argue that different proteins sharing the same epitope were being expressed at these different times and places. However, this pattern of staining is seen with both the MAbs and the antiserum against each purified protein; each antisera would be expected to recognize multiple epitopes on each protein.

The fasciclin I and II glycoproteins are good candidates for axon recognition molecules involved in the selective fasciculation choices of embryonic growth cones. However, whereas these molecules are good candidates for pathway labels, neither one alone is sufficient to account for the behavior of the growth cones we have previously observed. The complexity of fascicles and specificity of growth cone choices suggest that many labels may be involved and that the behavior of individual growth cones may often be based on the expression of several different molecules.

In the future we hope to determine the structure and function of fasciclin I and II, and to search for other related molecules. Both fasciclin I and II have been purified, characterized, and portions of them sequenced (P.M. Snow et al., unpublished data), and thus it should now be possible to clone the genes encoding them. Because the patterns of selective fasciculation in the early *Drosophila* embryo are nearly identical to those in the grasshopper embryo (Thomas et al., 1984; Goodman et al., 1984), there may be a concomitant molecular conservation as well. If this is the case, then perhaps the molecular probes for fasciclin I and II isolated from the grasshopper will be useful in finding the homologous genes and proteins in *Drosophila* and will thus allow a detailed genetic analysis of the problem.

CHAPTER EIGHT

Common Mechanisms of Growth Cone Guidance During Axonal Pathfinding

MONTE WESTERFIELD AND
JUDITH S. EISEN

Introduction

A central problem of neurobiology is understanding how specific neuronal connections are formed during development of the nervous system. Neurons must complete two steps while forming synaptic connections. The first is to grow a process into the region of the nervous system or body that contains an appropriate target, and the second is to form synapses with the correct type and number of target cells in that region. The first step, often termed pathfinding, usually occurs by growth of the axon directly to its target, although there are suggestions that in some systems there may be initial projections into inappropriate regions, followed by retraction. In either case, accurate pathfinding is absolutely required for the formation of correct connections, since it is known that many types of neurons will form synapses with incorrect targets if their axons are forced to grow to inappropriate regions. Thus, an understanding of the process of pathfinding and ultimately its cellular and molecular mechanisms is a prerequisite for understanding the origins of neuronal specificity.

At the heart of the pathfinding phenomenon is the growth cone. Recently much effort has been focused on the behavior of growth cones and on the morphology of growing axons in a variety of species. In this chapter we review some of these studies, emphasizing that many common features of growth cone navigation and axonal development are found during development of different animals and different parts of the nervous system. We suggest that this is because even in diverse systems, common mechanisms underlie axonal guidance.

Pathfinding by Invertebrate Growth Cones

Considerable advances in our understanding of axonal guidance have resulted from studies of the growth cones of identified neurons in invertebrates. These organisms have obvious advantages for cellular studies, since their nervous systems contain relatively few neurons, many of which can be uniquely identified in different individuals of the same species. Some of these neurons have been described during the earliest stages of development in the grasshopper (Bate and Grunewald, 1981). By following their detailed development we have learned that growth cones appear to display selective affinities for specific surfaces and that these affinities are cell specific (Bastiani et al., 1985b). For example, during development of the central nervous system (CNS) in the grasshopper, individual growth cones display cell-specific affinities (defined by filopodial contacts) for particular non-neuronal surfaces such as basement membrane or glial cell surfaces, whereas other growth cones display selective affinities for specific neuronal surfaces. Moreover, these selective affinities appear to be absolute and nonhierarchical, as demonstrated by alteration of the temporal appearance or ablation of the preferred surface (Bastiani et al., 1985a).

The formation of the first longitudinal axonal fascicles in the grasshopper CNS provides an example of this behavior. The first pioneer growth cones extend from the midline precursor neurons, MP1, dMP2, and vMP2 (Bate and Grunewald, 1981; Goodman et al., 1982; Taghert et al., 1982). They grow at essentially the same time, are confronted with the same environment, and yet make divergent pathway choices (Figure 1). Initially all three growth cones project to the dorsal basement membrane of the neuroepithelium; then the vMP2 growth cone turns anteriorly, while the MP1 and dMP2 growth cones turn posteriorly. The pathways diverge irrespective of the initial positions or time of arrival of the growth cones at this surface, suggesting that the growth cones are responding to guidance information present in their environment, rather than simply being channeled into the appropriate path by spatial or temporal constraints. This guidance information is most likely contained within the basal lamina itself, since electron-microscopic examination of the growth cones

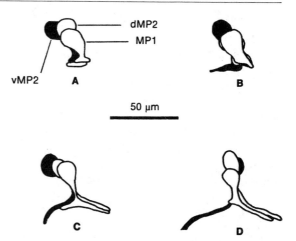

FIGURE 1. The growth cones of three neurons—MP1, dMP2, and vMP2—in the grasshopper CNS that pioneer central pathways. The cell bodies of these neurons are located close to one another. (A) Their growth cones initially extend to the basement membrane of the overlying epithelium. (B–D) Regardless of the time of arrival or position of the growth cones, the vMP2 growth cone then turns anteriorly (to the left in the figure) and extends along the ventral surface of the basement membrane, while the dMP2 and MP1 growth cones turn posteriorly (to the right in the figure) and extend along the dorsal surface of the basement membrane. Each drawing represents approximately 10 hr of development. (After Bastiani and Goodman, 1984.)

has demonstrated that the vMP2 growth cone extends along a relatively lateral part of the basement membrane, whereas the MP1 and dMP2 growth cones extend along the dorsal part of the basement membrane. Later-developing neurons form their projections by following specific pathways pioneered by these early growth cones. This apparently occurs due to a cell-specific selective affinity of growth cones for the axons of particular neurons (Goodman et al., 1982; Taghert et al., 1982).

Axonal pathways are formed in the appendages of insects by a set of peripheral pioneer sensory neurons (Bate, 1976). The growth cones of these sensory neurons extend along the basement membrane of the overlying epithelial sheet as they grow towards the central nervous system (Berlot and Goodman, 1984). Genetic (Palka, 1982; Palka et al., 1983) and experimental (Edwards et al., 1981; Bentley and Keshishian, 1982a,b; Blair et al., 1985) perturbations of the appendages have suggested that pioneer sensory growth cones are guided by polarity information, presumably in the form of a gradient along the epithelia (Ho and Goodman, 1982; Nardi, 1983; Berlot and Goodman, 1984; Blair et al., 1985) and by selective interactions with "guidepost" cells positioned along the axonal pathway (Bentley and Caudy, 1983). The growth cones of insect motoneurons extending from the CNS into developing limbs appear to select particular muscles to innervate by cell-specific interactions with particular mesodermal cells called muscle pioneers (Ho et al., 1983; Ball et al., 1985).

Pathfinding by Vertebrate Growth Cones

Considerably less detailed information exists about pathfinding by the growth cones of vertebrate neurons. To some extent this is because vertebrate systems are composed of a much larger number of cells and are therefore more complex than invertebrates. Moreover, most studies have focused on the projections of neurons and rearrangements of their projections, rather than on the nature of initial pathfinding. Recently, however, detailed descriptions of axonal morphology and growth cone pathfinding have been obtained from studies of developing vertebrate visual and motor systems.

Visual systems

In the mature vertebrate nervous system, the axons of retinal ganglion cells are found to project along a stereotyped pathway, the optic tract, from the eye to the brain (Shatz and Kliot, 1982; Thanos and Bonhoeffer, 1983; Holt and Harris, 1983). In the tectum the axons terminate in particular regions to produce topographic and eye-specific connections. The optic tract is probably established by the growth cones of retinal ganglion cells. Recent studies of individual axons have demonstrated that retinal ganglion cells usually project directly toward their appropriate targets, even from eyes transplanted to ectopic locations (Figure 2) (Harris, 1986). Thus, these growth cones can find and follow pathways to their targets by responding to environmental cues that may be widely distributed throughout the embryonic brain (Harris, 1986).

Motor systems

Vertebrate motor growth cones also appear to follow cell-specific pathways to their targets.

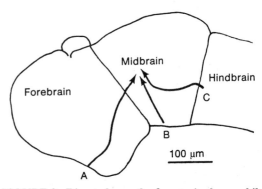

FIGURE 2. Directed growth of axons in the amphibian visual system. The initial projections of retinal ganglion cell axons to the tectum are diagrammed for three animals, one with a normal eye (A) and two with ectopically placed eyes (B, C). The arrows represent the pathways of axons traced from camera lucida drawings. The animals were near stage 39 of development. The brain is shown in lateral view with the divisions between the major subdivisions, forebrain, midbrain, and hindbrain, depicted. Axons from all three animals projected directly to the central neuropil region of the tectum, even though they entered different regions of the brain. (After Harris, 1986.)

Labeling of individual motor axons in the chick has demonstrated that growth cones from particular regions of the spinal cord sort out in the plexus region at the base of the limb bud (Figure 3) and project directly toward regions of the developing muscle appropriate for their mature function (Tosney and Landmesser, 1985a,b). The growth cones from a particular spinal segment sample a small fraction of the axis of the limb bud and at no time project diffusely into the limb (Landmesser, 1981). This directed growth appears to be due to active selection of pathways based on specific chemical cues, since the growth cones of motoneurons displaced a short distance from their normal location are able to traverse novel territories to reach their appropriate target muscles in the limb (Lance-Jones and Landmesser, 1980, 1981).

In our studies of the zebrafish, this analysis of pathfinding by vertebrate motor growth cones has recently been carried to the level of individual identified motoneurons. The motor system of the zebrafish is organized in a simple and stereotyped manner (Westerfield et al., 1986), and it has been possible to observe pathfinding by pioneer growth cones in live developing embryos (Eisen et al., 1986).

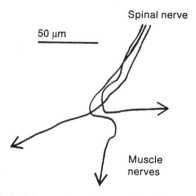

FIGURE 3. Divergent pathways of motor axons in the chick. Three individual motor axons sort out and cross over each other as they leave the spinal nerve (top) and enter different muscle nerves. The pathways were traced from camera lucida drawings. (After Tosney and Landmesser, 1985a.)

The axial muscles of adjacent body segments in zebrafish are separated from one another by myosepta that, in the adult, contain the ribs. Within a given segment the body muscles are separated into dorsal and ventral regions by a horizontal septum that spans the segment longitudinally from rib to rib. The bulk of this muscle is composed of white, twitch fibers (van Raamsdonk et al., 1980, 1982) that fire action potentials when activated (Westerfield et al., 1986). They are innervated, as in other lower vertebrates (Diamond, 1971; Blight, 1978; Roberts and Clarke, 1982; Forehand and Farel, 1982; Westerfield and Eisen, 1985), by two classes of motoneurons—primary and secondary (Myers, 1985). In the spinal cord, the primary motoneurons are distinguished by their relative medial, dorsal position and the large size of their cell bodies. Their large axons make a conspicuous loop around the medial aspect of the Mauthner axon before entering the ventral root. Secondary motoneurons have smaller cell bodies located farther ventrally and laterally. Their smaller axons pass lateral to the Mauthner axon and usually have no obvious contact with it.

These anatomical differences and inferences from other vertebrates (Diamond, 1971) have led to the proposition that primary motoneurons receive direct monosynaptic input from the Mauthner axon and mediate activation of body muscles during the escape response, whereas secondary motoneurons are used for other types of swimming (Myers, 1985; Westerfield et al., 1986). This concept is supported by observations of the peripheral innervation patterns of primary and secondary motoneurons. The axons of primary motoneurons are more than twice the diameter of secondary motor axons. They have very large terminal fields and as a group appear to innervate all the white muscle. The terminal field of a given primary motoneuron is confined to the body segment corresponding to the spinal segment containing its cell body. On the other hand, the axons of secondary motoneurons have smaller terminal fields and may innervate muscle fibers in more than one segment. Thus, activation of primary motoneurons during a Mauthner axon–mediated escape response (Eaton and Hackett, 1984) would effectively activate all the fast twitch fibers on one side of the body. Activation of subsets of secondary motoneurons would, presumably, produce contraction of subsets of muscle fibers, leading to a variety of other types of movements.

The primary motoneurons are uniquely identifiable as individual cells from animal to animal, based on the positions of their cell bodies and terminal fields. Each side of each spinal segment contains three primary motoneurons that have been named according to the longitudinal positions of their cell bodies (Westerfield et al., 1986). Each primary motoneuron innervates a stereotyped subset of muscle fibers in its corresponding body segment, as illustrated in Figure 4. The most rostral primary motoneuron in each segment, RoP, innervates muscle fibers located in the middle of the segment near the horizontal myoseptum; the most caudal cell, CaP, innervates fibers in the ventral part of its segment; and the middle cell, MiP, innervates the most dorsal fibers.

That the terminal fields of the three primary motoneurons are mutually exclusive was demonstrated by recording intracellularly from individual muscle fibers while stimulating the ventral roots (Figure 5). Each muscle fiber receives an input from a single primary motoneuron and from one or more secondary motoneurons. Inputs from primary motoneurons are distinguished from those of secondary motoneurons by their larger amplitudes and shorter latencies. Moreover, anatomical reconstructions demonstrate that each motor axon makes several synaptic terminals distributed along the length of each muscle fiber that it innervates. Thus muscle fibers in the zebrafish, as in other fishes (Hudson, 1969;

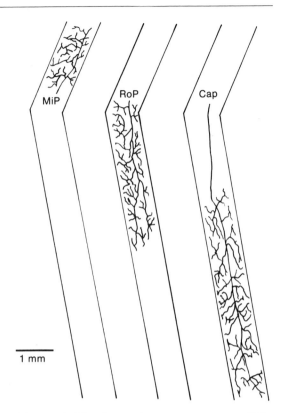

1 mm

FIGURE 4. The three primary motoneurons of the zebrafish (MiP, RoP, and CaP) establish mutually exclusive terminal fields. Side views of the fields as they appear in adults; rostral is to the left and dorsal, to the top. The slanted lines represent the myosepta that separate muscle fibers in adjacent segments. The MiP motoneuron innervates muscle fibers in the dorsal part of its segment; the RoP motoneuron innervates muscle fibers in the center of the segment, and the CaP motoneuron innervates muscle fibers in the ventral region of its segment. The terminal field of each motoneuron is shown innervating a different segment, although all three motoneurons actually are present on each side of each spinal segment. (After Westerfield et al., 1986.)

Bone, 1978; Ono, 1983), are polyneuronally and multiterminally innervated.

The stereotyped and cell-specific innervation patterns of the three primary motoneurons make them ideal subjects for studying the origins of neuronal specificity. How do these three neurons establish and maintain synaptic connections with mutually exclusive subsets of muscle fibers? At least two different mechanisms can be envisioned to be at work. At one extreme, primary motoneurons may initially overproduce projections to the developing muscle and may contact fibers in regions inappropriate for their adult function. Later, through selective cell death (Oppenheim, 1981) and/or synapse elimination and withdrawal of inappropriate projections (Purves and Lichtman, 1980), they would mold their specific adult fields as suggested by studies of the developing mammalian visual system (Hubel et al., 1977). At the other extreme, from the outset each primary motoneuron might project exclusively to the appropriate region of the developing muscle, perhaps due to cell-specific pathway recognition by its growth

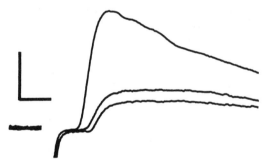

FIGURE 5. Zebrafish primary motoneurons establish stronger inputs to muscle fibers than do secondary motoneurons. Intracellular recordings from a muscle fiber that received inputs from three separate motoneurons. Responses were elicited at different, discrete thresholds by increasing the strength of the stimulus applied to the ventral root. The concentration of calcium in the bathing medium was reduced to block action potentials in the muscle fiber. The primary motoneuron that innervated this fiber produced the larger amplitude and shorter latency input. Activation of the axons of secondary motoneurons produced the two smaller and longer latency inputs. Resting potential, −83 mV; calibration, 5 mV for top trace and 2 mV for lower two traces, and 2 msec for all three traces. (From Westerfield et al., 1986.)

cone, as suggested from the studies of developing invertebrates described above.

These two possibilities have been examined during development of the zebrafish by direct observation of pathfinding by identified growth cones in the embryo (Eisen et al., 1986; Myers et al., 1986). Primary motoneurons were labeled with fluorescent lineage tracer dyes (Kimmel and Law, 1985) by injecting their early precursors in the blastula-stage embryo. The dyes were passed on to the daughters of each mitotic division; in animals chosen for study, a few neurons were labeled by 10 to 15 hours after fertilization of the egg, when the primary motoneurons were born (Myers et al., 1986). Individual primary motoneurons were identified by the characteristic positions of their cell bodies within the developing spinal cord. The motoneurons were watched with image-enhanced video techniques (Eisen et al., 1986) while they sprouted axons and while their fluorescently labeled growth cones navigated through the periphery.

Repeated observations of an individual neuron in a single developing embryo and comparisons of motoneurons in different embryos demonstrated that the growth cones of the three primary motoneurons in a given segment grow out of the spinal cord in a stereotyped sequence and that they follow cell-specific pathways into the region of muscle appropriate for their adult functions (Figure 6). The growth cone of the CaP motoneuron, whose cell body is located closest to the ventral root, invariably leaves the spinal cord first. It grows ventrally along the medial surface of the somite to the region of the horizontal septum where it pauses for a period of time up to two hours. After this characteristic pause, it continues ventrally to the bottom of the somite before turning laterally and dorsally to grow along the lateral surface of the somite near the rostral myoseptum of the segment containing its cell body. In the region of the horizontal septum and at intervals along its length, the CaP growth cone leaves varicosities that often

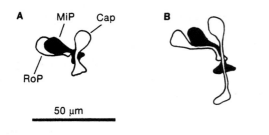

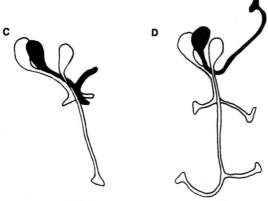

FIGURE 6. The growth cones of three motoneurons in the zebrafish that pioneer peripheral pathways. The cell body of each neuron is located in a characteristic rostral (RoP), middle (MiP), or caudal (CaP) position within the spinal segment. (A) All three growth cones initially extend out of the spinal cord to the region of the horizontal septum. (B-D) The CaP growth cone then continues ventrally, the MiP growth cone turns and extends dorsally, and the RoP growth cone extends laterally through the somite. The three motoneurons were drawn from different animals, at approximately 20 (A), 22 (B), 24 (C), and 28 (D) hr of development. The axons in (B-D) are shown with larger than normal diameters for clarity.

develop into side branches later in development.

The growth cone of the more rostral primary motoneuron, MiP, is the next growth cone to leave the spinal cord. It initially follows the CaP growth cone to the horizontal septum, where it, too, pauses. But then, unlike its predecessor, the MiP growth cone bifurcates and sends a branch into the dorsal region of the segment. Later, the ventral projection is withdrawn. The dorsal growth cone continues along the medial surface of the somite until it reaches the dorsal edge where, like the CaP growth cone, it turns laterally and enters the region of the rostral myoseptum of its own segment.

The growth cone of the most rostral primary motoneuron in each segment, RoP, is last to leave the spinal cord. It first grows caudally within the spinal cord until it reaches the level of the ventral root pioneered by the CaP growth cone. It then follows the CaP axon to the horizontal septum, where it turns laterally and follows a third pathway through the center of the somite to its lateral edge.

Common Mechanisms of Pathfinding

Our studies of pathfinding by identified motor growth cones in the zebrafish demonstrate that each neuron makes a cell-specific pathway choice from the onset of axonal growth and, with the exception of the ventral projection of the MiP growth cone, unerringly grows directly to the region of the somite appropriate for its adult function. This behavior clearly demonstrates that overproduction of axonal branches and withdrawal of inappropriate projections is not a major mechanism for ensuring synaptic specificity in the motor system of the zebrafish. In fact, the choice of divergent pathways at the horizontal septum is reminiscent of the cell-specific pathfinding of neurons in grasshoppers (Figure 7) (Raper et al., 1983a). This type of behavior suggests that the growth cones are recognizing environmental guidance cues (Raper et al., 1983b). We know from elegant in vitro studies that the growth cones of vertebrate neurons can make choices among variably adhesive substrates (Letourneau, 1982), further suggesting that divergent pathway choices may involve cell-specific differences in adhesion (Mason, 1985)

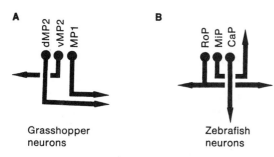

Grasshopper
neurons

Zebrafish
neurons

FIGURE 7. Comparison of pathfinding by identified central neurons in the grasshopper (A) and identified primary motoneurons in the zebrafish (B). The growth cones of invertebrate and vertebrate neurons make cell-specific pathway choices.

or affinity for environmental cues. In invertebrates, growth cones of neurons that pioneer peripheral pathways respond to a variety of cues, including cellular surfaces (Bentley and Keshishian, 1982a,b; Ho and Goodman, 1982; Bentley and Caudy, 1983) and noncellular components of the extracellular space (Ho and Goodman, 1982; Nardi, 1983; Berlot and Goodman, 1984; Blair et al., 1985; Caudy and Bentley, 1986). It seems very likely that the growth cones of primary motoneurons in the zebrafish use similar mechanisms to pioneer

FIGURE 8. The axons of fish and mammals show directed growth and limited remodeling. (A) The growth cones of zebrafish primary motoneurons initially grow directly to the appropriate region of muscle and later produce short side branches that project into inappropriate regions. During subsequent development these side branches are lost, while the terminal fields expand. The upper horizontal lines represent the horizontal septum; the lower horizontal lines indicate the ventral edge of the body. (B) The growth cones of retinal ganglion cells in the cat initially grow directly to eye-specific layers of the LGN, but also produce short side branches that project into inappropriate layers. Later the proximal side branches are eliminated, while the terminal fields expand by addition of more branches. The curved lines indicate the boundaries between layers in the LGN. (After Shatz and Sretavan, 1986.)

the peripheral pathways that ensure the formation of synaptic connections with appropriate target muscles. This hypothesis can be tested by the same methods used to study developing insects.

On the other hand, we also found that the varicosities left behind the primary motor growth cones often sprouted small side branches into regions of the muscle that would not be innervated in the adult. In Figure 8A, the bulk of the side branches in a young CaP motoneuron occupy the ventral part of the segment, although there are a few short side branches just ventral of the horizontal septum. This is a region containing muscle fibers that are not innervated by adult CaP motoneurons. The absence of projections in this region of the adult is due either to selective addition of new muscle fibers near the horizontal septum or more likely to the subsequent withdrawal of the proximal branches observed during the first two days of development. Nevertheless, it

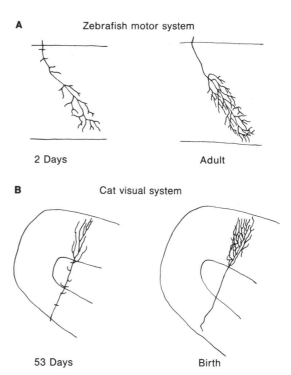

is clear that initially the CaP growth cone migrates unerringly along a path toward the region of muscle appropriate for its adult function and that the majority of side branch growth occurs in this area.

The two processes of directed growth to the appropriate region and limited remodeling by elimination of small side branches are reminiscent of the mode of development seen in retinal ganglion cell axons in the cat. Recent observations of individual ganglion cell axons have elucidated the morphological changes associated with the formation of the eye-specific layers in the dorsal lateral geniculate nucleus (LGN) during prenatal development of the cat's visual system (Sretavan and Shatz, 1986a). Although it has not been possible to trace the growth of individual axons, as was done for primary motor axons in the zebrafish, reconstructions of axons from fetuses of various ages suggest that axons initially project to distinct layers of the LGN yet sprout small side branches along their lengths, some of which project into inappropriate layers (Figure 8B). During subsequent development, the axons enlarge their terminal arbors and appear to withdraw the proximal side branches, so that by birth all axons have smooth trunks and elaborate terminal arbors confined to a single layer. This remodeling has been offered as an explanation for the segregation of inputs from the two eyes into distinct layers in the LGN (Sretavan and Shatz, 1986a). Similarities in the growth of zebrafish primary motor axons and cat retinal ganglion cell axons suggest common cellular mechanisms of development. Interestingly, directed growth and withdrawal of small side branches is also seen during development of pioneer neurons in the peripheral nervous system of insects (Bentley and Keshishian, 1982b; Bentley and Caudy, 1983; Caudy and Bentley, 1986) and during development of the central nervous system of the leech (Glover and Mason, 1986).

Other Mechanisms of Synaptic Specificity

What do these observations say about the cellular mechanisms actually used by growth cones for accurate pathfinding and the establishment of appropriate synaptic connections? Several alternatives to axonal guidance have been considered. Observations in the amphibian motor system (Lamb, 1976, 1977, 1979; McGrath and Bennett, 1979) have been interpreted to indicate that motoneurons failing to form functional or appropriate connections are eliminated during the period of naturally occurring cell death. Similar explanations have been offered for the formation of eye-specific layers in the primate retinogeniculate system (Rakic and Riley, 1983a,b). In its most extreme form (Oppenheim, 1981), this mechanism could increase synaptic specificity by elimination of incorrectly connected neurons. Cell death seems to be an unlikely mechanism to explain the specific patterns of motor innervation in the zebrafish, since there are only three primary motoneurons on each side of each spinal segment throughout development and there are no obvious signs of death among this population of neurons or in other early neurons that have been characterized (Mendelson, 1985). Moreover, studies of chick motoneurons before and during normal cell death have shown that at least some motoneurons fated to die have apparently formed appropriate functional projections into the developing limb (Landmesser and Morris, 1975; Landmesser, 1978a,b), and studies of invertebrates (White et al., 1983) and vertebrates (Lance-Jones and Landmesser, 1980) have revealed that neurons forced to innervate inappropriate targets are not eliminated in more than normal numbers.

Another mechanism used by neurons to modify innervation patterns is initial overproduction followed by elimination of synapses. This process is thought to be regulated by competition between axonal terminals and/or by trophic interactions between pre- and post-synaptic cells. Such competitive interactions have been suggested from studies of the developing nervous systems of vertebrates (Purves, 1980; Purves and Lichtman, 1980; Easter et al., 1985) and invertebrates (Blackshaw et al., 1982; Murphey and Lemere, 1984; Kramer and Stent, 1985; Murphey, 1985). For example, in the developing visual cortex of mammals, removing one eye or depriving one eye of patterned vision often decreases the number of cortical neurons driven by the deprived eye and reduces or eliminates the segregation of ocular dominance columns (Wiesel and Hubel, 1963, 1965; Hubel et al., 1977; LeVay et al., 1978). These observations suggest that segregation of terminal fields normally occurs due to competitive interactions between axons driven by different eyes. This interpretation recently has been tested at the level of individual axons (Sretavan and Shatz, 1986b). Removal of one eye during embryonic development of the cat's visual system resulted in a normal restriction of ganglion cell terminal arbors within the LGN, demonstrating that segregation of individual axonal terminals occurs in the absence of competitive interactions between neurons from the two eyes. However, it is still unclear whether the segregation is due to competitive interactions between ganglion cells within the remaining eye or is a result of axons recognizing guidance cues in the LGN. Competitive interactions in invertebrates appear to mold the central and the peripheral projections of leech neurons (Blackshaw et al., 1982; Kramer and Kuwada, 1983; Kramer et al., 1985; Kramer and Stent, 1985) and the branching of presynaptic axons (Murphey and Lemere, 1984; Murphey, 1985) and postsynaptic dendrites in insects (Shankland et al., 1982). In the leech, different mechanosensory neurons have receptive fields with stereotyped shapes and sizes (Kramer and Goldman, 1981). Each neuron's arbor is formed by precise and

stereotyped initial projections of its axon that are presumably guided by environmental cues (Kuwada, 1982; Kuwada and Kramer, 1983; Kramer and Kuwada, 1983). However, the exact locations and contours of the boundaries that separate adjacent fields vary significantly in different animals (Kramer et al., 1985) and are probably established by competitive interactions, since they can be altered by eliminating axonal branches (Kramer and Stent, 1985).

In the zebrafish, differences in the timing of axonal outgrowth among the three primary motoneurons could be an important factor in the establishment of cell-specific territories within the muscle. We consistently found that in any given segment the CaP growth cone pioneered the path to the horizontal septum and that it established an exclusive territory in the ventral muscle. Perhaps the small but consistent lag in the time of arrival of the MiP and RoP growth cones gave CaP an advantage in capturing and maintaining innervation of the ventral muscle by competitive exclusion. The role of competition in establishing axonal pathways and in molding terminal arbors could be tested by ablation of individual motoneurons or by delaying the outgrowth of individual identified growth cones.

Summary

The generation of specific neuronal connections requires accurate pathfinding and the formation of synapses with the appropriate number and type of target cells. While it is clear that quantitative matching of presynaptic and postsynaptic populations of cells depends upon competition between axonal terminals, trophic feedback between pre- and postsynaptic cells, and the modification of connections by the relative activity of different inputs, it is also apparent that neurons establish specific projections by axonal guidance along stereotyped pathways. Thus it appears that competitive interactions among axonal terminals serve mainly to refine the structure of arbors and to adjust the relative sizes and strengths of projections, whereas growth cone guidance during pathfinding serves to ensure the accuracy of axonal projections. Presumably both processes are required for synaptic specificity.

Studies of the initial innervation of body muscles by identified primary motoneurons in live zebrafish embryos have shown that individual growth cones make cell-specific, divergent pathway choices and grow directly to the region of muscle appropriate for their adult functions. Such directed growth has also been observed for developing axons in the leech (Kuwada and Kramer, 1983); insect central (Bastiani et al., 1985a,b) and peripheral (Edwards, 1977; Bate, 1978; Murphey et al., 1980; Palka et al., 1983; Keshishian and Bentley, 1983; Berlot and Goodman, 1984; Blair et al., 1985; Blair and Palka, 1985a,b) nervous systems; chick (Tosney and Landmesser, 1985b) and amphibian (Farel and Bemelmans, 1985; Westerfield and Eisen, 1985) motor systems; and *Daphnia* (Lopresti et al., 1973), amphibian (Harris, 1986), and mammalian (Sretavan and Shatz, 1986a) visual systems. Thus, despite opinions to the contrary (Easter et al., 1985), we suggest that vertebrate and invertebrate growth cones alike use common mechanisms to recognize and respond to environmental guidance cues that direct them to regions containing appropriate target cells. An understanding of the mechanisms used by neurons in pathfinding is a prerequisite for understanding the origins of neuronal specificity.

Peripheral Guidance Cues and the Formation of Specific Motor Projections in the Chick

LYNN LANDMESSER

Introduction

That the proper functioning of the nervous system depends at least in large part on the existence of highly ordered sets of connections between neurons and their targets is generally accepted. The mechanisms by which such arrays are formed during development is more controversial. Few would deny that some form of specific recognition, usually viewed as a variation of Sperry's chemoaffinity hypothesis (Sperry, 1963), is required. But there is now ample evidence that activity-dependent interactions between neurons can sharpen and possibly even create an orderly projection (Meyer, 1982; Schmidt and Edwards, 1983; Boss and Schmidt, 1984; Reh and Constantine-Paton, 1985; Sanes and Constantine-Paton, 1985), with only minimal information required in the form of chemical recognition molecules. The general consensus is that both of these mechanisms are used during neural

development and that some form of axonal guidance and target recognition sets up basic circuits, which are then made more precise by activity-dependent synaptic rearrangement.

More controversial has been the extent to which each of these two mechanisms is used in different species and in different parts of the nervous system. It has recently been proposed that although chemical recognition plays a dominant role in forming the simple and invariant nervous systems of invertebrates, it may play only a minor role in vertebrates, having been superseded by activity-dependent synaptic rearrangement (Easter et al., 1985). This in turn has led to the belief that the developmental mechanisms elucidated in invertebrates (Bentley and Caudy, 1984; Bastiani et al., 1985; Blair and Palka, 1985) are not relevant to vertebrate, and especially mammalian, neural development. I would contend, however, that the differences are not so much between phyla or species as between the tasks faced by the neurons in question. Thus the problem facing limb-innervating motoneurons is similar in both the grasshopper and the chick; they must grow from the central nervous system (CNS) to the limb and there make a series of divergent choices enabling them to synapse with appropriate muscles. One would predict that the mechanisms used would also be similar. Furthermore, since these basic motor circuits in vertebrates are formed early and can form in the absence of activity (Haverkamp, 1986; Landmesser and Szente, 1986), chemical recognition would be expected to play an important role. Real differences would exist even within a single species between the motor system just described and the visual system, where it is clear that the environment and patterned activity play a dominant role in circuit formation (Stryker and Harris, 1986).

Overall, activity-dependent synaptic arrangement is more widespread in vertebrates, but this is probably because it plays a dominant role in the newer parts of the brain, especially where there are large numbers of neurons in reiterated circuits, such as in the cortex and the cerebellum. This mechanism is best suited for partitioning targets among neurons of equivalent specificity, and would occur wherever there are many neurons of a given class. Thus it would also be expected to play a role in partitioning muscle fibers among the many neurons within a single vertebrate motoneuron pool, but not in establishing the match between that pool and its appropriate muscle.

In this chapter, I first provide evidence that the basic mechanisms employed to get motoneurons connected with appropriate limb muscles in a complex vertebrate such as the chick are quite similar to those used in invertebrates (Harrelson et al., Chapter 7) and in simpler vertebrate systems (Westerfield, Chapter 8), and second, that these mechanisms involve for the most part motoneuron growth

cones responding both to relatively nonspecific environmental cues that define major nerve pathways and to more specific target-derived cues. Next, I will present some evidence on the nature of these cues. Finally, since one problem with unambiguously demonstrating the existence of such cues in higher vertebrates has been the difficulty of making observations with the required spatial and temporal resolution in the complex in vivo nervous system, I will describe several simplified in vitro preparations that should prove amenable to this task.

How Specific is Motoneuron Outgrowth in the Vertebrate Limb?

In the grasshopper, where the behavior of growth cones arising from single identified motoneurons can be followed with good temporal resolution, motoneuron outgrowth has been shown to be highly precise. Motoneurons fasciculate with specific neuronal fascicles within the CNS, leave the CNS by cueing on a specific glial cell, and then grow along stereotyped pathways in the limb, until they exit the fascicle upon contacting specific muscle precursor cells (Ball et al., 1985).

In the chick limb, each muscle is innervated by hundreds of motoneurons, rather than a few, and these are distributed in elongated nuclei or motoneuron pools extending over several of the eight spinal segments that contribute to the limb (Landmesser, 1978a; Hollyday, 1980). By retrogradely staining the axons that project from a single muscle nerve, it has been shown that these axons are initially widely distributed within the segmental spinal nerves that contribute to that muscle, but as the spinal nerves converge in the two plexuses at the base of the limb, these axons sort out to form a compact cluster in a characteristic position (Lance-Jones and Landmesser, 1981a). The axons then maintain general topographical order as they grow down the stereotyped nerve pathways until diverging at a specific muscle nerve pathway. At that point their growth cones show highly individualistic behavior, often crossing over other axons to exit the nerve trunk (Tosney and Landmesser, 1985a). These observations led to the suggestion that axons must be responding to specific environmental cues in two places: in the plexus and at sites of muscle nerve emergence. The efficiency of this response could also be amplified by specific fasciculation among axons belonging to the same motoneuron pool, a suggestion for which there is as yet no evidence.

When the morphology and the trajectory of horseradish peroxidase- (HRP-) labeled growth cones were quantified, it was found that growth cones were larger and more lamellopodial, and made more abrupt turns at "decision" regions (the plexus and muscle nerve initiation points) than at other points along the axon's trajectory (growth from the cord to the plexus and then along major nerve trunks) (Tosney and Landmesser, 1985b). That axons were capable of responding to specific environmental cues had been previously demonstrated by rotating parts of the lumbosacral neural tube about the anterior-posterior axis, thereby displacing motoneurons (Lance-Jones and Landmesser, 1980, 1981b). Even when displaced by as many as five segments, axons would invariably synapse with their appropriate muscle if they entered their original plexus. They did this by first altering their trajectories in the plexus. It was concluded that motoneurons (1) must possess an identity, probably in the form of some specific molecule or subset of molecules on the cell surface, prior to axon outgrowth; (2) that there must exist environmental guidance cues in the limb; and (3) that displaced motoneurons are capable of responding to these guidance cues by altering their axonal trajectories (Lance-Jones and Landmesser, 1980, 1981a,b).

However, despite the demonstration some five years ago that specific guidance cues were necessary, we know nothing about their molecular nature and little about the way they might be distributed. For example, does each motoneuron pool bear a unique label or set of labels, or could a broad gradient of one or a few molecules explain the axonal guidance we have observed? Of some importance in distinguishing among these possibilities is knowing the precision with which axons from different motoneuron pools are able to make choices during outgrowth, ultimately synapsing with the appropriate muscle.

The earliest studies on the specificity of motoneuron outgrowth in this system, employing electrophysiological techniques (Landmesser and Morris, 1975), indicated that by stages 28 to 30, shortly after the onset of peripheral synapse formation and before the motoneuron cell death period, muscle nerves contained few, if any, segmentally inappropriate axons. Retrograde labeling of motoneuron pools at these stages confirmed that axons were making generally correct decisions, since the motoneurons were in essentially the same relative positions as the mature pools (Landmesser, 1978b). Since motoneurons are arranged topographically, the identities of motoneurons could be inferred from their positions within the spinal cord. In contrast, similar retrograde labeling studies in *Xenopus* indicated that a significant number of motoneurons were making projection errors and that these were corrected during the motoneuron cell death period (Lamb, 1976, 1979; but see Farel and Bemelmans, 1985; for the chick wing, see Pettigrew et al., 1979).

Retrograde HRP labeling suffers from the drawback that it is difficult to decide whether inappropriately positioned labeled cells have actually made errors in projection or have been labeled by diffusion of HRP away from the desired application site. This is especially a problem in developing systems, where the entire target structure is only a few millimeters in size. However, this drawback can be avoided by orthograde HRP labeling of axons through spinal nerve injection (Lance-Jones and Landmesser, 1981a). The axons whose segmental identity is obvious are stained to their growing tips, making it possible to chart with great precision the position of axons arising from a given segment throughout their growth into the limb. Confirming earlier observations on the chick hindlimb, these studies indicated that segmentally inappropriate axons could not be detected in muscle nerves at any developmental stage (Lance-Jones and Landmesser, 1981a; Tosney and Landmesser, 1985a; see for the chick wing, Hollyday, 1983). Thus, chick motoneuron outgrowth is highly precise at the segmental level.

It is necessary for axons to make more than segmentally appropriate decisions, however, since each segment of the cord contains a number of motoneuron pools, some with quite different physiological functions. As already mentioned, retrograde HRP labeling reveals that most motoneurons are also making motoneuron pool–appropriate decisions (Landmesser, 1978b; Farel and Bemelmans, 1985). However, relying on position alone to identify motoneurons presents a problem, since those motoneurons at the boundaries between pools cannot be unambiguously assigned to either pool. Thus, as many as 10 to 20% of lumbosacral motoneurons could make undetected projection errors. Fortunately,

motoneurons can be identified by another means: their functional activation patterns.

The isolated spinal cord–hindlimb preparation from chick embryos exhibits patterned motor output. A single shock to the thoracic cord activates lumbosacral circuits and elicits a series of kicks. Electromyogram (EMG) or muscle nerve recordings reveal that each motoneuron pool is activated in a characteristic manner, and many pools can be distinguished from one another by the duration of the EMG bursts as well as the length of the silent period at the start of each cycle (Landmesser and O'Donovan, 1984a). Thus, the sartorius motoneurons (Figure 1A top trace) exhibit a long silent period (approximately 500 msec) at the start of each cycle and burst durations of 2.5 to 3.5 seconds, whereas immediately adjacent extensor motoneurons in the same segment would exhibit minimal silent periods and burst durations not in excess of 1.5 seconds, similar to the caudilioflexorius, an extensor muscle shown in Figure 1A, bottom trace.

Therefore, motoneurons can also be identified to a large extent by their activation patterns. If, for example, any extensor motoneurons projected to the sartorius, this should be easily detectable, since they would fire during the sartorius silent period. Similarly, flexor motoneurons projecting to extensor muscles would produce abnormally long burst durations. However, for this form of motoneuron identification to be useful in developmental

FIGURE 1. Motoneuron activation patterns of embryonic chick extensor and flexor muscles in controls and ▶ following experimental manipulations. (A) Electromyographic recordings from stage 36 chick sartorius (flexor; top trace) and caudilioflexorius (extensor; bottom trace) muscle in an isolated spinal cord–hindlimb preparation during an electrically elicited movement sequence. The onset of several cycles is indicated by arrows. (B) An analysis of burst parameters of several experimental muscles innervated by foreign motoneurons following early anterior–posterior limb rotations, compared to control muscles. These histograms indicate when each muscle is active during a cycle, the zero time point being the onset of the extensor burst. Caudilio, caudilioflexorius; p. ITIB, posterior iliotibialis. (C) Similar histograms from sartorius and caudilioflexorius muscles in controls and embryos in which activity was chronically blocked during the cell death period. (B, from Landmesser and O'Donovan, 1984b; C, from Landmesser and Szente, 1986).

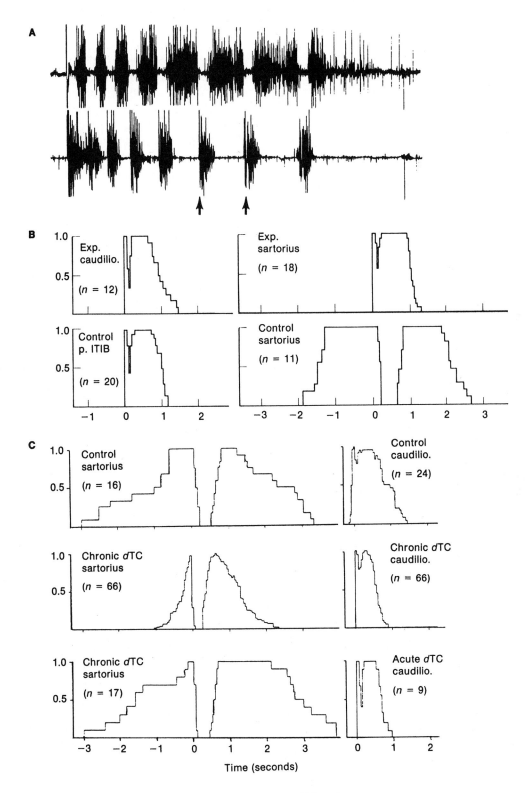

125

studies it must be shown that these motoneuron properties (duration of burst and silent period) are not altered when motoneurons project to inappropriate muscles.

This was demonstrated by causing motoneurons to project to foreign muscles following limb bud rotation prior to motoneuron outgrowth (Landmesser and O'Donovan, 1984b). It was found that motoneuron activation patterns developed autonomously and were not altered by innervation of an inappropriate muscle. Therefore, the identity of motoneurons could usually be inferred from the activation patterns and confirmed by retrograde HRP labeling. For example, the sartorius in the rotated limb (Figure 1B, top right) was activated as an extensor, with burst characteristics very similar to the control posterior iliotibialis (bottom left) and very different from the normal pattern in sartorius (bottom right). In fact, when this muscle was injected with HRP, all of the motoneurons were situated in the position of the posterior iliotibialis pool. In other cases, when hybrid activation patterns indicated that a muscle was innervated by several pools, this was confirmed by retrograde labeling.

Unfortunately, the characteristic activation patterns of motoneuron pools do not develop until the motoneuron cell death period, with possible removal of erroneous projections, is underway (O'Donovan and Landmesser, in press). However, all motoneurons can be rescued, including any that may have projected aberrantly, by chronic blockade of movement with in ovo injections of d-tubocurarine (dTC) or α-bungarotoxin (Pittman and Oppenheim, 1979). We repeated this study, using dTC, and at the end of the normal cell death period, isolated spinal cord preparations were made from treated embryos, the dTC was washed out, and motoneuron activation patterns were assayed (Landmesser and Szente, 1986).

We found that systemically applied dTC blocked not only the neuromuscular junction, but also cholinergic circuits within the cord, so that patterned bursts were not initially elicitable. However, after several hours of washing, typical burst sequences were obtained. Although burst durations and silent periods were shorter than control values for all pools assayed, flexor and extensor pools (sartorius and caudilioflexorius, respectively, in Figure 1C) were activated with distinctly different patterns, each similar to its own control pattern. The differences observed between experimental and control burst parameters appeared to be caused by a residual action of dTC on the spinal cord, since similar effects were produced by a single injection of dTC into untreated control embryos, 10 hours prior to sacrifice (compare the acute and chronic dTC-treated caudilioflexorius patterns shown in Figure 1C). In a few cases, this central effect was not as evident and burst parameters in chronically treated embryos did not differ from controls (Figure 1C, bottom left).

From these observations, as well as from anatomical observations made by Oppenheim (1981), we can conclude that chick lumbosacral motoneurons do not simply make segmentally appropriate choices, but make pool-specific choices with considerable precision. This strongly suggests that the motoneurons composing a pool may all be labeled in a similar manner that differs from those of immediately adjacent pools. Clearly, individual pools have dissimilar activation patterns, which probably result from differences in the interneuronal connections made onto them, as well as from differences in intrinsic membrane properties (M. O'Donovan, personal communication). It is therefore not unreasonable to propose that pool-specific differences in cell surface molecules could allow motoneurons to make specific and divergent choices during peripheral outgrowth similar to what has been described in arthropods (Bastiani et al., 1985). Since these choices are made in the chick limb by motoneuron growth cones as they grow through the plexus and muscle nerve decision regions, they must be due to recognition of specific environmental cues and cannot result from activity-driven synaptic rearrangement or from reggressive events such as axonal retraction and cell death.

The precise decisions that are made also appear to require more than the weak preference demonstrated by preganglionic autonomic neurons for ganglion cells and muscle fibers arising from similar segmental levels (Purves et al., 1981; Wigston and Sanes, 1985). For motoneurons, it has also been suggested that they might share labels in common with other cells arising from the same segmental level. They could then simply cue along and ultimately synapse with the somite-derived muscle precursor cells from the same segment that migrate into the limb ahead of them to form muscles (Landmesser, 1984). Individual limb muscles do derive their innervation and muscle precursor cells from the same segments (Beresford, 1983; Lance-Jones and Lagenauer, 1985). However, Lance-Jones and Lagenauer (in preparation and personal communication) have shown that following somite manipulations, motoneuron pools grow to and synapse with their appropriate muscles even though these are now composed of segmentally inappropriate muscle cells. Using chick–quail chimeras, Tanaka and Landmesser (1986) have also shown that motoneurons of one species synapse with the homologous muscles in limbs of the second species, even though these are in some cases derived from segmentally different sources. Thus, the rather pleasing and simple scenario described above is not consistent with available experimental data.

Could the functional activation of motoneuron pools in distinctive patterns play any role in axonal guidance or the clustering of like axons in the plexus? The activation of motoneuron pools in distinctive patterns does not arise until about stage 30, long after initial axonal outgrowth (O'Donovan and Landmesser, in press), and in isolated cord preparations from earlier stages, motoneurons fire only briefly and synchronously during "movement" sequences. Thus, a role for patterned neural activity seems unlikely, but until we have recordings from outgrowing axons, this possibility cannot be excluded.

General Guidance Features Responsible for the Position of Major Nerve Trunks

To explain the observations previously discussed, specific guidance cues must exist that are differentially responded to by subsets of motoneurons. Yet a variety of observations indicate the existence of an additional set of more general guidance features that are viewed as a preferred set of pathways for axonal growth or a highway system that corresponds to the gross anatomical nerve pattern (Lewis et al., 1983). This system is not specific, in that a variety of foreign axons will follow these same pathways when experimentally channeled into a given limb region (Hamburger, 1939; Lance-Jones and Landmesser, 1981b; Hollyday, 1981). However, this highway system may be important in constraining axons and also bringing them into limb regions where they can respond to specific guidance cues.

What do we know about this guidance system? During initial outgrowth, axons grow through a loose mesenchyme with considerable extracellular space. They do not grow along any preformed structures such as blood vessels or oriented arrays of mesenchyme cells; in addition, the mesenchyme cells immediately distal to the most advanced growth cones do not differ at the ultrastructural level from mesenchyme cells in regions where axons do not grow (Tosney and Landmesser, 1985c). Why, then, do axons prefer to grow through some regions and not others?

In the past, it was thought that axons could grow easily through most limb tissue and needed somehow to be constrained to stay on nerve pathways. However, a series of more recent observations has suggested that much of the developing limb may be rather inhospitable for axonal growth and that axons may require added factors to enable them to grow. To begin with, axons avoid regions rich in glycosaminoglycans (Tosney and Landmesser, 1985c). In addition, after growing rapidly to the plexus region at the base of the limb, they wait for several days (Hollyday, 1983; Tosney and Landmesser, 1985c), suggesting that the limb mesenchyme is initially not conducive to axon growth, a supposition supported by cultured slice experiments to be discussed later in this chapter (L. Landmesser, unpublished data). Experiments in which the limb was partially ablated (Tosney and Landmesser, 1984) showed that axons grew into a given limb region only if muscle fragments were present. This suggested that the muscle tissue may condition the intervening mesenchyme, allowing axonal growth through it. As in the plexus, motoneurons also wait for several days when they reach the differentiating muscles, until the developing myotubes begin to express neural cell adhesion molecule (N-CAM) (Tosney et al., 1986) and laminin (L. Dahm and L. Landmesser, unpublished data). Thus, in both the plexus and muscle regions, axons initially prefer to associate with other axons (Tosney and Landmesser, 1985a; Lewis et al., 1983), which suggests that the surrounding non-neural tissue is a poor substrate for axon elongation.

One possibility, then, is that nerve pathways are determined by the presence of molecules favorable for axonal growth, such as N-CAM and laminin. However, while present in low amounts in limb mesenchyme, N-CAM does not occur in greater amounts along presumptive neural pathways (Tosney et al., 1986). Laminin is present only in very low amounts distal to the growing nerve front and clearly does not define neural pathways (H. Tanaka and L. Landmesser, in preparation; but see Rogers et al., 1986). In fact, most motoneuron growth cones do not grow along the inner border of the myotome, which is very rich in both laminin and N-CAM, but through adjacent mesenchyme, where these molecules occur in much lower amounts (Tosney and Landmesser, 1985c; Tosney et al., 1986; H. Tanaka and L. Landmesser, in preparation).

Recent observations suggest another possibility. Using an anti-neural crest monoclonal antibody (Tanaka and Obata, 1984), we were surprised to see that many cells recognized by this antibody, and presumably of crest origin, formed a cloud around the distal nerve growth front (Figure 2) and appeared to extend somewhat beyond the growing tips of the axons (as visualized with a monoclonal directed against N-CAM or a motoneuronal cell surface antigen (H. Tanaka and L. Landmesser, in preparation). Thus, these cells are in a position to guide neurons and may be preferred to other non-neuronal cells. These "crest" cells have lower levels of N-CAM than surrounding mesenchyme cells and little, if any, laminin (H. Tanaka and L. Landmesser, in preparation). However, they have abundant Ng-CAM (L. Landmesser and U. Rutishauser, unpublished data). Thus, axon–crest interactions involving this as well as additional molecules may cause axons to follow certain pathways within the limb. This hypothesis must of course be experimentally tested. In addition, it will be necessary to determine why these "crest" cells follow the pathways and if in fact they are crest cells or some other cell type expressing a common epitope.

In Vitro Preparations to Study Guidance Cues

Based on the earlier described observations made on the developing chick motor system in vivo, a strong case can be made for

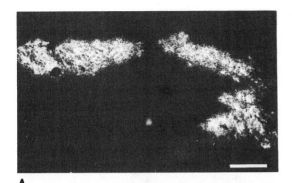

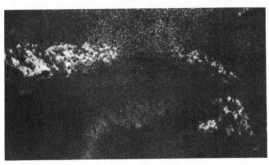

FIGURE 2. The distribution of cells expressing a neural crest epitope at the distal nerve growth front in a stage 23 1/2 embryo. (A) Frozen section, stained with a monoclonal antibody that recognizes neural crest, reveals a cloud of immunoreactive cells surrounding the nerve front (the crural trunk is on the left; the dorsal and ventral sciatic trunks, on the right). (B) The same section, stained with a monoclonal specific for N-CAM, reveals the axon fascicles. Bar = 100 μm.

the existence of specific guidance cues. However, it is impossible to elucidate the nature of these cues, especially their molecular basis, in the complex embryo. A major difficulty is inadequate temporal and spatial resolution. For example, it is usually not possible to know at what time and at exactly what position a given growth cone responded to a specific cue. This information is necessary in distinguishing among differing hypotheses of how guidance cues are distributed (i.e., as a broad gradient,

across the limb, or restricted to a more localized region, or restricted to a few specific cells, as in invertebrates). Another problem is that the embryo is not sufficiently accessible for experimental manipulation. Experimental surgery can be performed, but not with the precision needed to distinguish between proposed hypotheses. Also, in order to probe the molecular nature of guidance cues, it will be necessary to interfere with the function of specific molecules. Antibodies that block the function of specific molecules can be injected in vivo, but it is difficult to ensure their spatial extent and time of action. Other useful treatments might be toxic to the whole embryo. Therefore, it seems important to develop in vitro preparations in which more precise observations and manipulations can be carried out. Dissociated cell culture offers one possibility. Yet since the relevant cells participating in axonal guidance are not yet known, one runs the risk of studying an irrelevant cell type. More troubling is the possibility that a given growth cone response may require the participation of several cell types and/or molecules, and thus may not occur in the simplified culture situation. Therefore, I attempted to develop several preparations that, although simplified, conserve much of the in vivo environment encountered by growth cones.

The first consists of transverse slices through the entire embryo at the lumbosacral region. Embryos from stages 16 to 24 can be embedded in gelatin and sectioned on a vibratome to create slices of 200 to 350-μm thickness (Figure 3B). These are caused to adhere to tissue culture plastic dishes and placed in a defined N2 medium and cultured. Damage to the tissue appears minimal (compare the living slice in Figure 3B with the scanning electron micrograph specimen in 3A), and with transmitted light, considerable detail can be visualized, including individual cells and fascicles of axons. The morphology of the slices is maintained for up to 48 hours, although con-

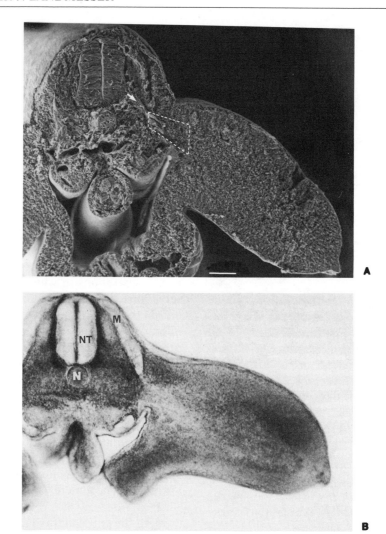

FIGURE 3. **A transverse view through the lumbosacral region of a Stage 20 embryo at the time of initial axonal outgrowth. Many of the same features can be seen in the scanning electron micrograph (A) and in a living 300-μm-thick cultured slice. NT, neural tube; N, notochord; M, myotome; in A, the plexus region is indicated by the region enclosed by the dashed line, and an arrow points to several outgrowing neural fascicles. Bar = 100 μm. (A, from Tosney and Landmesser, 1985c.)**

tinued growth results in an increase in slice thickness. At any point the neural tube can be injected with HRP to visualize the growing axons, including their growth cones (Figure 4A,B), and by comparing the slices with slices at the start of incubation, it can be shown that axonal outgrowth is similar to that in vivo. Although the general health of the slice persists, overgrowth of skin and disproportionate growth in some regions after 48 hours result in a sufficiently altered morphology that the preparation is no longer useful. Nevertheless, the 48-hour time window allows one to study several important time periods. Slices made at stages 18 to 19 can be used to study axon outgrowth from the cord to the plexus region. Slices made at stages 20 to 22 can be used to study axon behavior in the plexus and initial outgrowth into the limb. This later event encompasses the first important decision made by motoneurons: whether to project to the dorsal or the ventral nerve trunk. Finally, it is possible to label the outgrowing axons by injecting a lipid-soluble dye, DI-I (Honig and Hume, 1986), into the neural tube. Within two hours, the growing axons are labeled to their tips (Figure 4C,D), and since the dye is not toxic, multiple observations can be made

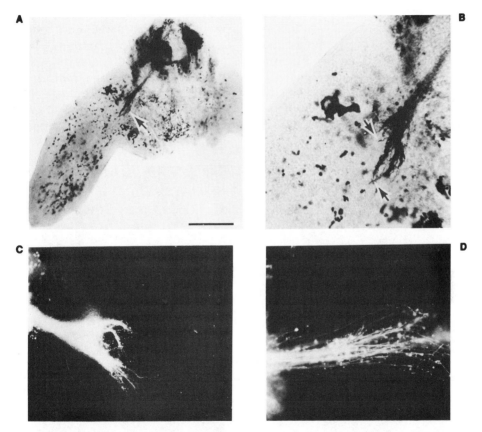

FIGURE 4. Labeling of outgrowing axons in cultured lumbosacral slices. (A) Low-power view of a slice in which the left lateral motor column was injected with HRP. After fixation and visualization of the reaction product, a mass of motor axons can be seen distal to the plexus (arrow) just beginning to diverge into dorsal and ventral nerve trunks. Other dark-staining profiles scattered in the tissue are red blood cells. (B) The distal growth front shown in A on higher magnification reveals individual axons and growth cones (arrows). (C) Visualization of living axons following injection of the lateral motor column with a solution of DI-I 10 hours earlier. The fluorescent dye has been incorporated into the membrane, allowing visualization of the growth front just distal to the plexus. (D) The plexus region, on higher magnification of another embryo, reveals individual axons and small fascicles. Bar = 250 μm in A; 280 μm in B, C; 50 μm in D.

without inducing any apparent damage. Therefore, it should be possible to observe growth cone behavior in real time. In summary, this preparation has numerous advantages for investigating factors involved in axon outgrowth and guidance; in addition, it should prove useful for studies on neural crest cell and somite cell migration.

If chemical guidance cues exist within the limb and/or plexus mesenchyme, additional information on their nature and distribution can be obtained by using axonal growth cones as a probe. Therefore, a second preparation was developed in which transverse limb bud slices are made and caused to adhere. In another embryo, spinal nerves are injected with DI-I to retrogradely label the motoneurons, the ventral neural tube is isolated,

and the cells are dissociated. The dissociated cells, greatly enriched for motoneurons, are then seeded onto the surface of the slices, where they adhere. Within 19 hours, axons are extended and, since they are fluorescent, their behavior can be observed with a standard fluorescence microscope (Figure 5). With this preparation it is possible to determine the trajectories taken by axons of known identity (anterior versus posterior, medial versus lateral, or even from a single motoneuron pool) at any position in the developing limb bud (L. Landmesser, in preparation). This should greatly facilitate the analysis of purported guidance cues and the manner in which motoneurons are labeled. It has already been possible to obtain evidence that the limb bud prior to stage 23 is not able to support axon el-

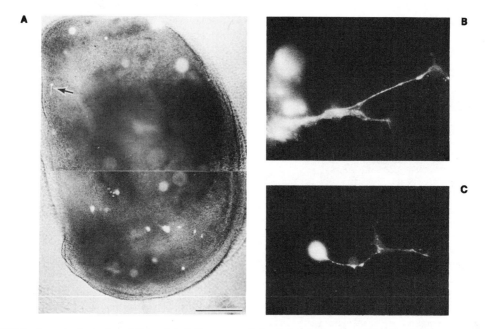

FIGURE 5. Axon trajectories of dissociated, fluorescently back-labeled cells cultured on transverse limb slices. **(A)** The anterior half of a 250 μm-thick transverse slice of a stage 24 limb bud cultured for 10 hours and then seeded with dissociated, DI-I–labeled motoneurons, and viewed with combined fluorescence and transmitted light 24 hours after plating the neurons. One motoneuron with axon is in focus (arrow); the other light splotches are motoneurons out of the focal plane. **(B, C)** Fluorescence optics showing labeled motoneurons and their axon trajectories at higher magnification. Bar = 250 μm in A, 50 μm in B, C.

ongation. When slices are made from limbs from stages 18 to 22, motoneurons adhere to them and remain viable, but none send out axons. On slices made from stages 24 to 25 limb buds, 84% of all DI-I-labeled motoneurons extend axons.

Summary

A considerable body of evidence reveals that axon outgrowth is highly precise in limb-innervating regions of the spinal cord in higher vertebrates. Segmentally identified motoneurons have been shown to make a series of divergent choices with little error (Landmesser and Morris, 1975; Lance-Jones and Landmesser, 1981a; Hollyday, 1983; Farel and Bemelmans, 1985; Tosney and Landmesser, 1985a). Thus, despite the much larger number of cells and the complexity of axons from many segments converging in the plexus, the behavior is similar to what has been described for zebrafish primary axial motoneurons (Myers et al., 1986) and in inverte- brates (Ball et al., 1985; Bastiani et al., 1985). It seems likely that axonal guidance via specific recognition molecules must play a dominant role in setting up this motor system. The available evidence suggests that the guidance features can be divided into two types. General guidance features that are capable of being followed by all motoneurons regardless of their identity create a limb "highway" system. Molecules such as NCAM and laminin do not seem to determine this highway system, although preliminary evidence indicates that neural crest-derived, probably presumptive glial, cells may be involved. Specific guidance cues are also required and these may involve selective fasciculation of axons as well as interactions between axons and other non-neuronal mesenchyme cells. Although the complexity of the in vivo system makes further analysis of these specific cues difficult, several in vitro slice preparations have been developed that should greatly facilitate the analysis of their distribution and ultimately their molecular basis.

CHAPTER TEN

Growth and Guidance of Axons in Two Pathways from the Mammalian Cerebral Cortex

KATHERINE KALIL

Introduction

Axons growing out in the developing central nervous system and those regenerating after injury must not only elongate but also find their targets by appropriate pathfinding. The mechanisms underlying these events, and the limited ability of damaged axons to regrow, are poorly understood in the mammalian central nervous system (CNS). In the present chapter, I describe efforts to understand how axons elongate during development and regeneration, and how they might be influenced to select certain pathways and arrive at proper destinations. Most of these studies were carried out in the pyramidal tract of hamsters, a long CNS pathway that arises from the cerebral cortex and extends the entire length of the spinal cord. An important feature of this system is that the axons grow out almost entirely after birth, so that they can be observed and manipulated in the postnatal animal. Another postnatally developing pathway arising from the cerebral cortex, the corpus callo-

sum, was also useful for studying axon pathfinding.

Precise knowledge of the time course of axon outgrowth in the pyramidal tract made it possible to correlate axon elongation with the elevated expression of a rapidly transported growth-associated protein, GAP-43. Moreover, elevated synthesis of this protein is a consistent feature of regenerating axons as well (Skene and Willard, 1981a,b), and the ability of young but not mature pyramidal tract axons to regrow after they are severed correlates closely with a developmental decline in the expression of GAP-43 (Kalil and Skene, 1986). The ability of axons to elongate during limited periods of development and for only circumscribed times after injury in young animals, when expression of GAP-43 is high, had led to the hypothesis that elevated synthesis of GAP-43 is directly involved in axon growth and that a failure to reinduce these proteins after injury may limit the axon's ability to carry out some steps in regeneration.

In an effort to elucidate some mechanisms

for guidance of growing axons, we focused on two aspects of pathfinding: fasciculation of neighboring axons, and the behavior of axonal growth cones in developing pathways. Pyramidal tract outgrowth was studied in vivo by injection of various anatomical tracers into the cortical neurons of origin or into the axons themselves. Axons in this pathway course parallel to one another in bundles, or fascicles, and appear to maintain consistent relationships with their neighbors as they course caudally. However, as axons cross within the pyramidal decussation to the contralateral side of the spinal cord, they defasciculate and undergo a dramatic reorganization. This observation has led us to conclude that axons from neighboring regions of the cortex actively rearrange to produce a new fiber order as axons enter the spinal cord. Further, during regeneration of the pyramidal tract, neither the original pathway nor the normal fiber order is preserved. Nevertheless, regenerating axons are able to establish functionally appropriate connections in the spinal cord. Axons in the developing corpus callosum also fail to preserve spatial order in axon trajectories from one set of neurons to their distant targets. Together, these observations argue that, at least in these pathways, fiber order must be relatively unimportant in the axon's ability to find proper sites of termination.

In recent years there has been renewed interest in the study of axonal growth cones in vivo, particularly as in vitro studies suggest that the morphology of the motile axon tip may reveal how axons navigate through their environment. The hamster corpus callosum, like the pyramidal tract, develops to a large degree postnatally and is therefore an ideal system in which to observe growth cones as they advance from one cortical hemisphere to the opposite side. We were especially interested in whether the shape and complexity of growth cones might reveal changes consistent with regions in which axons changed their behavior. Observations of several hundred

growth cones revealed a variety of growth cone morphologies from broad, delicate, veil-like lammelipodia to complex multibranched filopodia. Many of these spread over surprisingly large regions of the callosum, and the spatial extent of individual fibers was often accentuated further by multiple branching within the white matter. However, in contrast to some recent observations of growth cones in the peripheral nervous systems of vertebrate and invertebrate species, there were no consistent spatial or temporal relationships of growth cones to distal or proximal regions of their callosal trajectory that would suggest specific axonal functions such as pioneering, tracking, or decision-making behavior. As axons entered the contralateral cortex, however, the growth cones did become significantly smaller and simpler, suggesting possible changes in speed, adhesivity, or exploratory behavior when the growing tips neared their targets.

This brief overview suggests that despite an often discouraging complexity, the mammalian CNS provides opportunities for studying mechanisms of axon outgrowth and guidance. These may differ markedly from those of invertebrates. The present chapter will, when possible, attempt to underscore the significant differences between mammalian and non-mammalian systems in the growth and guidance of axons.

Development and Regeneration of the Pyramidal Tract

The hamster pyramidal tract arises from layer V of the sensorimotor cortex and extends the entire length of the nervous system, projecting to regions of the brainstem and to all segments of the spinal cord. In hamsters it is a completely crossed system, so that fibers from one cortical hemisphere terminate in the contralateral spinal cord after crossing in the medulla (Reh and Kalil, 1981). Corticospinal fibers project to the dorsal horn and to interneurons ventrally, but direct connections to motoneurons have not been demonstrated in hamsters. This pathway, as in other mammals, plays an important role in fine motor control, especially of the limbs (Reh and Kalil, 1982b; Keifer and Kalil, 1985).

The hamster CNS is extremely immature at birth, which occurs after a 16-day gestation period. Cortical neurons are still migratory and pyramidal tract axons have penetrated through the internal capsule only as far as the midbrain. As shown in Figure 1, the postnatal extension of the axons through the decussation does not occur until two days postnatal, and caudal growth into the spinal cord is not complete until almost two weeks after birth (Reh and Kalil, 1981).

In the adult mammalian CNS, severed axons typically do not regenerate. However, we found surprisingly vigorous regrowth of the hamster pyramidal tract if the axons were cut during stages of development when axons are actively elongating (Kalil and Reh, 1979, 1982; Kalil, 1984). Because the pyramidal tract forms a discrete pathway on the ventral surface of the brainstem, it was possible to sever the fibers on one side of the medulla above the decussation at various postnatal ages without damaging other regions of the brain. Injections of tritiated amino acid tracers into the appropriate region of the cortex ipsilateral to the lesion labeled the pyramidal tract and revealed massive regrowth of the severed axons. The regrowing fibers emanate from the pyramidal tract axons above the lesion but do not follow their normal route through the decussation and into the dorsal column of the spinal cord. Instead, the regenerating axons cross to the opposite side of the brainstem near the site of the injury several millimeters rostral to the normal decussation. The axons then coalesce into a compact bundle in a lateral region of the brainstem and spinal cord. Although the regenerating axons never resume a normal course in the dorsal column, they are

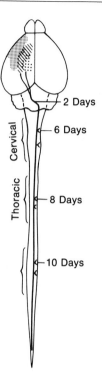

FIGURE 1. Schematic diagram of postnatal outgrowth of the hamster pyramidal tract, showing days when axons begin to innervate various levels of the spinal cord. Topography of corticospinal neurons is represented by large dots (lumbar area), crosshatching (cervical area), and small dots (trigeminal representation).

nevertheless able to project to their appropriate targets in the spinal cord. The establishment of synapses by the regenerating fibers was verified by electron microscopy (Kalil and Reh, 1982) and their functional role in fine motor control of the forelimb has been demonstrated by behavioral tests (Reh and Kalil, 1982b).

An important result of these experiments is that severed axons are only able to regrow for a limited period of time after an injury early in life. Regrowth is maximal when the pyramidal tract is cut between four and eight days of age. Thereafter, both the number of axons and the distances over which they regrow begin to decline, such that no regrowth occurs in animals more than three weeks of age (Kalil and Reh, 1979, 1982). The period of maximal regrowth corresponds to a period of extensive axon elongation. If axons are severed during this time, they are able to respond with further growth leading to formation of appropriate synapses. After this period, when axons have completed their normal growth, axotomy leads to retrograde degeneration rather than regrowth (Figure 2).

Pyramidal Tract Outgrowth and Expression of a Growth-Associated Protein

What can account for the failure of adult pyramidal tract axons to regenerate, whereas axons injured during development are capable of further elongation? Why is the mammalian peripheral nervous system able to respond to injury with functional regeneration, whereas the adult CNS does not normally do so? What factors permit pathways in the CNS of fishes and amphibians to regenerate but limit regeneration in the mammalian CNS? One hypothesis to account for these various dichotomies is that certain axonal proteins necessary for axon growth are expressed in much greater abundance in developing axons than in mature axons and that these proteins are not reinduced after injury to mature CNS axons that fail to regenerate (Skene and Willard, 1981b). In the last several years it has been shown that elevated synthesis and subsequent rapid transport of a small number of "growth-associated proteins" (GAPs) accompanies regeneration in the CNS of fishes and amphibians (Benowitz and Lewis, 1983; Benowitz et al., 1981; Heacock and Agranoff, 1982; Skene and Willard, 1981a) and in the mammalian periphery (Skene and Willard, 1981b). If elevated synthesis of GAPs is important for axon growth, one would also expect to find that during normal development, high

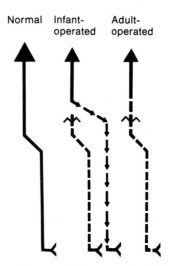

FIGURE 2. Schematic diagram showing pyramidal tract in normal, infant-operated, and adult-operated hamsters. After an infant lesion, axons below the cut degenerate (dashed line) but axons above the lesion regrow (arrows). After an adult lesion, fibers in the tract degenerate (dashed line) above and below the lesion.

levels of GAPs are expressed in neurons of the mammalian CNS. This is true for at least one of them (GAP-43), which does appear to be developmentally expressed in many regions of the mammalian CNS (Jacobson et al., 1986).

It was important, however, to show a precise temporal correlation between GAP-43 synthesis and axonal outgrowth. Moreover, a correlation between failure of adult CNS regeneration, as we had found in the pyramidal tract, and failure of axotomy to reinduce GAP synthesis would lend further support to the hypothesis that synthesis of these proteins is important for regenerative axon growth. Because we knew the exact time course for normal and regenerative axon outgrowth, the hamster pyramidal tract proved to be an ideal system for studying the relationship between axon outgrowth and GAPs in the mammalian CNS (Kalil and Skene, 1986). At various ages we injected the hamster sensorimotor cortex with ^{35}S methionine. Two-dimensional gel electrophoresis of ^{35}S methionine-labeled proteins synthesized in the cortical neurons and subsequently rapidly transported into the pyramidal tract revealed, first, that a very acidic protein with an apparent molecular weight of 43,000Da is expressed at a much higher level in 12-day-old hamsters than in four-week-old animals (Figure 3). Densito-metry showed that labeling of this protein relative to the total radioactivity on each gel declines 10 to 12-fold between 12 and 28 days of age. Second, this developmentally regulated 43kDa protein transported in hamster pyramidal tract comigrates on these gels with GAP-43, a protein induced during regeneration of toad optic nerve (Skene and Willard, 1981a). The behavior of this protein on gels, and the observation that an antiserum raised

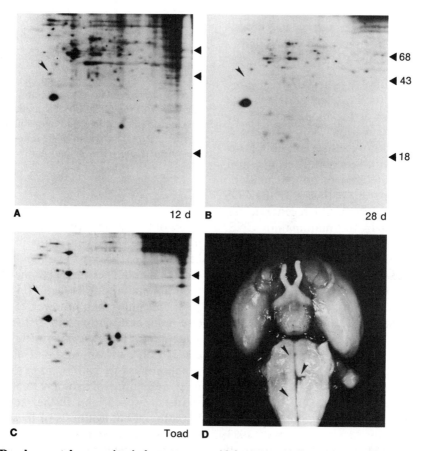

FIGURE 3. **Developmental expression in hamster pyramidal tract at 12 days (A) and 28 days (B) of a protein (arrows) similar to GAP-43 from regenerating toad optic nerves (C). Numbers at right indicate the positions of molecular weight markers: BSA (68,000), ovalbumin (43,000), and β-lactoglobulin (18,000). Acid side is to the left in all fluorographs. (D) Ventral view of an adult hamster brain, illustrating the segment of the pyramidal tract (between arrows on left side) used in two-dimensional gel electrophoresis. The arrow at right shows the position of a typical pyramidal tract lesion. (From Kalil and Skene, 1986.)**

against a developmentally regulated protein from rat brain recognizes both toad GAP-43 and the hamster protein (Jacobson et al., 1986), led us to conclude that this protein is indeed GAP-43. It was also possible to show a precise correlation between outgrowth of the pyramidal tract during development and decline in synthesis of GAP-43. Synthesis of GAP-43 is maximal during the first week of life but, as shown in Figure 4, begins to decline after eight days. This corresponds to a time when many of the axons in the pyramidal tract destined for the rostral spinal cord have stopped elongating. During the next week, the further decline in GAP-43 corresponds to the cessation of growth by axons arriving in the caudal spinal cord. The synthesis of GAP-43 declines toward adult levels by the end of the third postnatal week, when innervation of the spinal cord by the pyramidal tract is virtually complete (Reh and Kalil, 1981).

Not only is normal developmental outgrowth of the pyramidal tract well correlated with GAP-43 synthesis, but the declining ability of the severed axons to regenerate is accompanied by declining levels of GAP-43 synthesis. Injury to the pyramidal tract during development neither reverses nor delays the normal developmental decline of this protein, as shown in the vignettes in Figure 4, nor does axotomy of the adult pyramidal tract reinduce GAP-43 synthesis at any time after the injury. The failure of pyramidal tract injury to reinduce GAP-43 synthesis is similar to failure of GAP induction after injury to adult rabbit optic nerves (Skene and Willard, 1981b), which suggests that this may be a failure common to injured mammalian CNS pathways and that synthesis of GAPs may be one limiting event in the growth and regeneration of mammalian CNS axons.

At present, the mechanisms for the developmental regulation of GAP-43 are unknown. One possibility is that GAP-43 synthesis is regulated by axon–target interactions. Negative evidence for this is provided by the work of Benowitz et al. (1983), who showed that synthesis of a GAP-43–like protein increases greatly during early stages of regeneration of the goldfish optic nerve but declines whether or not the tectum is present. In the hamster pyramidal tract, the observation that axotomy neither prevents nor delays the decline of GAP-43 synthesis is also consistent with a regulation of GAP-43 that is independent of the target in neonatal hamsters.

A mounting body of evidence suggests that adult mammalian CNS neurons can respond to the local non-neuronal environment in dramatic ways. Whereas damaged axons do not typically regenerate within the adult CNS, various CNS axons, both injured and intact, have the ability to extend into peripheral nerve bridges implanted into the CNS (David and Aguayo, 1981; Richardson and Issa, 1984; So and Aguayo, 1985). In many cases regenerative axon elongation into the peripheral graft is more extensive than normal, even though growth back into the CNS is limited. These results emphasize the importance of the environment through which the axons grow and suggest the possibility that GAP-43 may be environmentally regulated. One form of environmental regulation consistent with our observations is that the maturing CNS environment might produce an inhibitor of GAP-43 synthesis. If this were the case, an inhibitor along the pathways of developing pyramidal tract axons would bring about the developmental decline of GAP-43, or, in the case of neonatal injury, a decline in GAP-43 in the portion of the axon rostral to the lesion.

Further evidence that GAP-43 participates in some aspects of axon growth comes from several recent reports showing that GAP-43 is a major component of growth cone membranes (Meiri et al., 1986; Skene et al., 1986). Although antibodies to GAP-43 react only weakly in adult tissue (Meiri et al., 1986), and GAP-43 is more highly concentrated in

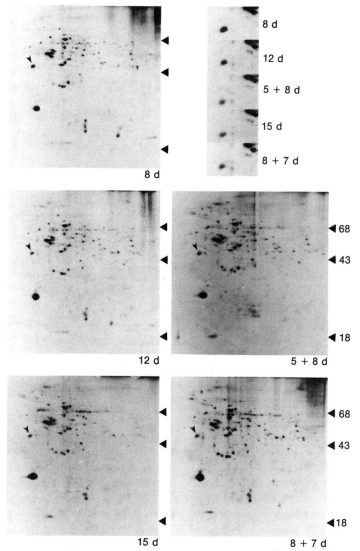

FIGURE 4. Early time course of decline in GAP-43 in normal and axotomized neurons. Pyramidal tract samples were from animals in which the cortex was labeled with ^{35}S methionine at 8, 12, and 15 days postnatal (left) or 7 to 8 days after a lesion at 5 or 8 days (right). Fluorographs were exposed so that the product of the exposure time multiplied by the total amount of radioactivity applied to the gel was constant for all samples. Acidic end is to the left; position of GAP-43 is indicated by arrows. Vignettes at upper right are of regions of the fluorographs containing labeled GAP-43. Other numbers at right as in Figure 3. (From Kalil and Skene, 1986.)

growth cones than in mature synaptic terminals (Skene et al., 1986), similarities have also been found between GAP-43 and several other proteins associated with functions requiring plasticity in mature synapses (Chan et al., 1986). These observations have led to suggestions that such proteins are important not only for growth cone mediated axon extension, but also for plastic changes at adult synapses such as those underlying learning and memory (Meiri et al., 1986; Jacobson et al., 1986). Thus, in the future it will be of great interest to understand how GAPs are regulated and whether they might mediate changes in cytoskeletal and membrane organization underlying axon outgrowth during development and regeneration, as well as synaptic plasticity in adulthood.

Pathfinding by Pyramidal Tract Axons

Regenerating pyramidal tract axons do not follow their original pathway in the CNS, yet they still terminate in appropriate regions of the spinal cord to mediate normal motor behavior (Kalil and Reh, 1982; Reh and Kalil, 1982b). Evidence from a number of studies on central and peripheral avian and amphibian neural pathways (Lance-Jones and Landmesser, 1981; Harris, 1984, 1986; Constantine-Paton and Capranica, 1976) has shown that axons arising from embryonic implants that leave the CNS at abnormal positions or that enter the CNS from regions of displaced neural tissue can travel over anomalous trajectories and yet terminate in appropriate targets. This suggests that stereotyped pathways may not be absolutely necessary in achieving correct target innervation. One possibility is that guidance cues in the vertebrate brain are not restricted to the particular pathways that are normally followed. A more broadly distributed guidance mechanism perhaps based on gradients of adhesivity between axons and their substrates would help to explain why

regenerating pyramidal tract axons growing in anomalous regions of the brainstem nevertheless always grow in the caudal direction toward their targets in the spinal cord.

Despite the adaptability of growing axons to pathfind in foreign territory, it is clear that in normal development axons take highly stereotyped routes toward their targets. In the central and peripheral nervous systems of grasshoppers it has been possible to analyze the stereotypical pathfinding behavior of axons at the level of single identified neurons. Studies of "pioneering" axons have shown that these fibers forge the first neural pathways by orienting toward specific environmental cues that, at certain choice points, may be localized to "guide post" cells (Bentley and Caudy, 1983; Caudy and Bentley, 1986). Repeated observations in both fly and grasshopper embryos have shown that pioneering neurons establish stereotyped pathways by orienting toward neuronal and non-neuronal cues and that later axons selectively fasciculate with earlier fibers by making specific choices via their growth cones (for a review see Bastiani et al., 1985; Harrelson et al., Chapter 7). These observations have led to the "labeled pathways" hypothesis, according to which growth cones in the CNS are guided by selective fasciculation with specific axons.

Can these insights be applied to complex vertebrate pathways, in which axon numbers far exceed a handful of identifiable fibers and, in the case of the mammalian CNS, may number hundreds of thousands of axons? To address this question we decided to look at the patterns of fasciculation in two major mammalian CNS pathways: the pyramidal tract and the corpus callosum. Both of these systems originate from the cerebral cortex and develop postnatally in the hamster. Pyramidal tract neurons in the sensorimotor cortex are topographically arranged in a map of the animal's body from the earliest ages at which the neurons can be retrogradely labeled. As illustrated in Figure 1, axons arising from neurons in different regions of the cortical map must travel over long pathways to match up with spinal cord segments related to appropriate regions of the body. Similarly, callosal fibers must match up with homotypic regions of the contralateral cerebral hemisphere (Wise and Jones, 1976).

A parsimonious model for reducing the amount of information to match up these thousands of axons with their targets is one in which unique identities are not conferred upon single fibers, but upon bundles or fascicles of axons that travel together and behave in similar ways. To some extent this resembles the passive spatial model invoked by Horder and Martin (1978), in which populations of axons are passively channeled into pathways that maintain nearest neighbor relationships among the fibers, which in turn reflect the topographic positions of their neurons of origin. To test this hypothesis we injected various anterogradely transported dyes such as horseradish peroxidase (HRP) and rhodamine into the cerebral cortex of newborn hamsters so as to label efferent axons in the

pyramidal tract and corpus callosum (Kalil and Norris, 1985, 1986). In some instances, HRP was injected directly into the axons. As shown in Figure 5, the initial segment of the pyramidal tract in the internal capsule shows a remarkable degree of spatial order. Axons are grouped into stereotyped fascicles that appear in consistent patterns from one animal to the next and represent a spatial map of the cortical cells (Kalil and Norris, 1985). As axons begin to extend through the brainstem, fiber order is preserved, but as they approach the decussation, fiber order becomes less precise. At the decussation, the axons completely reorganize. Filling of axons by local injection of HRP into the pyramidal tract (Figure 6) shows that individual axons leave their parent fascicles to seek out new partners. In many in-

FIGURE 6. Camera lucida drawing of pyramidal tract axons at the decussation labeled by local injection of HRP into the pyramidal tract of a 12-day-old animal. Coronal section shows highly individual axonal trajectories, as axons leave their original fascicles. Bar = 0.05 mm.

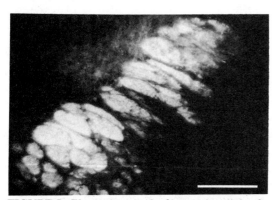

FIGURE 5. Photomicrograph of internal capsule of a hamster whose sensorimotor cortex was injected with rhodamine B isothiocyanate (RITC) at 2 days. Survival time was 2 days. Fluorescence microscopy shows characteristic fascicles formed by RITC-labeled efferent cortical axons. Bar = 1 mm.

stances, axons weave in and out, taking highly individual trajectories among hundreds of other axons. Even within a given fascicle, axons maintain neighboring relationships for only very short distances. As axons leave the decussation to enter the spinal cord, they form new fascicles that are composed of fibers different from those observed rostral to the decussation (Figure 7). The composition of the fascicles within the spinal cord is not known, but anterograde labeling from the cortex as well as retrograde labeling from the spinal cord, shows that the fascicles do not represent axons from contiguous neurons in the cortex or fibers destined for a given segment of the spinal cord. Similar results were obtained in the corpus callosum. Initially, axons arising from a small region of the cortex enter the white matter as a bundle. However, as axons progress toward the midline, they begin to separate from their original neighbors and undulate widely over broad regions of the callosum (Figure 8). Although axons from a given cortical region stay in a narrow anterior-posterior plane, individual axons make

numerous turns, which sometimes extend from the dorsal to the ventral borders of the callosum.

Thus, passive models based on fasciculation of spatially contiguous axons in which fiber order into the target region is maintained cannot account for the active rearrangement of fibers at the pyramidal decussation or within the callosum. A passive, nonselective adhesion of fibers that leave the cortex close to one another could explain the grouping of axons into bundles in the initial segments of these pathways. However, the subsequent local fiber rearrangements in which individual axons show unique, and often tortuous, trajectories seem to suggest that fibers are responding to environmental cues as if they had a distinct and separate identity. One possibility is that even in mammalian pathways containing thousands of axons, single fibers obey directional cues in the environment much as individual axons in the simpler nervous system of the grasshopper embryo pathfind by making decisions at certain choice points about which other specific axons to follow. The pres-

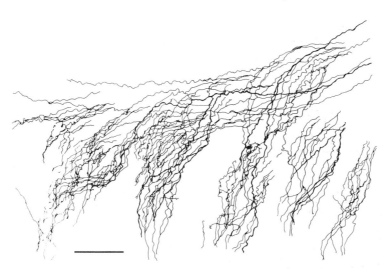

FIGURE 7. Camera lucida drawing of pyramidal tract axons at the decussation labeled by local injection of HRP into the pyramidal tract of an 18-day-old animal. Parasagittal section shows axons leaving their original fascicles within the decussation to join new fascicles as they enter the spinal cord. Bar = 0.05 mm.

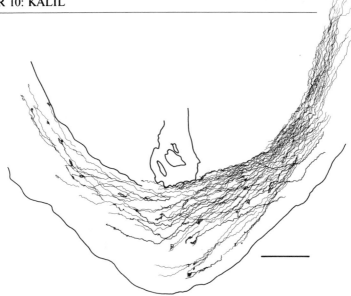

FIGURE 8. Camera lucida drawing of callosal axons and their growth cones labeled by HRP injected into the sensorimotor cortex of a hamster 5 hours postnatal. These represent some of the earliest callosal fibers. Note the wide divergence of axons as they approach the midline. Bar = 1 mm.

ent results, as well as those from other complex vertebrate systems, suggest that during development fibers may be making decisions about their trajectories, but that these decisions need not necessarily involve tracking along other specific fibers. For example, within the developing plexus region of the peripheral nerves innervating the chick hindlimb, there is an active sorting out of spinal nerves (Lance-Jones and Landmesser, 1981). Previously closely fasciculated fibers diverge, making it unlikely that axons navigate in this region by fasciculating with early pioneering fibers (Tosney and Landmesser, 1985). In the embryonic optic nerves of the monkey and cat, growth cones lose their original neighboring fibers over very short distances, and individual axons move freely between fiber fascicles (Williams and Rakic, 1985; Williams et al., 1986). Moreover, preliminary evidence in the corpus callosum (K. Kalil and C. Norris, unpublished data) does not suggest a model of

axon pathfinding based on fasciculation of later fibers onto earlier pioneering axons, since the earliest fibers to traverse the callosum are widely divergent. However, more work is needed to be certain that the axons labeled with the HRP techniques represent the very first fibers to enter this pathway.

Growth Cone Morphology in the Corpus Callosum

At present it is not known what environmental influences could account for the highly individualistic trajectories of single axons in certain regions of the pyramidal and callosal pathways. The complex turns taken by axons in both systems, particularly when they cross the midline, suggest a series of decisions by single fibers in response to molecular cues in the environment. Evidence from observations both in tissue culture (Letourneau, 1975, 1983) and in the embryonic grasshopper central and peripheral nervous systems (Caudy and Bentley, 1986; for a review, see Bastiani et al., 1985) suggests that growth cones respond differentially to surfaces based on their degree of adhesion to the substrate. When insect growth cones are faced with decisions at choice points, they become broad and complex, whereas growth along connectives is characterized by long, tapered growth cones (Raper et al., 1983). Since the decisions made by growth cones determine the pathways taken by the axons trailing behind them, there has been a recent surge of interest in describing the morphology of growth cones in the vertebrate nervous system, particularly in regions where growth cone behavior might reflect directional "decisions." For example, in the peripheral nerves of the lumbosacral region of the chick embryo, Tosney and Landmesser (1985) differentiate between nondecision regions such as nerve trunks, where axons maintain a constant fiber order as they grow out, and decision regions such as the plexus region, where axons are confronted with several possible directional decisions. They found that in decision regions, axons diverge widely and their growth cones are large and more likely to be lamellipodial, whereas in nondecision areas such as spinal nerves or trunks, small varicose forms are observed. However, the first growth cones in the plexus region do not seem to suggest, by their larger size, any specific pioneering role. Similarly, a study of growth cone morphology in the embryonic mouse retinal axon pathway showed that simple forms predominate when axons follow a well-defined pathway, but at choice points such as the optic chiasm, growth cones appear broader and more complex (Boloventa and Mason, in press).

The corpus callosum has several advantages as a mammalian system in which to correlate growth cone morphology with possible modes of axon guidance. Fiber outgrowth across the callosum is largely postnatal, and with injections of HRP crystals into the cortex, growth cones can be well labeled because of the short distances from the cortical cell bodies to the ends of the growing tips. Furthermore, HRP-labeled axons and their growth cones tend to remain in a narrow anterior-posterior plane, making it possible to observe most of the processes of a given growth cone within the thickness of a single 100-μm coronal section. Following injections of HRP crystals into the sensorimotor cortex of animals several hours to four days postnatal, perfusion-fixed brains were sectioned on a vibratome and reacted with the DAB method (Figures 9 to 12). Camera lucida drawings and measurements of over 400 growth cones in the developing corpus callosum revealed a wide variety of elaborate growth cone forms, averaging about 20 μm in length and 5 μm in width. Lamellipodia were by far the most common growth cone type, accounting for about 80% of all those studied in the callosum. As shown in Figure 9, lamellipodia closely resemble the ruffled veil-like structures observed in vitro, and frequently show diverging branches. It is not known how growth cones branch in vivo, but in vitro branching can occur by exertion of tensions along the lateral margins of the growth cone body (Letourneau, 1986). In some cases, a miniature growth cone, probably signifying the beginning of a branch point, arises from the base of a fully formed growth cone (see Figure 10B). In other cases, collateral branches arise directly from the axon (see Figure 11). Figure 10 shows typical examples of numerous filopodia extending from the body of the growth cone (A, B) and several long filopodia branching from the growing tip (C, D). Blunt club-shaped endings described in other vertebrate pathways (Tosney and Landmesser, 1985) are extremely rare in the callosum. We conclude that most of the growth cones in the callosum at the ages examined have complex morphologies (Figure 13), independent of whether they are early or

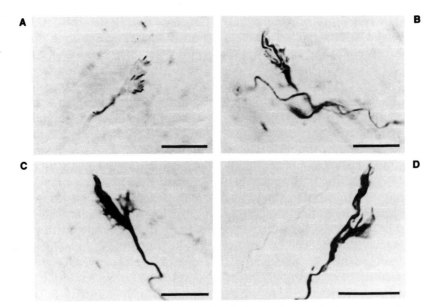

FIGURE 9. Photomicrographs of HRP-labeled lamellipodial growth cones in the corpus callosum. (A) and (B) (5 hours and 2.75 days postnatal, respectively) show typical veil-like lamellipodia without filopodia. (C) and (D) (5 hours and 20 hours postnatal, respectively) show ruffled lamellipodia with diverging branches. Bar = 10 μm.

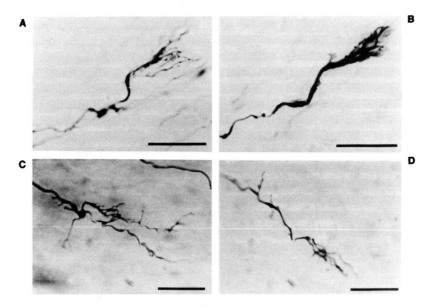

FIGURE 10. Photomicrographs of HRP-labeled filopodial growth cones in the corpus callosum. A and B (20 hours postnatal) are typical examples of numerous filopodia arising from the body of the growth cone. C and D (24 hours postnatal) show several long, branching filopodia extending from the growing tip. Bar = 10 μm.

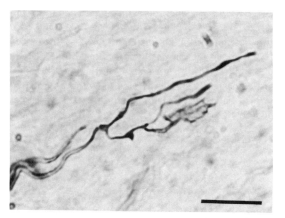

FIGURE 11. Photomicrograph of HRP-labeled callosal axon with branches arising directly from axon (5 days postnatal). Note different growth cone morphologies on each ending. Bar = 10 μm.

late. Our results also imply that growth cone morphologies are transitory shapes expressed at any point along the callosal trajectory. Further, in this system axons that might play a pioneering role need not have growth cones that are larger or more complex than those that follow. It is nevertheless interesting that in a pathway consisting of axons that do not appear highly fasciculated but have widely diverging trajectories, the growth cones are predominately complex, suggesting that the axons might be relatively independent pathfinders.

In contrast to the uniformity of growth cone size in the callosal white matter, growing tips entering the cortex three to four days postnatal are consistently smaller (approximately one-third the length of those in the callosum). As shown in Figure 12, growth cones in the cortex appear simpler and less spread out. In tissue culture, rapidly moving growth cones have large, spread-out lamellipodia, whereas those pausing or moving more

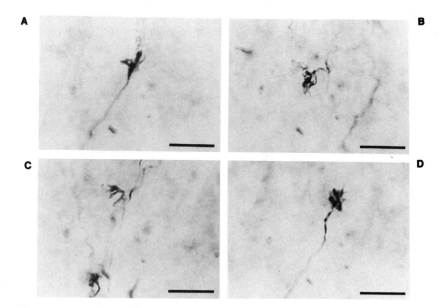

FIGURE 12. Photomicrographs of HRP-labeled growth cones entering cortex (2.75 days postnatal) show lamellipodia and filopodia that are smaller than those in callosal white matter. Bar = 10 μm.

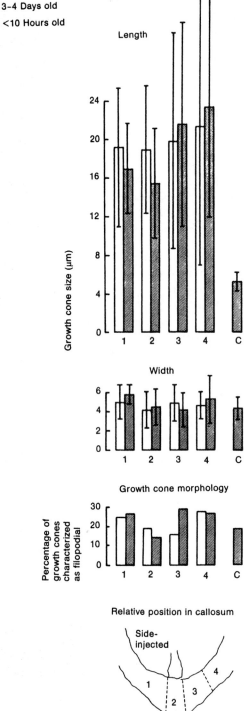

FIGURE 13. Histograms showing size and morphology of 312 growth cones with respect to age and relative position in the corpus callosum (1–4) or cortex (C). Neither age nor position in callosal white matter determines growth cone size. Growth cones in cortex are consistently smaller. At all ages the majority of growth cones are lamellipodial; about 20% are filopodial. Bar = 10 μm.

slowly are smaller and pared away (Argiro et al., 1984). If size and speed can be similarly related to the situation in vivo, these results imply that growth cones move rapidly along the callosal pathway but pause or slow down to seek out neuronal targets in the cortex. In vitro, spread-out growth cones reflect a greater adhesivity to the substrate (Letourneau, 1975). It is not clear whether differences in growth cone morphology in the callosal white matter versus the cortical gray matter also reflect differences in growth cone adhesivity for these two regions of the brain.

The present observations in the callosum are consistent with a model in which axons respond to a broadly distributed set of cues. The exact trajectory of any single fiber might be independently exploratory, much like the behavior of a hunting dog sniffing the scent of its quarry. This view contrasts with the conclusions of a recent study by Kuwada (1986) that pathways in the fish spinal cord develop using guidance mechanisms similar to those in grasshoppers.

In the future, it will be important to develop preparations such as a slice through the mammalian brain that could permit direct observations of growth cone behavior in real time under a variety of experimental conditions. Such observations might help to clarify the similarities and differences between axon pathfinding in tissue culture, in invertebrate pathways, and in more complex vertebrate systems.

Summary

In the present chapter I have emphasized several differences among neural systems in nonmammalian species and the often more complex pathways of mammals with respect to axon outgrowth and guidance. Clearly, one important difference is that CNS axons of lower vertebrates regenerate, whereas severed axons in the adult CNS of mammals do not generally do so. Developmental axon outgrowth in a variety of vertebrate species is accompanied by elevated synthesis of GAPs; reinduction of at least one of them (GAP-43) occurs in axons that do regenerate. Thus, an important goal for understanding mechanisms in regenerative and developmental outgrowth is to define at the molecular level how proteins associated with growth are regulated in axons that do regenerate versus axons that do not regrow after injury.

With respect to guidance mechanisms, an important difference between simpler invertebrate pathways and the complex brains of vertebrates is the sheer numbers of axons. Even at early times in development, growth cones are rarely confronted with only a few other axons among which they must choose. For example, early callosal axons have growth cones whose filopodia are capable of contacting hundreds of other developing axons. For this reason it has not been possible to identify a specific identified axon as a pioneer in the vertebrate CNS, as has been accomplished in insects. Nor can the problem of numbers of vertebrate axons be made more manageable by supposing that neighboring axons simply fasciculate and are passively conveyed to their targets. Rather, in many regions axons reach choice points and must actively reorient. Results from the two hamster CNS pathways considered in this chapter suggest a possible model for vertebrate axon guidance that may depend less on identified axons taking stereotyped trajectories along which other fibers follow by selective fasciculation and more on axons responding individually to molecular gradients widely distributed in the CNS. This could account for the continuous series of turns observed in axons along the callosal and the pyramidal pathways. Continuous decision making guided by molecular cues in the environment might also account for the complexity of growth cones along the entire callosal trajectory. Certainly the ability of regenerating axons or those experimentally displaced during development to find their targets by following alternative pathways is consistent with such a model.

The function of the complex mammalian brain requires that neural connections be appropriately formed during development, but at present the degree of specificity necessary for higher brain function is not known. Thus, in the future it will be important to address the daunting complexity in development of synaptic connections of the mammalian brain. To do this will require techniques using simplified preparations that bridge the gaps between our understanding of how growth cones behave in the defined conditions of tissue culture, how growth cones guide axons in the relatively simple nervous system of invertebrates, where choices are limited, and how axon growth and guidance can occur in the complex brains of vertebrates, where the choices of synaptic partners are myriad.

IV

INTERCELLULAR CONTACTS

STEPHEN S. EASTER, JR.

Once a growth cone has reached its target, it slows down, branches a bit, and makes synapses. This decision to terminate is a crucial one in the development of the nervous system. Motor axons must be in the appropriate muscle. Retinal axons must be in the correct region of the brain, such as the tectum, and in the correct subregion of that. Organized behavior depends on the highly ordered set of connections between neurons and between neurons and non-neuronal target cells. If the connections are not precise, and if they are not maintained, the resultant behavior will be accordingly disorganized.

The determinants of specific connections remain elusive, but the literature is rich with provocative hypotheses and experiments. The prevailing idea, traceable to Ramon y Cajal (1928) and Langley (1895), but formalized and promoted most actively by Sperry (1963), is widely known as the chemoaffinity theory. According to this idea, the future pre- and postsynaptic neurons acquire labels, presumed to be chemical, which identify them. When the growth cone reaches its proper target, it recognizes it by the chemical label there. Perhaps the two labels are also the agents that join the cells' processes together, akin to an antigen–antibody reaction, or lock-and-key. Whatever the detailed mechanism, the general idea of chemoaffinity is that cells acquire a chemical identity early, and their final

151

connections are strongly influenced by these chemicals. The limitations of this idea have been discussed in a recent review (Easter et al., 1985).

Two of the three chapters in this section, those by Hatten and Edmondson (Chapter 11) and Frank et al. (Chapter 13), deal directly with central issues of chemoaffinity. Chapter 12 by Lichtman et al. concerns another issue, that of the stability and lability of synapses.

Hatten and Edmondson investigate the molecular basis of a specific intercellular association—the one between cerebellar granule cells and the radial Bergmann astroglia. Normally, the former migrate exclusively along the latter, a case of specific cell recognition. But in the *weaver* mutant mouse, migration is defective. Hatten and Edmondson have studied this migration in vitro, under conditions that allowed them to identify a molecule, astrotactin, that probably mediates this specific interaction. Astrotactin is different from several of the other, apparently more general purpose, cell adhesion molecules that Edelman and his collaborators investigated (Edelman and Crossin, Chapter 1). It seems certain that other specific intracellular interactions must be mediated by molecules as unique as astrotactin appears to be. Identification of these molecules will contribute significantly toward putting some "chemo" in "chemoaffinity."

In Chapter 13, Frank et al. resuscitate an old theory and provide elegant and persuasive supportive evidence. The theory in question is that of "end organ specification" of the central nervous system. In the 1930s and 1940s, two rather antipodal views of neuronal development were in competition with one another. One, a clear forerunner of chemoaffinity, was referred to by Sperry (1943) as the theory of "autonomous central self differentiation of coordination patterns," which he associated with Paul Weiss. Sperry favored the theory of "end organ specification," which, roughly stated, holds that the peripheral structures (both sense organs and muscles) specify central neurons, which in turn specify more central ones, and so forth until the central connections are all specified. (A cell is "specified," in the context of chemoaffinity, when it acquires its chemical label.) Thus, muscles would specify motor neurons, retinal ganglion cells would specify tectal neurons, etc. The validity of this idea remained uncertain for many years, but it surfaced again years later when Miner (1956) exchanged back and belly skin on tadpoles and found that after metamorphosis, the frog mistook back for belly. That is, when an irritant was applied to the back (on the skin that had originated on the tadpole's belly), the frog tried to sweep it off its belly. This was interpreted as evidence for the skin having labeled sensory neurons in accord with the address of the skin's origin rather than its final location. Subsequent work in the retinotectal system led to the conclusion that retinal axons specified tectum (Schmidt, 1978).

Interesting as they are, none of these earlier experiments are as compelling as those of Frank et al, who show that dorsal root ganglion cells are specified by the end organ (skin or muscle) that they innervate. Although one ought not to suppose that all specification proceeds from the outside in, the experiments described in Chapter 13 make clear that it can proceed in that direction in this one case, and for that reason, the work represents a milestone.

Chapter 12, by Lichtman et al., investigates, admirably directly, synaptic stability and lability. It would be folly to assume that once an array of synapses has formed, it will not change. But until very recently, it was not possible to study, in vivo, the same synapse more than once. Lichtman and his collaborators have developed ingenious methods to do exactly that, and they have used these methods to provide novel insights on interaxonal competition. The new method makes possible a new class of experiment, and in doing so opens a new area of research.

These chapters offer three divergent looks at intercellular association. The techniques are more powerful than those that were available to the originators of this field of study, but the clarity and elegance of the originators' ideas are no less evident today than when they were first stated.

CHAPTER ELEVEN

Mechanisms of Neuron–Glia Interactions In Vitro

MARY E. HATTEN AND
JAMES C. EDMONDSON

Introduction

The developing mammalian nervous system presents many remarkable examples of specific cell-cell interactions. Among these, neuron-glia interactions are extremely important because they establish the architectonics of most brain regions (Sidman and Rakic, 1973). In the embryonic brain, the processes of radial astroglial cells guide the migration of postmitotic neurons outward from the ventricular surfaces where they originate to the cell layer where they will form synapses. When the period of neuronal migration ends, stellate astroglia replace the radial astroglia and provide cell contacts that stabilize the position of the neurons. Although neuroanatomical evidence suggests that the radial glia might transform into stellate cells (Schmechel and Rakic, 1979), the cues for and mechanisms of this morphological transformation are unknown.

Our laboratory has developed a model system with which to study neuron-glial relationships. The objectives of the study were to provide a culture system where highly differentiated forms of astroglial cells—forms associated with the glial support of migration and compartmentalization—are seen and where complex neuronal behaviors—especially neuronal migration—are also seen in real time. Having defined culture conditions where astroglial cells undergo morphological differentiation, we used cellular antigen markers (Raff et al., 1979) to identify astroglia and define their cellular relationships with neurons.

All of our in vitro studies were done with cells dissociated from developing mouse cerebellum, because the cerebellum presents one of the most striking paradigms of glial-guided migration, that of the migration of granule neurons on Bergmann cells, and because of the existence of a defect in the mutant mouse *weaver,* in which a failure of granule neuron migration occurs in concert with abnormalities in neuronal contacts with Bergmann cells (Rakic and Sidman, 1973; Sotelo and Changeaux, 1974).

A key technical feature of the microculture system is the very small volume of the cultures, generally 10 to 30 μl. A second feature is the culture substratum, generally either a microtiter test plate or a glass coverslip affixed as a false bottom to a 5 mm hole in the center of a Petri dish, which is pretreated with a low concentration of polylysine (25 to 100 µg/ml) to generate a hierarchy of cell substrate adhesion. Astroglia bind to the culture surface, and neurons adhere to the astroglia rather than to the substratum. The vast majority (80 to 90%) of the cells in the microcultures are granule neurons, and the remaining 10 to 15% are highly differentiated forms of astroglia (Hatten and Liem, 1981). Very few oligodendroglia survive, because horse serum is used to supplement the culture medium.

Neuron-Glia Interactions in Microcultures

Glial cells and their processes can be visualized when the cells in the cultures are stained with antisera raised against the glial filament protein, a cellular antigen marker for astroglia (Figure 1) (Liem et al., 1978).

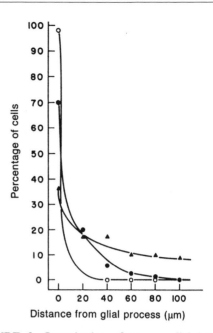

FIGURE 2. Quantitation of neuron–glial interactions in microcultures by measuring the distribution of neurons dissociated from E14 (▲), P1 (●), or P7 (○) cerebellum along astroglial processes. Cells were plated at low density (10^4 cells/ml) in microcultures and stained with AbGFP after 48 hr in vitro. At this density, the area within 20 µm of stained astroglial cells and their processes was less than 20% of the total area of the culture dish. The videotape recordings of phase and fluorescence images of the cultures were replayed for identical fields, and the distance between the center of a stained astroglial fiber and the center of the nucleus of phase-bright, unstained neurons was measured on the video screen with a ruler. Distances on the screen were calibrated with a Leitz stage micrometer projected through the microscope onto the video screen. (From Hatten and Liem, 1981.)

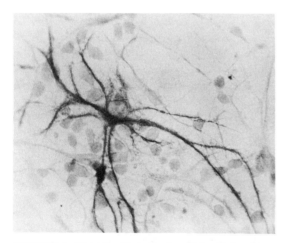

FIGURE 1. Visualization of astroglial cells in microcultures by staining with antisera raised against the glial filament protein (AbGFP). Cerebellar cells were harvested on postnatal day 7 and plated in microcultures for 48 hr prior to staining. The stellate processes of a stained astroglial cell are evident. Numerous unstained small neurons are harbored among the arms of the glial cell.

Using the identified glial processes as landmarks, we analyzed the cellular relationships of neurons with astroglia by measuring the distance of each neuron to the nearest stained glial process. When the number of neurons at a given distance from a stained glial arm was plotted, three findings emerged (Figure 2) (Hatten and Liem, 1981; Hatten,

1984). First, when cells are harvested at embryonic periods, prior to the time when young granule neurons migrate down along Bergmann astroglia or interact with astrocytes in the deeper layers of the cerebellar cortex, the distribution of neurons relative to the glial cells is random (Hatten and Liem, 1981; Hatten, 1984). Just after birth, a time when cell migration is beginning, almost 70% of the cells in the culture are within 20 μm of a stained process. By the end of the first postnatal week, when neuronal migration is in full force, nearly all of the neurons contact glial arms. These findings suggest that specific neuron-glia contacts can occur in vitro, and that these contacts are developmentally regulated in the developing mouse cerebellum.

When the morphological characteristics of astroglial cells dissociated from early postnatal mouse cerebellum were analyzed in detail, two different forms of glial cells were seen

(Figure 3). The predominant form accounted for 80 to 90% of the identified astroglia and had a stellate shape. A large number of neurons generally nestled among the arms of the stellate glial cells. Another form of glial cell was also seen, this one having much longer processes, a shape more like the Bergmann astroglia than astrocytes (Hatten et al., 1984).

To test whether neurons behaved differently on these two forms of astroglial cells in the cultures, we made time-lapse video recordings of living cultures, after which we fixed and stained the cultures with AbGFP to identify astroglial processes. These studies showed that neurons migrate on the highly elongated arms

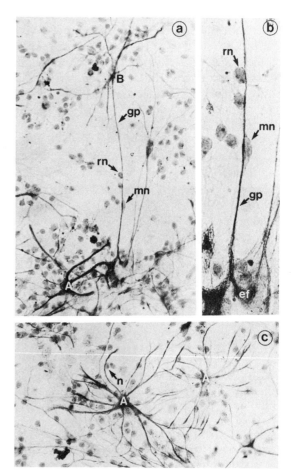

FIGURE 3. Specific associations of cerebellar neurons with two different forms of astroglia. After 48 hr in vitro, glass coverslip microcultures of cerebellar cells harvested from mouse cerebellum at postnatal day 7 were stained with AbGFP. (A) Several dozen unstained cells nestle at or near branch points of the processes of stained glia with a stellate morphology (the A label). Other stained glial cells (the B label) extend longer (150 μm) glial processes (gp) against which several unstained, phase-bright cells are apposed; one unstained cell resembles a resting neuron (rn), another a migrating neuron (mn). Magnification, ×348. (B) The glial process of glial cell B shown in (A) at higher magnification. The thickened, leading process of the cell resembling a migrating neuron (mn) contacts the stained astroglial process (gp). The end foot (ef) of the astroglial cell, densely packed with stained glial filaments, is adjacent to the processes of the stained, stellate astrocyte (A), and appears to end on an unstained, flat cell. Magnification, ×868. (C) Stained astroglia with stellate morphology (A label)— the majority of the glial cells seen in the culture— harbor a number of neurons among their processes. Magnification, ×348. (From Hatten et al., 1984.)

of Bergmann-like astroglia, but anchor on the arms of stellate astroglial cells (Hatten et al., 1984).

Thus one form of astroglial cell, the Bergmann-like cell with highly elongated processes, supports neuronal migration, whereas the stellate astrocyte appears to inhibit neuronal movement (Hatten et al., 1984). Neurons bound to stellate astrocytes are highly active cells, extending processes down along the glial process or out onto the culture substratum, but they do not move about on the glial arms. Occasionally the astrocytes move on the culture substratum, carrying the neurons stationed along their arms as they go.

Neuronal Migration In Vitro

Neuronal migration in vivo takes the postmitotic neuron out from the zone where it originates to its position in the cortex, a distance that often spans hundreds or thousands of micrometers (Sidman and Rakic, 1973; Rakic, 1972; Rakic et al., 1974). By reconstructing serial electron micrographs of migrating cells, Rakic and colleagues have shown that migrating neurons assume a characteristic posture on the glial process (Rakic et al., 1974). The migrating neuron closely apposes its cell soma against the radial glial arm, extends a thickened leading process in the direction of migration, and scales the glial process by wending around it. Once in the proper layer of the cortex the neuron retracts the leading process, and rounds up and detaches from the radial glial process to assume its position and establish synaptic relationships.

The features of neuronal migration, seen in vitro in real time, provide a striking image of the relationship between the migrating neuron and the glial cell (Hatten et al., 1984). Glial processes that support migrating neurons are 1 to 2 µm in diameter, thicker than most neurites, and stretch across a distance of 100 to 200 µm, attaching to the substratum at the glial

cell body and at the end foot of the elongated process. The migrating neuron perches on the glial process by closely apposing its soma against the glial arm, in much the same posture that Rakic has described for cells seen in fixed material in vivo. Out in front of the cell soma, the neuron extends a thick leading process that seems to balance the moving neuron on the glial arm. An axon trails behind the cell soma of the migrating neuron, sometimes, but not always, tracking along the glial process.

In the culture setting, neuronal migration generally occurs at speeds of 10 to 60 µm/hr. When two neurons are seen migrating on the same glial arm, they often both proceed in the same direction. Intermittently the neurons stop migrating by retracting their leading process and rounding up on the glial process. The cues that arrest migration are not apparent. Migrating neurons often pause for periods of minutes or hours, remaining in a rounded configuration on the glial arm, and then begin migrating again after re-extending the leading process.

A question that has often arisen about the inside-out model of neuronal migration in the developing brain (Sidman and Rakic, 1973), a model that proposes that later-generated neurons will migrate out beyond their earlier counterparts, is whether a migrating neuron can move past a stationary cell without either cell losing contact with the glial arm, thereby losing its course along the glial guiding process.

This situation is frequently observed in vitro, and a remarkable series of events keeps both cells attached to the glial arm. The stationary cell poises on the glial process in a rounded configuration, having withdrawn its leading process. As a migrating neuron approaches the stationary neuron, it shovels its leading process underneath the edge of the motionless cell and lifts the immobile cell away from the glial arm by burrowing its leading process underneath it. During the passage of the migrating cell, processes of the stationary cell retain contact with the glial fiber. After the migrating cell passes by the stationary cell, the latter resumes its position on the glial process (Hatten et al., 1984).

Recently we have augmented phase-contrast microscopy with video-enhanced differential interference contrast (Nomarski) microscopy (Allen et al., 1981; Edmondson and Hatten, 1987) to allow us to resolve fine details of neuronal migration, especially the behavior of the leading process (Figure 4). With this microscopy, the leading process can be clearly distinguished from the glial process, allowing a detailed view of the moving neuron. Two features are of special interest. First, the surface of the leading process is highly ruffled, and filopodia are numerous from the tip of the leading edge to the cell soma. Second, the neuron moves with a saltatory cadence (i.e., short intervals of movement punctuated by regular pauses) that is reminiscent of fibroblast motion (Abercrombie et al., 1970). Throughout the period of migration, the neuron retains an extended posture along the glial arm, the cell soma snugly apposing the glial process and the axon trailing out behind the cell (Edmondson and Hatten, 1987).

In one video sequence, an example of the hierarchical adhesion of neurons to glia, fibroblasts, and a polylysine-coated culture dish was seen. A neuron migrates along a glial arm as described, during which time a fibroblast moves into the field and approaches the migrating neuron. A filopodium from the neuronal leading process contacts the fibroblast momentarily, then withdraws, and the neuron continues its migration along the glial process. This sequence suggests that astroglia interact with fibroblasts more avidly than they do with the culture substratum, but that they will choose the glial arm over either a fibroblast or the substratum (Noble et al., 1984; Fallon, 1985).

The time-lapse recordings vividly support the view of migration that Rakic reconstructed

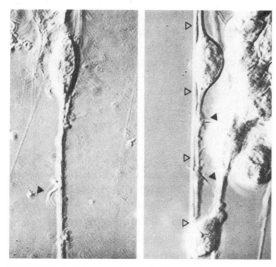

FIGURE 4. Video-enhanced Nomarski microscopy of migrating granule neuron and its relationship with an astroglial process. (Left) Ruffles of cytoplasm (▲) extend up from the leading process. (Right) The glial process (△) extends vertically across the field and a granule neuron migrates down along the process by extending a leading process (▲) in the direction of migration and apposing its cell soma against the glial arm. An axon trails the migrating cell. The leading process of a migrating neuron is a highly active expanse along the glial process. The distance across the screen is 60 μm.

from serial electron micrographs (Rakic et al., 1974). In the microculture system, however, the migration occurs in real time, allowing experimental manipulation and the use of the cultures as a screen for molecules that regulate migration. Thus two lines of research have emerged from the culture system: first, the use of the cultures as an assay for molecules that regulate neuron-glia interactions, and second, the biological study of the cellular aspects of neuronal migration along astroglial processes.

Glial Process Outgrowth and the Establishment of Neuron-Glia Relationships in the Cultures

To understand how neuron-glia interactions come about in the microcultures, we have analyzed the time course of their formation after the cells are plated (C.A. Mason et al., in preparation). These studies have shown that astroglial cells attach to the dish very rapidly after the cells are plated and thereafter extend processes that interact with neurons. The emerging processes of glial cells are tipped by a large growth-cone–like structure that actively interacts with neurons and forms stable contacts with them upon contact. Thus astroglial cells seem to attain their complex shapes in much the same fashion as do neurons, by extending a process that is guided by a motile growth cone.

The primary distinctions of the growing glial processes are that the process emerges as a thick arm of cytoplasm tipped by a growth cone (instead of a thin neurite tipped by an expansive growth cone, as is the case for neuronal processes) and that the glial growth cone forms stable contacts with neurons.

Different aspects of neuron-glia interactions are visible at different times after the cells are plated in the cultures (Table 1). In the first 12 hours after plating, glial process extension occurs in concert with the establishment of contact relationships with neurons. Over the next 12 hours, glial cells continue their morphological differentiation into one of two forms: a stellate form that organizes dozens of neurons among its arms, and a Bergmann-like form that binds very few neurons along its

TABLE 1. Time Course of Neuron-Glial Interactions In Vitro

Time In Vitro	Neuron-Glial Interaction
0–12 hr	Neurons contact astroglia Astroglia extend processes
12–24 hr	Glial processes organize arrangement of neurons
24–48 hr	Complex neuronal behaviors are seen: Migration on elongated glia Anchorage on stellate glia

highly elongated processes. Twenty-four to 48 hours after plating, neuronal migration can be seen along the elongated glia. After two to three days, neuronal migration ceases and neurite outgrowth predominates.

The observations on glial process outgrowth suggest that glial form is established as cell-cell contacts are made with neurons. To test whether emerging glial forms depend on neuronal contacts, we have used three different experimental systems. First, we have separated neurons and astroglia and studied the morphological characteristics of astroglia in the presence and absence of neurons. Second, we have analyzed cells from the neurological mutant mouse *weaver,* because in *weaver* cerebellar granule neuron migration along Bergmann glia fails and the astroglia have poorly differentiated forms. Third, we have used immunological methods to raise antisera that block neuron-glia interactions in vitro.

Neuronal Regulation of Astroglial Differentiation

To study the influence of granule neurons on astroglial morphology, we developed methods to separate the two cell types physically prior to plating (Hatten, 1985). When orphaned astroglia were cultured, they failed to express the highly differentiated forms normally seen in cultures of dissociated cerebellar cells and proliferated rapidly (Figure 5). When highly purified granule neurons were added to the isolated astroglial cells, neurons bound to the glia rapidly and glial process extension commenced within an hour. Concomitant with the transformation of astroglial morphology from a highly flattened form to a complex shape, thymidine incorporation dropped two- to fivefold. Thus, within six hours of the addition of the neurons, neuron-glial contacts formed, thymidine incorporation dropped, and glial process outgrowth commenced.

Recently we have studied the mechanism

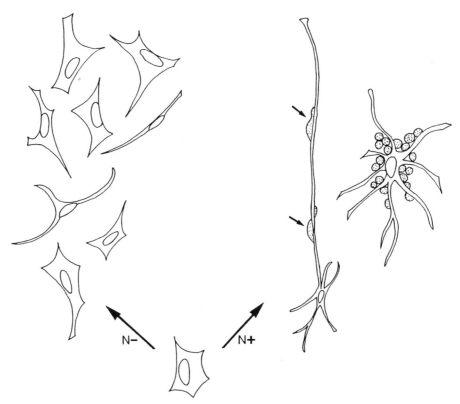

FIGURE 5. Model of the influence of neurons on astroglial proliferation and morphological differentiation. In the absence of granule neurons (N−), cerebellar astroglia assume a flat morphology and proliferate rapidly. However, when neurons are added to the astroglia (N+), the neurons rapidly bind to the glial cells, after which glial proliferation ceases and the astroglia differentiate into complex forms. (From Hatten, 1987.)

of the neuronal influence on glial proliferation and morphological differentiation by testing the effects of conditioned media and fixed neurons or neuronal membranes on orphaned astroglial cells (Hatten, 1987).

Although we have tested a range of concentrations of spent media from neurons or astroglia, no effects of these media have been seen for astroglial cell number of morphology. Similarly, co-culturing the neurons and glia in separate chambers with intermixed culture media does not influence glial number or morphology. These results suggest that trophic effects alone do not influence astroglial differentiation.

In contrast, when granule neurons that have been lightly fixed are added to cultures of purified astroglia, the fixed cells rapidly bind to the astroglia and thymidine incorporation drops within six hours, as was seen with living cells. The same results were seen when a plasma membrane fraction purified from granule cells was added to the astroglial cells, suggesting that a receptor-ligand mechanism mediates neuronal regulation of astroglial proliferation.

Unlike living cells, however, neither fixed neurons nor plasma membranes harvested from fixed granule neurons induced astroglial differentiation. Thus, neuron-glial contacts appear to regulate glial differentiation, but living neurons are required to induce astroglial process outgrowth in the microcultures (Hatten, 1987).

Weaver Granule Neurons Fail to Bind to Normal Cerebellar Astroglia

As an extension of our studies on the influence of neurons on astroglial form, we have separated cerebellar cells from the neurological mutant *weaver* into highly purified fractions of granule neurons or astroglial cells and then mixed and matched the mutant cells with normal cerebellar neurons or astroglia. The aim of these experiments was twofold. First, we wished to analyze the site of action of the weaver gene. Does it act on the granule neuron to prevent binding to the glial cell, or does it act on the astroglial cell to hamper glial differentiation and thereby prevent neuronal migration? Second, we wished to provide more information on the general question of the cellular regulation of neuron-glial contacts.

When weaver neurons were co-cultured with normal astroglial cells, they failed to make the contacts typical of migrating neurons, that is, they did not appose their cell soma against the glial process or extend a leading process out along the glial arm (Hatten et al., 1986). A few weaver neurons touched glial processes, but these contacts were generally at the tip of the neurite only. In addition, many normal astroglial cells did not undergo morphological differentiation in the presence of weaver granule neurons. Most of the glia had flat forms, and those with elongated processes often had enlarged endings, similar to those seen on astroglia cultured alone.

In contrast, normal neurons induced weaver astroglia to differentiate into forms resembling those seen in the normal mouse cerebellum, and numerous granule neurons were seen migrating along their processes. Whereas fewer than 10 migrating neurons per culture were seen when weaver neurons were

co-cultured with normal astroglia, more than 600 were seen when normal neurons were co-cultured with weaver astroglial cells. The details of normal granule neuron migration on weaver astroglial processes closely resembled those described for normal neurons on normal astroglial processes. The neuron apposed its cell soma against the glial arm and extended a thickened leading process in the direction of migration.

The experiments with weaver and normal cells suggest that the granule neuron is a primary site of action of the weaver gene. This study therefore supports the findings of Goldowitz and Mullen (1984), who showed with chimeric mice that the granule neuron is a primary site of action of the weaver gene. In turn, the experiments underscore the importance of the cell-cell contacts between neurons and astroglia in astroglial differentiation. When neuron-glia contacts fail, as in the case of the weaver mouse, astroglial differentiation is impaired.

Thus, in *weaver,* the defect in the weaver granule neuron likely sets in motion a cascade of events. The granule neuron fails to bind to the astroglial cell and impairs astroglial differentiation, and subsequently granule neuron migration fails. In turn, ectopic granule neurons die in superficial layers of the cortex, and a number of errors in synaptic connectivity follow.

The findings with weaver cells are consistent with our studies on normal cells. Cell-cell contacts with neurons appear to play a role in astroglial process outgrowth and in subsequent roles of the astroglial cells as a framework for neuronal positioning (Figure 6).

Astrotactin: A Novel Cell Surface Antigen that Mediates Neuron-Glia Interactions In Vitro

To identify the molecules involved in neuron-glia relationships, we injected rabbits with whole cerebellar cells that had been dissociated from early postnatal cerebellum by

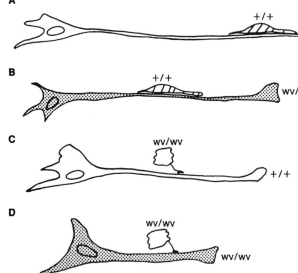

FIGURE 6. Model of cell-cell interactions of normal (+/+) and weaver (*wv/wv*) granule neurons with normal and weaver cerebellar astroglial cells. (A, B) Normal granule neurons form migration appositions with either normal or weaver astroglial cells and induce the morphological differentiation of both normal and mutant astroglia. (C, D) Weaver granule neurons fail to bind to either normal or weaver astroglial cells, and astroglial differentiation is poor in their presence. (From Hatten et al., 1986.)

gentle trypsinization and then allowed to recover from the enzymatic treatment by a brief period in cell culture. Immune sera were then introduced into living cultures to test for their ability to block neuron-glia interactions. To assay for blocking activity, the distribution of neurons relative to glial processes and the amount of glial process outgrowth was measured.

One of the sera tested had potent blocking activity for both of these events (Edmondson et al., in press). As shown in Figure 7, in the presence of Fab fragments of this serum, the distribution of neurons relative to glial processes was identical to a modeled random distribution (assuming no specific interactions) after 24 hours in culture; in the control, more than 90% of the neurons were within 20 μm of a stained glial arm. The total area within 20 μm of a stained glial process was only 20% of the culture surface, and so the cells were sparse enough to quantitate their associations. Thus the antibody prevented the neuron-glia interactions normally seen in cultures of dissociated early postnatal cerebellar cells.

When antibody-treated cultures were stained with AbGFP to visualize astroglial

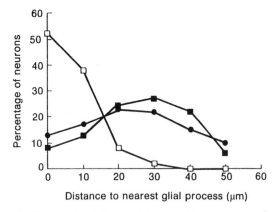

FIGURE 7. Distribution of cerebellar neurons relative to astroglial processes in the presence of anti-astrotactin antisera. The distance of neurons from stained astroglial processes was measured by microscopy and plotted as the number of neurons at a given distance from the glial arm. At the time of plating (■), the position of neurons is random relative to astroglial cells. After 24 hr in the absence of the antibody (□), the vast majority of the neurons are within 20 μm of a stained glial process. When the antibody is added (●), the distribution of the neurons remains random relative to position of stained astroglial process over 24 hr. In all cases, the area < 20 μm from a glial process is between 10–20% of the entire culture dish surface. Thus the differences are due to both less glial outgrowth and fewer neuron-glial associations. (From Edmondson et al., in press.

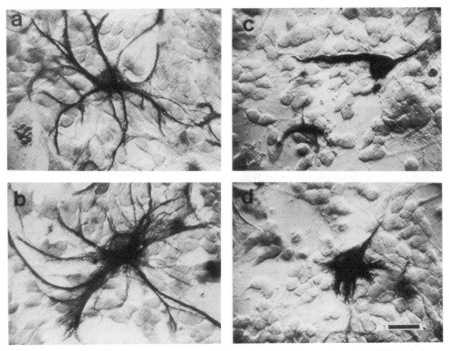

FIGURE 8. **Effect of anti-astrotactin antibody on astroglial morphology. Cultures of dissociated cerebellar cells were treated with anti-astrotactin antisera at plating and stained with AbGFP after 48 hr in vitro. In untreated cultures (A, B), stained astroglia have the characteristic stellate form seen in the cultures and numerous neurons are seen around the glial cell. In cultures treated with anti-astrotactin antisera (C, D), stained astroglial cells have stunted, thickened processes with enlarged endings. Neurons are positioned randomly. (From Edmondson et al., in press.)**

morphology, several changes were noted from untreated cultures (Figure 8). First, in the presence of the antibody, glial processes were shorter and thicker than in untreated cultures, and resembled astroglial forms seen when astroglia were cultured in the absence of neurons. In addition, some astroglia in Fab-treated cultures had a single, thick process several micrometers wider than any seen in untreated cultures and many had markedly thickened terminals (Edmondson et al., in press).

Immunostaining of cerebellar cells in microcultures with the whole rabbit antiserum revealed that the antigens recognized by the antiserum are on the cell surface of both neurons and astroglia. In tissue sections of

cerebellum, a developmentally regulated pattern of staining was seen. At embryonic periods prior to migration and positioning of granule neurons, the only staining seen is of the pia and blood vessels. After birth, the antiserum stains the molecular layer where migration occurs along Bergmann astroglia and the internal granular layer. The pattern of staining with the antiserum suggests a diffuse localization rather than a discrete cellular localization.

To characterize the antigens recognized by the antibody, we metabolically labeled cerebellar cell glycoproteins in vitro, immuno-precipitated the labeled glycoproteins with the antibody, separated the antigens by SDS-PAGE, and used autoradiography to visualize the bands. In material labeled with either [³H]-fucose or [³⁵S]-methionine, a large number of bands (more than a dozen) were recognized by the antibody (Figure 9) (Edmondson et al., in press).

To purify the blocking activity from the polyspecific antiserum, we absorbed the antibodies with whole cells, tested for blocking activity, and immunoprecipitated labeled cerebellar glycoproteins to identify the bands that were removed. The most useful immunoabsorbent was the pheochromocytoma (PC12) cell (Greene and Tischler, 1976), a cell line that expresses most of the described neuronal antigens, including neural cell adhesion molecule (N-CAM) and the nerve growth factor–inducible large external (NILE) glycoprotein (L1 antigen, Ng-CAM) (Thiery et al., 1977; Hirn et al., 1981; Salton et al., 1983; Lindner et al., 1985; Grumet and Edelman, 1984).

Absorption with PC12 cells did not remove the blocking activity in the microwell culture assay; that is, the PC12-absorbed immune serum still disrupted neuron-glia relationships, and stunted astroglia were seen. Immunoprecipitation showed that this absorption removed all but a few bands from the starting polyspecific antiserum, including BSP-2 (N-CAM), NILE (L1, Ng-CAM), and Thy-1. The remaining bands had molecular weights of 100 kD and 80 to 90 kD (Figure 9).

When the PC12-absorbed antiserum was then absorbed with purified granule neurons, the blocking activity was lost and immunoprecipitation showed that a prominent glycoprotein of molecular weight 100 kD had been removed (Figure 9). These experiments suggested that a novel cell surface antigen of molecular weight 100 kD is involved in neuron-glial contacts and in glial process outgrowth in vitro. We have named the antigen astrotactin because it mediates contact relationships between astroglia and neurons. By immunoprecipitation, material bound by anti-

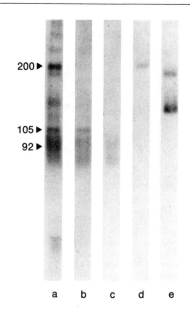

FIGURE 9. Identification of antigen that blocks neuron-glia interactions in microcultures. Cerebellar cells were metabolically labeled with [³H]-fucose for 4 days, after which they were removed from the dish by scraping and glycoproteins were solubilized. Immunoprecipitation followed by SDS-PAGE and autoradiography of (A) immune serum, (B) immune serum preabsorbed with PC12 neurons, and (C) immune serum preabsorbed with PC12 and then granule neurons; (D) immunoprecipitation with anti-NILE antiserum; and (E) immunoprecipitation with anti–BSP-2 antiserum.

Absorption of the antiserum with PC12 cells removes activities of the molecular weights at which antigens recognized by BSP-2 and NILE are seen. Subsequent absorption with granule neurons removes a glycoprotein of molecular weight 100 kD and also removes the blocking activity.

NILE (anti–Ng-CAM, L1) or anti–BSP-2 antibodies (N-CAM) were removed by PC12 absorption, suggesting they are not involved in the blocking activity. Experiments are in progress to analyze the effects of anti-astrotactin antisera on granule neuron migration along single, identified astroglial processes in vitro.

To prepare an affinity-purified monospecific antiserum against astrotactin, the

PC12-absorbed immune serum was passed over an affinity column of granule neuron membranes. When added to microcultures, Fab fragments of affinity-purified antibodies blocked neuron-glia relationships in the cultures at concentrations 500-fold less than those of the unpurified Fab fragments. In addition to its effects on the distribution of neurons relative to astroglial cells, the antibody induced an altered astroglial morphology as seen in unpurified Fab fragments (Figure 10).

To directly test the effects of antigens removed by absorption with PC12 cells, we added Fab fragments prepared from BSP-2, L1, and NILE antisera (Hirn et al., 1981; Salton et al., 1983; Lindner et al., 1985). BSP-2

recognizes the N-CAM antigen; L1 and NILE recognize the antigens recognized by Ng-CAM. None of the antisera tested affected neuron-glial interactions or cell migration in the microculture system (Figure 10).

To visualize the influence of anti-astrotactin antibodies on astroglial process outgrowth and the interaction of the glial growth cone with neurons, we used high-resolution Nomarski microscopy with video-enhanced contrast, as described by Allen and coworkers (1981), and recorded the images with a laser disk recorder (Figures 11 and 12). In untreated cultures, glial processes extended as a thick arm of cytoplasm tipped by a large, motile growth cone and formed highly stable contacts with neurons upon their initial interac-

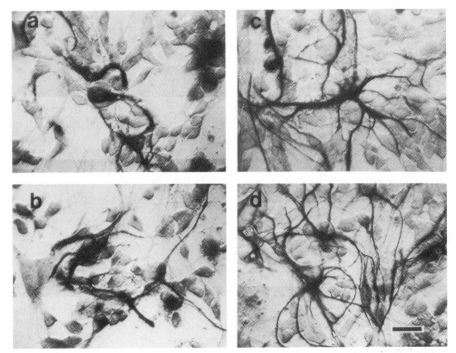

FIGURE 10. Effect of affinity-purifed anti-astrotactin antiserum and anti-NILE Fab fragments on astroglial morphology. In (A) and (B), cultures were treated with affinity-purified anti-astrotactin antisera for 24 hr, after which they were stained with AbGFP to visualize astroglial morphology. Treated cultures have astroglial cells with shortened, thickened arms. In (C) and (D), cultures treated with anti-NILE Fab fragments, glial cells grew extensive processes and interacted with neurons as in untreated cultures. (From Edmondson et al., in press.)

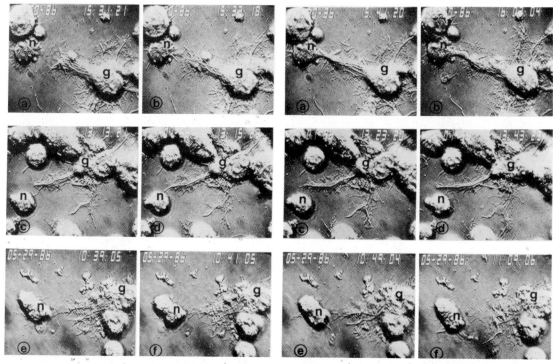

FIGURE 11. Time-lapse video microscopy of astroglial process outgrowth and interactions of the glial growth cone with neurons, at 0 minutes (A, C, and E) and 2 minutes (B, D, and F). In (A) and (B), a normal astroglial cell (g) extends a thickened process tipped by a large growth cone. The growth cone encounters a neuron (n) and immediately forms a stable contact. In (C) and (D), in the presence of anti-astrotactin Fab fragments, glial processes (g) are more flattened and the growth cone contacts neurons (n), but fails to make a stable bond. In (E) and (F), weaver astroglial cells (g) also extend elongated processes that fail to form stable contacts with weaver neurons (n). (From Edmondson et al., in press.)

FIGURE 12. Time-lapse video microscopy of astroglial process outgrowth and interactions of the glial growth cone with neurons, at 10 minutes (A, C, and E) and 30 minutes (B, D, and F). In (A) and (B), the neuron (n) continues to extend a growth cone–tipped neurite on the surface of the glial cell (g). In (C) and (D), in the presence of anti-astrotactin Fab fragments, the process of the glial cell (g) fails to interact with the neuron (n). In (E) and (F), weaver neurons (n) and glia (g) also fail to form stable contacts. (From Edmondson et al., in press.)

tion (Figures 11a,b; 12a,b). In contrast, anti-astrotactin antibody-treated astroglial cells had a less organized growing process and did not form stable contacts with neurons. Instead, the astroglial process contacted and then withdrew from the neuron repeatedly. In concert with this failure to form contacts with neurons, the astroglial growth cone did not ad-

vance, and a stunted glial arm resulted (Figures 11c,d; 12c,d) (Edmondson et al., in press).

Since studies on the mutant mouse *weaver* showed that weaver granule neurons fail to bind to astroglial processes, we videotaped glial process outgrowth of weaver cells in vitro and compared the results with anti-astrotactin antibody–treated cells. Weaver astroglia showed a striking similarity to the antibody-treated cells. Weaver astroglial processes were more stunted and flattened than their normal counterparts, and they failed to establish contacts with neurons (Figures 11e,f; 12e,f).

Two other lines of evidence suggest that astrotactin is missing or defective in *weaver*. First, in contrast to normal granule neurons, weaver granule neurons do not remove the blocking activity of anti-astrotactin serum. Second, immunoprecipitation of [^{3}H]-fucose labeled glycoproteins from weaver granule cells showed that the astrotactin band, but not the BSP-2 (N-CAM) or NILE (Ng-CAM, L1), was greatly reduced, especially in cells taken from the midline of the weaver cerebellum. Thus, astrotactin is missing or defective on weaver granule neurons.

Summary

The experiments described in this chapter suggest that neuron-glia relationships and complex neuronal behaviors such as neuronal migration on Bergmann-like astroglia and anchorage on stellate astroglia can be studied in vitro. The differentiation of the astroglial cells into forms that support these neuronal behaviors, in turn, depends on cell-cell contacts with neurons. Contacts with neurons are required for glial cells to leave the cell cycle, to extend processes, and to maintain processes.

Antisera raised against the cell surface molecule astrotactin, block neuron-glia interactions involved in astroglial process extension and in neuronal positioning on astroglia in the cultures. It remains to be seen whether astrotactin is involved in neuronal migration, whether it is present on weaver granule neurons, and whether it is a neuronal or glial membrane molecule involved in astroglial growth regulation.

These experiments illustrate the usefulness of culture systems or culture conditions that promote complex cellular behaviors and cell-cell interactions important to brain histogenesis. Unlike cell adhesion assays, they provide a functional assay for highly specific cellular relationships in the developing nervous system.

It will be of interest to determine the relationship of molecules such as astrotactin, identified by the microculture system or other systems where complex behaviors occur, to adhesive glycoproteins such as N-CAM, Ng-CAM, laminin, or fibronectin, and extracellular glycosaminoglycans.

CHAPTER TWELVE

The Development and Maintenance of Synaptic Connections at the Neuromuscular Junction

JEFF W. LICHTMAN
MARK M. RICH
ROBERT S. WILKINSON

Introduction

In mammals and other vertebrates, synaptic connections undergo a major reorganization during the final stages of neural development. Throughout the nervous system of young animals, synaptic connections are being both established and eliminated in postnatal life, when the nervous system is already functioning. These synaptic adjustments seem to be more quantitative than qualitative. Axons have already established synaptic connections with appropriate targets, but the numbers of synaptic connections need to be modified. In parts of the nervous system where synaptic reorganization has been studied, axons initially innervate a larger number of target cells than is apparently necessary. The reorganization that ensues pares down the number of target cells contacted by each

axon through the elimination of some synapses, while at the same time new synapses are added, so as to strengthen the connections that axons maintain. In essence, then, this process *redistributes* the synaptic endings of individual neurons.

Synaptic reorganization is of interest for several reasons. First, it is arguably the final event in neurogenesis. As such, it occurs when it is technically easy to study neuronal function: animals are accessible (frequently ex-utero) and relatively mature, axons can be easily stimulated, and postsynaptic cells can be recorded from even with intracellular electrodes. A second reason for interest is the implication that postnatal neural activity (i.e., experience and behavior) profoundly affects the outcome of synaptic reorganization, and thus permanently alters the structure of the nervous system. Thus, the phenomenon of developmental synaptic reorganization is intriguing because it parallels another phenomenon in which neural activity leads to permanent changes in the nervous system—memory.

The major difficulties in studying synaptic reorganization are technical. First, counting the number of synapses at various developmental stages is not particularly revealing because synapses are being established and eliminated concurrently. Because the reorganization changes the numbers of axons innervating target cells, what is important is which axon's synapses are being eliminated and which are being established. Recently, new techniques have become available to permit such identification (see discussion later in chapter). Second, synaptic reorganization is a dynamic process and will not be well understood until it becomes possible to observe the changes as they occur over time.

With the aim of better understanding synaptic reorganization, we have attempted to develop new methods and find ideal biological preparations for the study of synapse elimination. In particular we wish to circumvent the two technical limitations mentioned above: the inability to separately identify the synaptic endings of competing axons, and the inability to view the process of synapse elimination as it occurs over time.

Viewing Axonal Competition

We have concentrated our efforts on the transversus abdominis, an extremely thin snake muscle (Wilkinson and Lichtman, 1985). This ventral muscle is segmentally repeating; 80 to 100 fibers course from each rib to the ventral midline surrounding the abdominal cavity. Each segmental component of the muscle contains three fiber types—fast twitch, tonic, and slow twitch—arranged in an alternating pattern (Wilkinson and Lichtman, 1985). Two features make this muscle especially useful. First, the fibers fan out into a

monolayer shortly after leaving the ribs. Thus every neuromuscular junction is visible in the intact muscle without sectioning. Second, the nerves from adjacent segments often join up, giving rise to a continuous endplate band along the two segments supplied by two segmental nerves. This feature permits us to separately activate and label two populations of axons whose synapses are in close proximity.

In order to visualize the terminals of different axons separately we have taken advantage of "vesicle recycling" at active nerve terminals (Heuser and Reese, 1973). After trying a number of fluorescent molecules, we found that sulfonic acid derivatives of rhodamine, fluorescein, and 8-hydroxypyrene were internalized by nerve terminals after a short (3- to 5-minute) period of nerve stimulation (Lichtman et al., 1985). Because each of these probes has a different emission spectrum, in principle we could label the synaptic endings of different axons with different colors by stimulating them separately. Thus, in an adult, supramaximal segmental nerve stimulation in the presence of bath-applied fluorescent probes will label all of the nerve terminals within the segmental component of the muscle. If two adjacent segmental components are stimulated in the presence of different colored probes, then the terminals originating from each nerve can be easily distinguished. In adult animals where the endplate band continues without interruption between segments, this technique demonstrates the fact that the axons from the two segments intermingle near the segmental boundaries.

Thus, nerve terminals on muscle fibers at the segmental borders labeled green (with sulfofluorescein) can be seen in between muscle fiber endplates that are labeled red (with sulforhodamine). As the endplate band continues farther into each segment's muscle, the terminals become exclusively red in one muscle and exclusively green in the other. We were surprised to find that the innervation of twitch and tonic muscle fibers at the segmental boundaries was different (Lichtman et al., 1985). Synaptic boutons innervating each twitch fiber endplate were exclusively one color, indicating that no twitch fibers were innervated by motor axons originating in both segments. Indeed, electrophysiological and other anatomical studies in the snake show that adult twitch muscle fibers are innervated by one axon only. This of course means that although at the segmental boundary the axons from the two segments will intermingle, each twitch fiber endplate will only be innervated by one axon and thus all the boutons will be one color or the other. The tonic endplates near the segmental boundaries, however, were dramatically different, in that most contained some boutons labeled red and other boutons labeled green (Figure 1). Thus tonic muscle fiber endplates near the segmental border are often multiply innervated by axons arising from different spinal segments. The continued presence of multiple innervation may mean that synapse elimination is less vigorous on tonic fibers than on twitch fibers, where all the endplates are singly innervated. This is interesting because a major difference between twitch and tonic fibers is the absence of action potentials in tonic muscle. Postsynaptic neural activity has been postulated to play an important part in the mechanism of synaptic reorganization (Lichtman and Purves, 1981; Purves and Lichtman, 1985).

Inhibition of synapse elimination is, however, not the only explanation for maintained multiple innervation of tonic fibers. It is possible that tonic fibers near the segmental boundaries are multiply innervated due to an overabundance of tonic axons near these border endplates. To see whether tonic fibers generally are multiply innervated, we labeled all of the terminals of one tonic axon one color (e.g., red) and the remaining terminals in the muscle another color (e.g., blue) (for technical details, see Lichtman et al., 1985). This technique permits us to look at an entire tonic motor unit and ask which of the endplates contacted by that axon (containing red boutons) are also innervated by other axons (containing blue boutons). We found that each tonic motor axon contacts approximately 100 endplates and that typically well over half of these are innervated by more than one axon. Thus, tonic endplates are generally multiply innervated.

The anatomical arrangement of the multiple axons that converge on the same tonic muscle fiber endplate makes several points. First, the inputs are quite intermingled. Individual boutons from one axon are often completely surrounded by boutons from another axon. This intermingling suggests that little in the way of competitive interactions occur directly between the converging axons. Second, the number of boutons supplied by an axon to a multiply innervated endplate ranges widely. One axon may innervate tonic fiber endplates to which it contributes roughly half the boutons, all but one of the boutons, or sometimes only one bouton. Thus, a tonic axon does not seem to have to make a certain-sized synaptic contribution to be maintained at a tonic fiber endplate. Finally, despite the multiple axonal innervation, the total number of boutons at each tonic fiber endplate is always many fewer than the number of boutons supplied by a single twitch motor axon to each twitch fiber endplate. This may also be related to less vigorous competitive forces at work that encourage neither the elimination nor the proliferation of synapses.

To show whether or not the difference in innervation pattern between twitch and tonic fibers is related to the efficacy of synapse elimination, we needed to see if twitch fibers in snake muscle are in fact temporarily multiply innervated during development. To assess multiple innervation during development, the same strategy of stimulating adjacent segments in the presence of different colors was

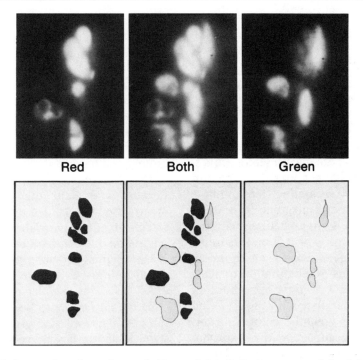

FIGURE 1. Multiple innervation of a tonic muscle fiber endplate in the transversus abdominis muscle of the garter snake. The three upper panels show the same endplate viewed using epi-illumination with excitation and emission wavelengths for red fluorescing rhodamine (left), green fluorescing fluorescein (right), and a double exposure showing both of these images (center). This endplate, located at the segmental boundary, was filled by stimulating two adjacent segmental nerves in the presence of different fluorescent probes (see text; also Lichtman et al., 1985). As shown in the tracings (below), the two segmental nerves contribute synaptic boutons to this endplate that are quite interspersed. The persistence of multiple innervation is due to an inhibition of synapse elimination on tonic muscle fibers. Magnification ×2000.

used. At birth, in several species of snake, the stimulus-induced uptake technique works very well and shows nerve terminals to be much smaller and less well differentiated. Despite the immaturity of the nerve terminals at birth, the twitch fiber endplates are all singly innervated. Two weeks before birth, however, the same labeling strategy shows the nerve terminals to be even smaller, but importantly, at this stage, the twitch endplates near the segmental boundaries are multiply innervated. We are currently following the elimination of multiple axons in the embryonic stages of the egg-laying black rat snake (J.W. Licht-

man, unpublished data). Our studies thus far suggest that the transition from multiple to single innervation of twitch fiber endplates is gradual. At the earliest stages, the multiply innervated endplates of twitch fibers are round structures diffusely occupied with the terminals of the different axons. The endplates at this stage in reptiles (as well as in mammals and birds) are essentially round structures in which nerve terminal membrane completely overlies a round plaque of acetylcholine receptors (R. Balice-Gordon and J.W. Lichtman, unpublished data). As development proceeds, the nerve terminals and receptor clusters coalesce into small regions surrounded by un-innervated gaps; at about the same time, multiple innervation is eliminated (Steinbach, 1981; Slater, 1982).

Our labeling of multiple axons with different colors suggests that as synapse elimination proceeds, the inputs become segregated within the endplate region. Thus, one axon is seen to occupy a contiguous large area of the endplate, whereas another axon seems sequestered. Interestingly, there appear to be regions of noninnervation separating the multiple axons. These regions are likely to increase in size as innervation is eliminated, perhaps accounting for the large gaps that appear in the nerve terminal innervation as development proceeds. Thus, synapse elimination appears to occur by a gradual regional segregation of inputs within an endplate with areas of "no man's land" in between the competing terminals. The area occupied by one axon enlarges with time, while the area occupied by axons to be eliminated shrinks.

In contrast, tonic motor axons are clearly multiply innervated during development and maintain their multiple innervation throughout life. The most salient change in tonic fiber innervation is an *increase* in the number of boutons at the tonic endplates without any change in the number of converging inputs. Thus the different adult patterns of twitch and tonic fiber innervation are due, at least in part, to the developmental difference during the period of synapse elimination.

Viewing Living Nerve Terminals Over Time

A major caveat in the interpretation of images obtained by activity-dependent labeling is that particular endplates are viewed only once, making the conclusions about gradual segregation inferential. To circumvent these problems we have also begun to study the same endplate sites at multiple times. In order to directly view changes in nerve terminals over time, we have attempted to develop vital

staining methods. As a first step we began with the important finding of Yoshikami and Okun (1984) that motor nerve terminals are well stained by dyes that have been developed as mitochondrial markers. We found, however, that the particular dyes used by Yoshikami and Okun were not ideal for mammalian endplates (where we could view the same nerve terminal over time) because of high background and reports of toxicity (Lampidis et al., 1984; Mai and Allison, 1983; Gear, 1974). Because the dyes that they used ranged widely in structure, we decided to make a systematic search of mitochondrial probes to see if we might find one that was better (Lichtman et al., 1986; Magrassi et al., 1987). One dye, 4-(4-diethylaminostyryl)-*N*-methyl pyridinium iodide (4-Di-2-ASP; Molecular Probes, Eugene, Oregon), turned out to be a remarkably good stain. A 30-μM solution of this pyridinium dye stains all the surface nerve terminals in mouse muscle within three minutes. The fluorescence is bright and, even with high exciting light levels, not toxic to the endplates (Magrassi et al., 1987; Lichtman et al., 1987).

To begin, we looked at the same endplates stained with this dye at multiple time points in adult mice (Lichtman et al., 1987). We chose to study nerve terminals in the mouse sternomastoid muscle because of its accessibility. Mice were anesthetized and the muscle exposed through a ventral midline incision. The muscle was stabilized by placing a support under the muscle that is used to lift its midportion (where the endplates are) off the body. A coverslip attached to a spring was used to flatten the surface of the muscle and was necessary because our water-immersion microscope objectives are corrected for coverslips. Endplates located on surface fibers are arranged in a band roughly midway along the length of the fibers. Because the exact positions of the endplates are unique, it is quite straightforward to find the same endplate on subsequent viewing.

We found that in adult animals, nerve terminal branching patterns did not seem to change much over time (Figure 2) (Lichtman et al., 1987). Thus even at intervals of six months, the same endplates were quite recognizable, the only obvious change being one of size. As animals grow their nerve terminals also enlarge (Hopkins et al., 1985). At the neuromuscular junction, this enlargement does not seem to increase the complexity of the arbor but only the length of the branches (Figure 3). Interestingly, the total increase in area occupied by an adult endplate over five to six months ($+370$ μm $\pm$ 82) is roughly the same irrespective of the size of the endplate. This means that a small endplate may even double in total size, whereas the percent change of size of the initially larger endplates is much less (Figure 4). Only one nerve terminal out of approximately 120 studied showed any obvious new sprouting. Thus, contrary to expectations (Wernig et al., 1984; Anzil et al., 1984), mammalian neuromuscular junctions in the sternomastoid muscle do not generally appear to be undergoing continual sprouting and retraction. They do, however, enlarge and are thus not entirely stable entities. It should be pointed out, however, that this enlargement may not be attributable to actual growth but rather to stretching as the underlying muscle fiber hypertrophies with age.

A number of reports have indicated that synaptic competition as occurs in developmental synapse elimination can be recapitulated at adult muscle during nerve reinnervation (Tate and Westerman, 1973; McArdle, 1975; Gorio et al., 1983; Werle and Herrera, 1986). Thus we have begun to study synaptic reorganization during the reinnervation of adult mouse muscles (Rich and Lichtman,

1986; M.M. Rich and J.W. Lichtman, unpublished data). The advantages of using an adult muscle to study synapse elimination include the larger size of the structures to be visualized and the greater ease of doing the microscopic studies in a living adult.

Nerve crush within 2 to 5 mm of the mouse sternomastoid muscle rapidly and completely eliminates nerve terminals in the muscle. Staining with 4-Di-2-ASP shows the nerve terminals to be entirely absent three days after crush. By five days after crush, however, the first signs of newly growing axons can be appreciated and by ten days after, axons have reoccupied most of the original endplate regions as well as giving rise to numerous sprouts that mostly recede over the following several weeks. We know that endplate sites have been reoccupied by staining both nerve terminals (with 4-Di-2-ASP) and postsynaptic receptor clusters (with rhodaminated α-bungarotoxin or rhodaminated cobratoxin) before crush and at various times afterwards. The particular postsynaptic receptor distribution revealed by rhodaminated α-bungarotoxin is a very reliable marker of each particular endplate. Even when the nerve is prevented from reinnervating the muscle by cutting and ligating the nerve, the receptor distribution at each endplate site remains the same, although with time, denervation-induced atrophy causes the widths of the endplates (but not their lengths) to shorten considerably. Using receptor staining to mark the site of endplates, we could thus see what sort of changes occur when axons reinnervate muscle. In the nerve crush experiments, it was clear that the old endplate site was very attractive to reinnervating axons. At more than half

(Overleaf)

FIGURE 2. Endplates viewed at multiple time points in the living mouse sternomastoid muscle. Endplates visualized with 4-Di-2-ASP were photographed in anesthetized mice and rephotographed 10 days (A), 27 days (B), 90 days (C), and 148 days (D) later. Although endplates increased in size over larger durations (e.g., D; see also Figure 3), few signs of remodeling were seen. (Data from Lichtman et al., 1987.)

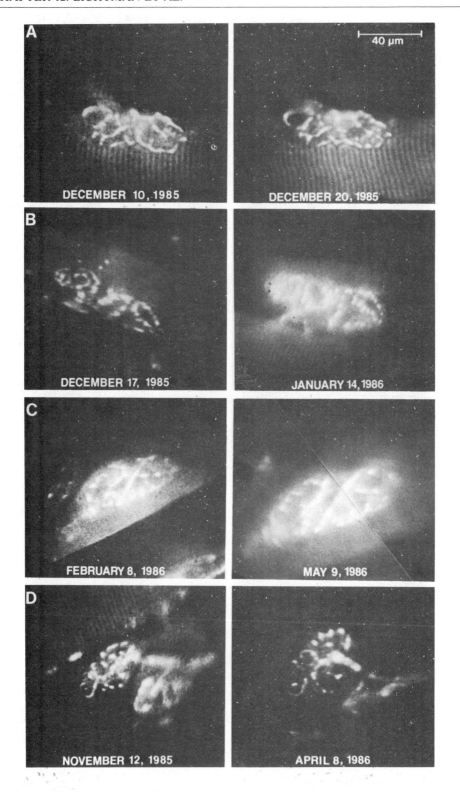

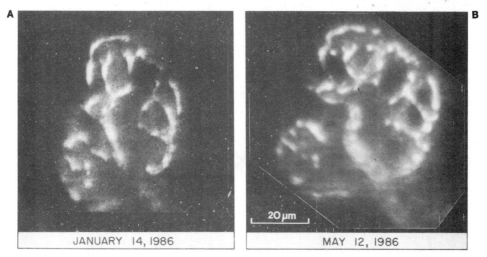

FIGURE 3. Nerve terminal viewed at higher power using a 100× water-immersion objective (primary magnification ×1250; N.A. = 1.2) initially (A) and after 4 months (B). Apart from overall growth, the changes in terminal configuration over this time interval are small. The total length of the terminal branches at this endplate increased from 387 to 522 μm. Much of the enlargement was due to lengthening of existing branches. There were, however, some slight changes (e.g., the bifurcation at the left end of the top branch). In virtually all cases, such refinements were minor at the intervals we studied (<6 months).

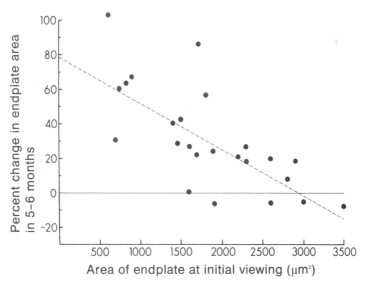

FIGURE 4. Relative growth of small endplates is greater than that of large endplates. The percent change in area of endplates over 5 to 6 months is graphed as a function of the size of the endplate at initial viewing. It can be seen that the smallest endplates increased their size by a much greater proportion than progressively larger endplates, which may mean that the linear growth of endplate branches is the same in all endplates once adulthood is reached. Dotted line is linear regression ($r = .71$, slope $- .028$, $p < .001$). (Data from Lichtman et al., 1987.)

the endplates, reinnervating axons seemed to reoccupy the old site with virtually no change in the branching pattern of the axon. This re-creation of the original endplate is all the more remarkable because it is unlikely that the same motor axon is involved each time. Thus, as also shown in the frog (Letinsky et al., 1976), the old endplate is preferentially reinnervated.

We did, however, also see obvious changes in the endplates of approximately 30% of the reinnervated muscle fibers. The changes were of two types. On some muscle fibers, regions of the original endplate were missing (as judged by the absence of both nerve and acetyl-choline receptors from regions initially stained), and on some fibers, new synaptic regions were present in the immediate vicinity of the old endplate and contained new sites of nerve contact and new acetylcholine receptor clusters. The colocalization of new additions to the nerve terminal and acetylcholine recep-tors was quite interesting. Of the profusion of sprouts that are seen following nerve regrowth, only those that induced receptors in the subja-cent muscle membrane were maintained. This is one of several pieces of evidence suggesting an important link between synapse main-tenance and the underlying receptors.

The newly formed synaptic regions con-tinue to change over at least one month. The synaptic additions become larger as both the nerve terminal and the size of the receptor cluster increases. Similarly, the deletions con-tinue, in some instances erasing large regions of the original endplate site. On some fibers, both sorts of changes were seen to occur so that the total area of neuromuscular contact did not change very much.

The remodeling during reinnervation is particularly interesting because it is nerve de-pendent. Thus, as mentioned previously, in a chronically denervated muscle the acetylcho-line receptors near the endplate do not re-model. It is true that new large plaques of acetylcholine receptor become visible on chronically denervated fibers (Ko et al., 1977).

We have found that after several weeks of denervation in the mouse sternomastoid mus-cle, these new clusters are rarely near the old endplate sites and are usually round, in con-trast to the more irregularly shaped acetylcho-line receptor additions seen at the region of the old endplates. Thus, we do not believe that the denervation-induced acetylcholine receptor clusters are related to the changes we have observed.

An obvious question is why some end-plates change during reinnervation, while others remain unchanged. Our anatomical studies suggest an important difference be-tween situations when endplate sites are sim-ply reoccupied and situations when they are transformed. Endplates that are reoccupied without change generally seem to be reinner-vated by a single motor axon. Following nerve crush, the old Schwann cell tubes provide rapid access to the old sites. Often the endplate is completely reoccupied by one axon. When single innervation of the site occurs, our find-ings suggest that only very subtle changes take place in the structure of the endplate. On the other hand, in a certain number of cases, more than one axon finds the same endplate and each begins to reinnervate it. During develop-ment, such multiple innervation of endplate sites on twitch fibers is unstable, leading to the elimination of all but one axon. In adult rein-nervation, several studies suggest the same oc-currence (Gorio et al., 1983; McArdle, 1975; Werle and Herrera, 1986). Our results in the mouse show that multiply innervated end-plates are often remodeled, suggesting that ax-onal competition causes these changes. For example, occasionally we can clearly see two axons entering an endplate from opposite sides; in such situations there are often signs of substantial reorganization. The appearance of these endplates suggests that the region of the endplate at the boundary between the areas occupied by different axons is largely el-iminated. Neither acetylcholine receptors nor nerve are present in a "no man's land" between the two axonal territories (Figure 5). This is very reminiscent of the reorganization seen in the developing snake neuromuscular junction as multiple innervation is eliminated, as mentioned earlier. A surprising feature is that not only are the nerve terminals of the two axons separated from each other by an uninnervated territory, but the acetylcholine receptor clusters that used to be located in the intervening spots are also missing. Thus, receptors have been either selectively degraded or pulled away from the region between the two competing axons. This loss of receptors from uninnervated territories is another piece of evidence suggesting that receptors play an important role in the competitive process.

Summary

The ability to view axonal interactions during synapse formation and elimination should aid in the understanding of several outstand-

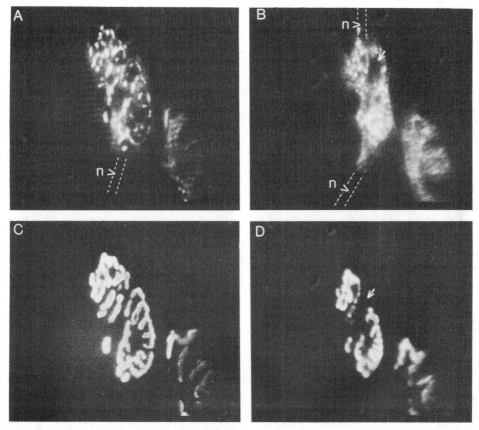

FIGURE 5. Remodeled endplate following reinnervation. On the left are shown the nerve terminal (A) and acetylcholine receptors (C) of an endplate. On the right the same nerve terminal (B) and acetylcholine receptors (D) were rephotographed 18 days later (after nerve crush). The most significant change is a gap in the center of the nerve terminal (shown by arrows) that was initially occupied by both acetylcholine receptors and nerve terminal (compare A, C with B, D). Interestingly, anatomical evidence of two axons entering this endplate site is seen after reinnervation (n indicates the location of axon trajectory to endplate).

ing problems in neurobiology. For example, axonal competition—an attractive explanation for a number of phenomena both in development and maturity—will not be subject to scientific scrutiny until there is a better description of the cell biology underlying axonal interactions. We are trying to discover the mechanisms underlying synaptic competition by devising techniques permitting us to view the process with greater clarity than was previously possible. In the snake neuromuscular junction, we have labeled different axons' terminals separately in order to view the way in which one axon becomes dominant on each muscle fiber during synapse elimination. We viewed the same nerve terminals and receptor patches at multiple times in living mouse muscle to better understand the dynamic aspects of synaptogenesis and elimination. Our work thus far argues for competition in which one axon gradually acquires territory within an endplate region as the other axons' contributions become progressively smaller. In both development and reinnervation, several axons appear segregated from one another as competition proceeds, suggesting that the area between them is rendered unattractive to innervation. Work in mouse muscle argues for changes in receptor distribution that occur in concert with the interactions between nerves.

It is well known that axons can cause postsynaptic receptors to move in the plane of the membrane and aggregate under their terminals (for a recent review, see Peng and Poo, 1986). The present work may offer some insight into the functional significance of this movement. During the synaptic competition in development and reinnervation, axons seem to be vying for the receptors or some closely associated component. The significance of this competition is that our results argue that axon terminals are not maintained unless receptors are juxtaposed on the postsynaptic membrane. This may mean that an axon can cause another axonal ending to be eliminated if it can pull away its underlying membrane components.

At first sight this interpretation seems to contradict the well-known experiments of McMahan, Sanes, and their colleagues (Marshall et al., 1977; Sanes et al., 1978) showing that axons will apparently make functional synapses on basal lamina ghosts in which it has generally been assumed that no receptors or other postsynaptic components are likely to be present because the muscle fiber is absent. However, the basal lamina sheaths never empty completely and small fragments of muscle fiber plasma membrane remain attached to the basal lamina for weeks (Sanes et al., 1978). Interestingly, components of the muscle fiber plasma membrane appear to remain longer in the junctional region than elsewhere, and one of these components, the acetylcholine receptor, is still demonstrable with fluorescent probes for at least a week following muscle degeneration (Slater et al., 1985; M.M. Rich and J.W. Lichtman, unpublished data). Because acetylcholine receptors and presumably other muscle membrane components are still likely to reside at this original endplate site in close apposition to the basal lamina ghosts, it is possible that even in the absence of muscle, postsynaptic membrane components (e.g., acetylcholine receptors) cause nerve axons to return and differentiate at the old endplate site. If the presence of receptors or some closely linked component is responsible for maintaining nerve terminals, one would expect that endplates would undergo some decrease in size when the muscle fiber is damaged, since only membrane fragments remain at the endplate site. In fact, within two to three days of muscle fiber damage (produced by freezing or cutting near the origin and insertion of the mouse sternomastoid muscle), the nerve terminals (which are undamaged by this procedure) undergo a substantial change in shape (M.M. Rich and J.W. Lichtman, unpublished). The nerve terminal contact, which initially consists of a series of long branches, becomes a few small spots of contact. Even when the muscle fiber regenerates, the end-plates and receptors can be seen to be, at least for a time, far less extensive than initially. Thus in mouse muscle, postsynaptic components *are* necessary to maintain a large terminal. The fact that there seems to be a point-to-point correspondence between nerve ter-minal and receptor patches may argue that the postsynaptic components necessary for terminal maintenance are closely associated with (or perhaps even part of) the acetylcholine receptor. If this were the case, then interaxonal competition for these components could lead to the elimination of axons that are no longer adjacent to areas rich in these components.

CHAPTER THIRTEEN

Strategies for Selective Synapse Formation Between Muscle Sensory and Motor Neurons in the Spinal Cord

ERIC FRANK
CAROLYN SMITH
BRUCE MENDELSON

Introduction

One hallmark of the mature nervous system is the great precision of interconnections between neurons. A major area of interest in developmental neurobiology, well demonstrated in the chapters of this volume, is how specific connections are established during development. It seems likely that a number of different mechanisms operate in concert to produce the final specific pattern. Certain aspects of neuronal connectivity result from the precise lineage of individual neurons (Sulston et al., 1983). A neuron's position within the developing embryo can also influence its choice of synaptic partners (Weisblat, Chapter 14) as well as its synaptic chemistry (Le Dourain, 1982). In some cases, these two determinants seem to be sufficient for specifying a neuron's synaptic relations with other cells completely, as, for example, for motoneurons innervating the various muscles of limbs in chickens (Landmesser, 1980). In other cases, however, the identity of a neuron's targets in one region may have a dramatic influence on the pattern of synaptic connections it makes in a second region; a particular example of such a case will be discussed in this chapter. Finally, temporally correlated patterns of electrical impulse activity in groups of neurons have been found to have a profound influence on their connections.

The thesis of this chapter is that all of these phenomena contribute, to varying degrees in different systems, to the establishment of synaptic specificity. Different strategies are probably emphasized in different systems, depending on the particular requirements of that system. One approach to understanding why particular strategies are used in a given system is to compare the mechanisms by which synaptic specificity is established in different systems. In this chapter we examine the strategies used in the development of the monosynaptic stretch reflex and then compare and contrast this system to others.

The system we have used for these studies is the set of synaptic connections in the spinal cord between sensory and motor neurons innervating muscles. These connections form the neuronal basis of the well-known monosynaptic stretch reflex, and they act to help maintain a muscle at a chosen length. Peripherally, the sensory neurons innervate stretch receptors in muscles; centrally, they make excitatory synaptic connections with motoneurons projecting back to the same muscle. When the muscle is stretched or slackened, the firing frequency of the sensory afferents is increased or decreased accordingly, thereby modulating the activity of the motoneurons. These changes in motor activity produce corresponding changes in the force of muscle contraction and hence compensate for the initial change in muscle length.

Because of its simplicity and relative accessibility, this reflex has been extensively studied at the cellular level (Eccles et al., 1957; Mendell and Henneman, 1971), and it has proven useful for developmental studies as well (Eccles et al., 1962b). The axons of both the presynaptic (sensory) and the postsynaptic (motor) neurons are located in peripheral nerves, where they can be selectively stimulated. The patterns of connections between sensory and motor neurons can be assessed by recording intracellularly from motoneurons, and the connections are known to be highly specific (for a review of this work in cats, see Burke and Rudomín, 1977). In frogs, these connections are made relatively late during larval development, well after the sensory and motor neurons have innervated their target muscles (Frank and Westerfield, 1983). This makes it possible to monitor the functional

formation of these synapses with electrophysiological recording techniques, something that has been difficult to do for synaptic connections in other parts of the central nervous system. And because the tadpole develops as a free-swimming larva, surgical manipulations can be made before the period of synaptogenesis, so that the influence of novel peripheral targets of sensory and motor cells on the synaptic connections between these neurons can be studied.

Normal Pattern and Development of Sensory-Motor Synapses

Anatomy of the Stretch Reflex

The anatomy and physiology of the monosynaptic stretch reflex in frogs is briefly described here to facilitate a description of our experimental results. It should be noted at the outset that the cellular basis of this reflex is very similar to that found in cats, and virtually all of the description of the normal adult reflex pathway was first made in the cat. The analogous system in frogs is described here only because it relates directly to the developmental and manipulative experiments that will be described subsequently.

A schematic drawing of the reflex is shown in Figure 1. The afferent side of the reflex is composed of large-diameter (Ia in mammals) sensory fibers whose somata are located in the dorsal root ganglia (DRGs). Their myelinated axons bifurcate within the DRGs; one branch extends peripherally, where it joins motor axons and forms a peripheral nerve that supplies an individual muscle. These fibers terminate in muscle spindles, and they are highly sensitive to muscle stretch. Centrally, these sensory axons enter the spinal cord through a dorsal root and bifurcate again to run rostrally and caudally in the dorsal funiculus. At intervals of a few hundred micrometers, they send out collaterals that penetrate without branch-

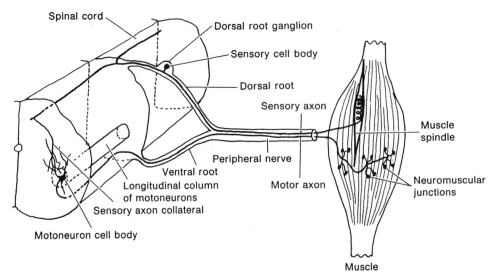

FIGURE 1. Anatomy of a stretch reflex. Impulses are generated in the sensory axon when the muscle is stretched. These impulses are conducted into the spinal cord through the dorsal roots, and they evoke excitatory postsynaptic potentials (EPSPs) in the motoneurons contacted by the sensory axon. The resulting increase in activation of the motoneuron produces greater tension in the muscle, thereby resisting the initial stretch.

ing through the dorsal horn to arborize and terminate in a ventral neuropil containing the dendrites of motoneurons. The cell bodies of these motoneurons are located in the ventral horn of the spinal cord, and their axons project out to the periphery through the ventral roots. Figure 2 shows the projection of triceps brachii muscle sensory neurons onto triceps brachii motoneurons, as revealed by labeling the peripheral axons of these neurons with horseradish peroxidase (HRP) and allowing sufficient time for the HRP to be transported into the spinal cord.

Normal Pattern of Connections Between Muscle Sensory and Motor Neurons

The basic functional pattern of spinal reflexes in the cat was first described in detail by Sherrington early in this century. In the 1950s, with the advent of intracellular recording from motoneurons, Eccles and collaborators mea-

sured the synaptic input from muscle sensory axons onto various classes of motoneurons directly (Eccles et al., 1957). They found that stretch-sensitive axons from one muscle make their strongest connections with motoneurons projecting back to the same or synergistic muscles, and they rarely contact motoneurons supplying antagonistic muscles. In fact, the sensory input from antagonistic muscles usually evokes a strong disynaptic inhibition of the motoneurons.

A similar pattern is found in frogs. Cruce (1974) and Tamarova (1977) suggested that, in the frog, homonymous inputs (i.e., from the same muscle) are stronger than inputs from functionally unrelated muscles. We documented this suggestion for the triceps brachii muscle system in the forelimb of bullfrogs (Frank and Westerfield, 1982a). Triceps sensory fibers make relatively strong monosynaptic projections to triceps motoneurons, but only weak projections to motoneurons inner-

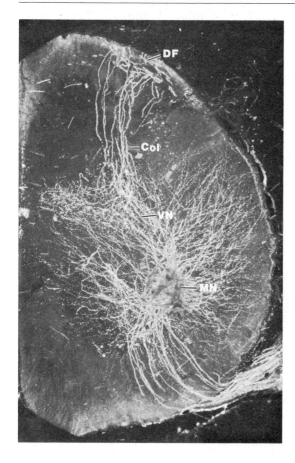

FIGURE 2. Triceps muscle sensory axons and motoneurons in the brachial spinal cord of a bullfrog. The triceps nerve was labeled with HRP and lysolecithin in a cuff placed in the arm 6 days before the frog was perfused with fixative, and transverse sections were reacted with tetramethyl benzidene (Frank et al., 1980). The sensory fibers course longitudinally within the dorsal funiculus (DF) and periodically send collaterals (Col) that descend to the region of the ventral neuropil (VN), where they arborize extensively. In this region, they contact the dorsomedial dendrites of brachial motoneurons (MN). In this and subsequent figures, dorsal is up and lateral is to the right. Magnification ×91.

vating several other, functionally unrelated muscles, such as the subscapularis and pectoralis muscles. Figure 3 shows intracellular recordings made from three kinds of motoneurons; in each case the triceps muscle nerve was stimulated to activate triceps sensory neurons. A large excitatory postsynaptic potential (EPSP) was recorded in the triceps motoneuron, but not in the other two, nontriceps motoneurons. By making recordings from many motoneurons, one finds that this result is highly reproducible; the histograms in Figure 4 show that triceps EPSPs in triceps motoneurons are, on average, about eight to ten times larger than in the two kinds of nontriceps motoneurons.

An important advantage of this system for experimental work is that the synaptic specificity can be studied at the level of connections

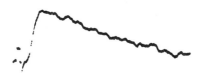

Medial triceps →
Internal-external triceps

Medial triceps → Subscapularis

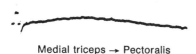

Medial triceps → Pectoralis

FIGURE 3. Intracellular electrical recordings from three types of brachial motoneurons illustrating the inputs they receive from medial triceps sensory axons. The strongest triceps sensory projections are made onto triceps motoneurons (top recording), while the projections to subscapularis and pectoralis motoneurons (middle and bottom recordings) are weak and often nonexistent. Each record is the average of several individual traces. Calibration pulses are 0.5 mV and 2 msec.

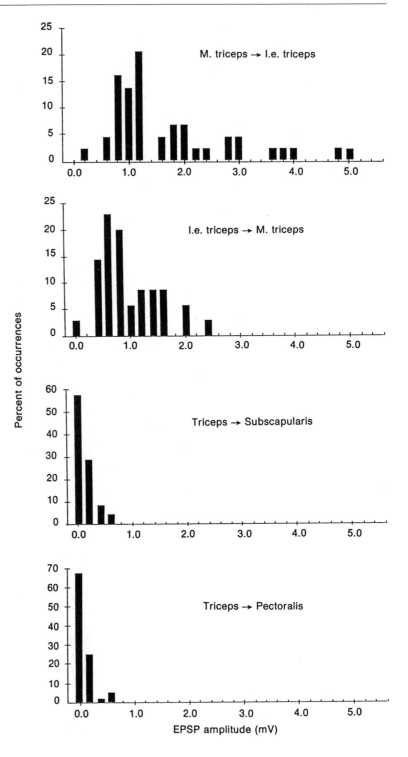

◀ FIGURE 4. Amplitude histograms of triceps sensory EPSPs in four types of brachial motoneurons. Only those synaptic potentials with latencies less than 5.5 msec, and therefore mediated monosynaptically, are included (for details, see Frank and Westerfield, 1982a). EPSPs elicited from medial (M.) or internal-external (I.e.) triceps sensory afferents are larger in synergistic (I.e. or M.) triceps motoneurons than in subscapular or pectoralis motoneurons, even though all four types are located in the same region of the spinal cord. (After Frank and Westerfield, 1982a.)

between individual presynaptic and postsynaptic cells, something that is normally possible only at the neuromuscular junction or in invertebrates. The method involves activating a single stretch receptor and recording the resulting synaptic potential from a motoneuron. Figure 5 is a schematic diagram of this method. Single stretch receptors are activated by tapping the muscle with a small probe, and the sensory impulses are recorded from the intact muscle nerve. By careful positioning of the probe, one can activate a single sensory axon. The synaptic potentials are recorded intracellularly from motoneurons. Unfortunately, these potentials, called unitary potentials, are quite small (often less than 0.1 mV) and are difficult to distinguish from other spontaneous synaptic potentials and noise. This difficulty can be overcome by averaging many individual responses, using the extracellularly recorded sensory impulse as a trigger for a single averager. Synaptic potentials evoked by these impulses occur at a fixed latency and so summate with each other in the average, while spontaneously occurring synaptic potentials or noise are not time-locked to

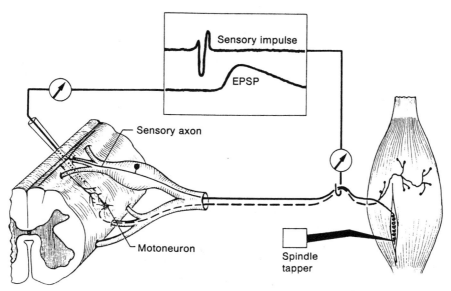

FIGURE 5. Diagram of preparation used to measure unitary EPSPs in motoneurons. The hemisected spinal cord is removed in continuity with the brachial nerve and the medial triceps muscle. Tapping a stretch-sensitive sensory ending in the muscle elicits an impulse that can be recorded en passant in the peripheral nerve. A simultaneous intracellular recording from a triceps motoneuron shows an EPSP elicited by the sensory impulse. Repetitive taps of the same sensory ending are used to measure the average amplitude of the individual (unitary) EPSP. (From Lichtman and Frank, 1984.)

the sensory impulse and are "averaged out."

This technique, called spike-triggered averaging, was developed by Mendell and Henneman (1971) for their studies of the stretch reflex in cats. In their original studies, Mendell and Henneman found that each muscle afferent from the triceps muscle in the hindlimb contacted nearly 90% of the homonymous motoneurons and only 50 to 60% of the synergistic motoneurons. In later studies, the absence of significant input to unrelated motoneurons was confirmed using the technique of spike-triggered averaging. We obtained similar results for the triceps brachii system in bullfrogs (Lichtman and Frank, 1984). As shown in Figure 6, each individual stretch afferent from the medial triceps muscle contacts over 95% of the medial triceps motoneurons and about half of the synergistic internal-external triceps motoneurons. Virtually no contacts are made with the unrelated subscapular or pectoralis motoneurons.

These unitary EPSPs are quite small. Impulses in a single medial triceps afferent evoked, on average, an EPSP of only 195 μV in homonymous medial triceps motoneurons and only 78 μV in internal-external triceps motoneurons. Similarly, the reciprocal projection from single internal or external triceps afferents onto their own motoneurons was 140 μV and onto medial triceps motoneurons, 55 μV. Despite their small size, the synaptic contacts are highly specific. Evidence will be presented below that both the specificity and the strength of these contacts are rather rigidly specified during development and are not dependent on competitive interactions among the sensory neurons.

Pathway Selection and Synaptic Specificity

One strategy used by developing neurons to innervate their correct targets is to extend their axons along particular pathways. In the grasshopper, the growth cones of individually

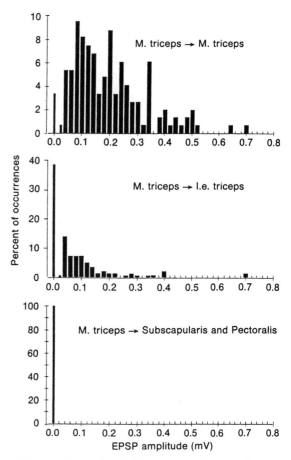

FIGURE 6. Amplitude histograms of averaged unitary EPSPs from individual medial (M.) triceps sensory axons in four types of brachial motoneurons. The first two bins in each histogram have been subdivided into two bins, each 10 μV wide, to illustrate that the smallest nonzero projections are above the noise level. Homonymous projections (top histogram) are more frequent (>96%) and larger (mean of 195 μV) than those to synergistic internal-external (I.e.) triceps motoneurons (62% and a mean of 78 μV). No projections to subscapularis or pectoralis motoneurons (bottom histogram) were seen in this series of experiments. (After Frank and Lichtman, 1984.)

recognizable neurons make highly reproducible decisions about which pathways to grow along and when to leave one pathway and join another (Goodman et al., 1984; Harrelson et al., Chapter 7.) Similar experiments have recently become possible in developing fish. Three identified segmental motoneurons follow reproducible pathways to innervate different regions of the body wall musculature (Eisen et al., 1986; Westerfield and Eisen, Chapter 8). The earliest neurons ("pioneers") appear to follow pathways determined by nonneuronal cells, while neurons developing later follow the axonal pathways laid down by the pioneer neurons (Kuwada, 1986). Lance-Jones and Landmesser (1981) have also demonstrated that in chick embryos, the axons of hindlimb motoneurons destined to innervate different muscles follow distinct pathways through mesenchymal tissue en route to their final targets. To what extent could particular pathways within the spinal cord serve to guide muscle sensory neurons to their appropriate synaptic partners?

Developing sensory neurons send their axons into the appropriate area of the spinal cord from the very beginning. Collaterals are formed at the appropriate rostrocaudal levels and grow into the appropriate regions within the spinal gray matter (Smith, 1983; Jackson and Frank, 1987; Smith and Frank, in press). These axon trajectories may well be determined by specific pathways within the spinal cord; the pathway for muscle sensory neurons could be responsible for bringing them into the vicinity of appropriate motoneurons, while different pathways might guide cutaneous afferents to more dorsal regions. The choice of which motoneuronal targets are innervated might also be controlled by pathways (Wyman, 1973). In many vertebrates motoneurons innervating different muscles are widely separated along the rostrocaudal axis. Lüscher and coworkers (1980) have demonstrated that single muscle sensory afferents

make stronger projections to motoneurons in their spinal segment of entry than elsewhere, and Zengel and coworkers (1983) have reported a similar result. At least part of the specificity of muscle sensory–motor connections could therefore result simply from the spinal level at which the muscle afferent entered the spinal cord.

These topographical features cannot explain the specificity of the synaptic connections between muscle afferents and particular types of motoneurons in the brachial spinal cord of the frog, however. All sensory afferents supplying the forelimb enter the spinal cord through a single dorsal root (see the left sides of Figures 9A and B). Moreover, the positions of triceps, subscapular, and pectoralis motoneurons overlap extensively, even though these three types of motoneurons receive very different amount of triceps sensory input. In fact, as illustrated in Figure 7, the terminal axons of triceps sensory fibers ramify within the dendritic arbors of subscapularis and pectoralis motoneurons, which they *do not* innervate, as extensively as they do within the arbors of triceps motoneurons, which they *do* innervate (Lichtman et al., 1984). Similarly, Mendell and coworkers (Scott and Mendell, 1976; Nelson and Mendell, 1978) found that sensory afferents from the soleus muscle in cats projected differentially to motoneurons supplying different muscles even when these motoneurons were adjacent to one another in the spinal cord. These observations are most consistent with a view of neural specificity in which there is a cellular recognition between appropriate synaptic partners rather than one in which specificity is the result of guidance of sensory axons to a particular place where their preferred targets are waiting.

Differential Timing of Synapse Formation

Another mechanism that contributes to the establishment of specific synaptic patterns in-

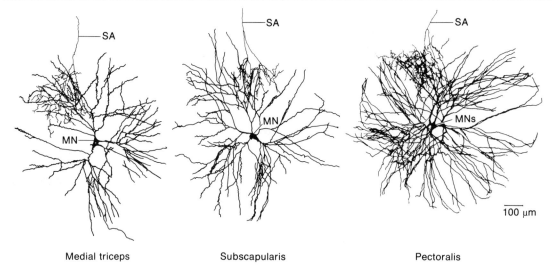

Medial triceps   Subscapularis   Pectoralis

FIGURE 7. Relative positions of triceps sensory axon arbors and dendrites of three types of brachial motoneurons. In each camera lucida drawing, one medial triceps sensory axon (SA) and one or more motoneurons (MN) were injected intracellularly with HRP, and the cells were reconstructed from transverse sections. Triceps sensory arbors overlap extensively with the dendrites of all three types of motoneurons, despite the fact that only triceps motoneurons receive significant triceps input. (For details, see Lichtman et al., 1984.) Bar = 100 μm; magnification ×52.

volves precise and differential timing of synaptic formation between different pairs of synaptic partners (Jacobson, 1978; Macagno, 1978). In the present context, one could imagine that nontriceps motoneurons are refractory to innervation by any muscle sensory neurons at the time that triceps sensory afferents are innervating triceps motoneurons. These nontriceps motoneurons could be innervated by sensory afferents from their own muscles at a different stage in development.

To test this idea, we compared the developmental stages at which several different classes of sensory-motor connections were established (Frank and Westerfield, 1983). As shown in Figure 8, subscapularis motoneurons were innervated by sensory afferents from the pectoralis muscle at the same time that triceps sensory afferents were innervating triceps motoneurons (and not innervating subscapularis motoneurons, as explained

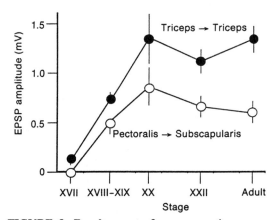

FIGURE 8. Development of monosynaptic connections between muscle sensory and motor neurons. EPSPs were judged to be monosynaptic if they had central latencies of 3 msec or less (for details, see Frank and Westerfield, 1983). These connections are first detectable at stage XVII (Taylor and Kollros, 1946), and the different classes of sensory-motor connections appear during the same developmental stages. (After Frank and Westerfield, 1983.)

more fully in a subsequent section). At least for these connections, then, the differential timing of synapse formation cannot explain the specificity.

Effect of Novel Peripheral Targets on the Development of Sensory Neurons

The results presented so far demonstrate that the synaptic connections between muscle sensory and motor cells are specific but they do not address the issue of how that specificity is achieved. One approach to this question is to study to what extent these connections within the spinal cord are determined by the peripheral tissues that the sensory neurons contact.

The idea that peripheral targets might specify synaptic connections within the central nervous system (CNS) stems from the theory of myotypic modulation proposed by Paul Weiss in the 1930s. According to this theory, the synaptic inputs a motoneuron receives are specified by the particular muscle that motoneuron happens to innervate. The evidence for this theory is not compelling because more recent experiments have shown that the initial projection patterns of embryonic spinal motoneurons are highly specific; small groups of motoneurons located in specific regions of the developing spinal cord grow out unerringly to innervate specific muscles in the leg (reviewed in Landmesser, 1980). Even when motoneurons are forced to innervate novel muscles, their central inputs, as revealed by their patterns of spontaneous electrical activity, remain unaltered (Landmesser and O'Donovan, 1984).

However, a number of studies have suggested that the peripheral targets of sensory neurons might be important in specifying their central projections. During development, many of the central connections of sensory cells are made after they have contacted their targets in the periphery (Ramon y Cajal, 1929;

Windle, 1934; Vaughn and Grieshaber, 1973; Frank and Westerfield, 1983; Smith, 1983). A developing sensory neuron might initially be capable of innervating a variety of peripheral targets; the particular target it contacted could then specify which central neurons it should innervate. This hypothesis has been tested by forcing developing sensory neurons to innervate novel peripheral targets and examining the central connections they subsequently form (Weiss, 1942; Kollros, 1943; Sperry and Miner, 1949; Miner, 1956; Baker et al., 1978). The reflex behavior of the adult animals often suggested that the sensory neurons had made central connections that were functionally appropriate for their novel targets.

The highly stereotyped set of synaptic connections between muscle sensory and motor neurons can be used to test this hypothesis more directly. In tadpoles, sensory neurons from ganglia that would normally supply muscles of skin of the trunk are made to innervate the forelimb instead, and their central projections are then studied in adult frogs by intracellular recordings from forelimb motoneurons. We have used two kinds of manipulations for these studies. In the first (shown schematically in Figure 9A), DRG2, the sensory ganglion that provides the normal innervation of the forelimb, is removed from early-stage tadpoles, inducing DRG3, which normally supplies the rostral trunk, to innervate the forelimb (Frank and Westerfield, 1982b). In the second series of experiments (Figure 9B), DRGs 2 and 3 are removed and sensory ganglia from midthoracic levels are transplanted to the brachial level (Smith and Frank, in press). In both types of experiments, if the surgical manipulation is made relatively early during larval development (by stage IX; see Taylor and Kollros, 1946), sensory neurons from thoracic levels supply various targets in the forelimb. Tactile stimulation often results in withdrawal of the limb, and the stretching of individual muscles evokes electrical im-

FIGURE 9. Schematic diagrams of preparations used to study the influence of peripheral targets on central projections of sensory neurons. (A) DRG2 removals: The second DRG (DRG2), which provides all of the normal sensory innervation of the forelimb, is removed at early larval stages. In the juvenile frog (shown), DRG2 and the second dorsal root (DR2) are absent; DR3 is abnormally large and the third spinal nerve (SN3) forms a plexus with SN2. Sensory neurons in DRG3 now innervate the forelimb. (B) DRG transplants: DRGs 2 and 3 are removed at larval stages and DRGs 4 and 5 from the same animal are transplanted to the brachial level. The transplanted sensory neurons innervate the forelimb and project centrally into the brachial spinal cord. (A, after Frank and Westerfield, 1982b.)

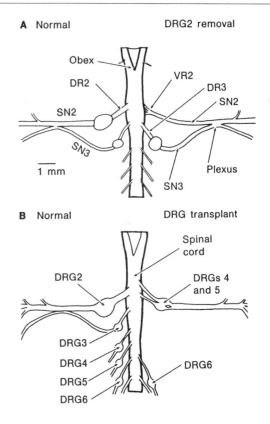

pulse activity in the dorsal roots of the foreign sensory ganglia.

Peripheral Influence on Central Projections of Sensory Neurons

In addition to innervating targets in the forelimb, the thoracic ganglia in both types of experimental animals projected anatomically into the brachial region of the spinal cord, as illustrated in Figure 10. For the animals in which only DRG2 had been removed, this meant that DRG3 sensory axons had grown rostrally from their entry point into the spinal cord to arborize in the region of forelimb motoneurons, something they normally do not do (Frank and Westerfield, 1982b). These central projections placed the foreign sensory afferents in a position where they could potentially contact what were now functionally appropriate synaptic targets (i.e., forelimb motoneurons), so it was possible to examine the influence that their novel peripheral targets had on their choice of synaptic partners within the brachial spinal cord.

The results were clear-cut. Foreign sensory axons innervating muscles made monosynaptic excitatory connections with forelimb moto-neurons, whereas sensory axons in cutaneous nerves did not (Frank and Westerfield, 1982b; Smith and Frank, 1987). Moreover, as illustrated in Figure 11, sensory afferents from an individual triceps muscle head innervated triceps motoneurons supplying both the same head and the synergistic triceps heads, and they made much weaker projections to non-triceps motoneurons (subscapularis and pectoralis) that were located in the same region of the spinal cord. Figure 12 shows the combined results from a number of frogs in which DRG3 innervated the arm (DRG2 was removed at larval stages III to VIII); the projection of foreign triceps sensory afferents, as measured by the amplitude of the monosynaptic EPSP, was an average of eight times stronger to triceps versus nontriceps motoneurons. Sim-

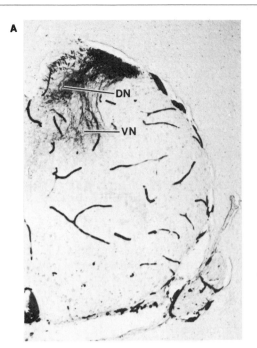
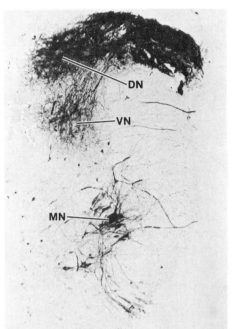

FIGURE 10. Anatomical projections to the brachial spinal cord of thoracic sensory axons innervating the forelimb. (A) Projections of neurons in DRG3 in a frog whose DRG2 was removed 6 months earlier, at stage II. (B) Projections of thoracic neurons in a frog whose DRGs 4 and 5 were transplanted to the brachial level at stage VIII. For both frogs, HRP was applied to the cut dorsal root. In A, the sensory axons have grown rostrally and arborized in the brachial spinal cord (shown in this section), where they usually do not project. A few brachial motoneurons (MN) are also labeled in B. In both A and B, the thoracic sensory neurons have formed dorsal and ventral neuropils (DN and VN), the sites of termination of cutaneous and muscle sensory axons, respectively (Jhaveri and Frank, 1983). Magnification ×64.

ilar results were seen in frogs with transplanted ganglia (Smith and Frank, 1987). These results imply that when sensory neurons are made to innervate novel peripheral targets, they establish central connections with precisely the same spinal cord neurons that would have been contacted by the normal sensory innervation of those targets. Apparently sensory cells can match their peripheral targets with functionally appropriate central targets.

A straightforward interpretation of these findings is that the developmental fate of sensory neurons is *instructively* influenced by their peripheral targets. Sensory neurons could initially be pluripotent with respect to precisely which target structures they will innervate. If a sensory neuron happens to innervate a particular forelimb muscle, it will make central connections with the appropriate brachial motoneurons, whereas if it innervates muscles or skin of the trunk, it will not. This means that specific and orderly sensory-motor connections could be established during development without the need for precise prespecification of sensory neurons for particular peripheral and central targets.

Alternatively, sensory neurons could al-

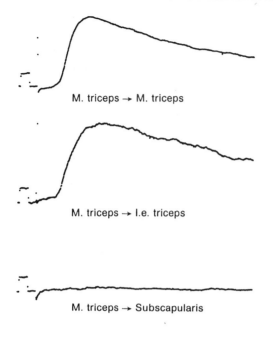

M. triceps → M. triceps

M. triceps → I.e. triceps

M. triceps → Subscapularis

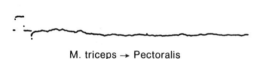

M. triceps → Pectoralis

FIGURE 11. Synaptic responses evoked in motoneurons by transplanted sensory neurons. Stimulation of medial triceps muscle afferents produced EPSPs in homonymous medial triceps motoneurons (top trace) and in synergistic internal-external triceps motoneurons (second trace). The projections to subscapularis and pectoralis motoneurons are relatively weak, as in normal frogs. Calibration pulses are 0.5 mV and 2.0 msec.

ready be determined to innervate particular targets at the time they begin to grow axons, much as motoneurons seem to be. Those neurons that failed to innervate their correct peripheral targets would then die during development. There are some difficulties with this hypothesis, however. To explain our findings with transplanted thoracic ganglia one must postulate that these ganglia initially con-

tain all the neurons needed for their normal thoracic targets as well as neurons specified for forelimb targets such as the triceps muscles. Moreover, thoracic ganglia probably also contain sensory neurons that can innervate the hindlimb and form the corresponding appropriate central connections in the lumbar spinal cord, since supernumerary hindlimbs innervated by thoracic sensory neurons show appropriate reflex behavior (Miner, 1956; Mendell and Hollyday, 1976). Although we do not know how many sensory neurons are generated in thoracic ganglia, we think it unlikely that the number is sufficiently large to include neurons prespecified for all the targets in the trunk and both limbs as well. Therefore it is probable that the peripheral targets of developing sensory neurons have an instructive influence on their choice of synaptic partners within the CNS.

Comparison with Similar Experiments in Other Animals

Other investigators have examined the influence of novel peripheral targets on central projections of muscle sensory neurons. In general, these studies have found central projection patterns to be little changed by the manipulations. Since the results differ from our own, it is appropriate to describe them and attempt to locate the source of the discrepancy.

Eccles and collaborators (Eccles et al., 1960, 1962a,b) crossed pairs of antagonistic muscle nerves in the hindlimbs of neonatal kittens to see if there would be an appropriate rearrangement of sensory-motor synapses in the spinal cord. For the most part, muscle afferents continued to innervate the same subsets of motoneurons as they normally would. In one type of nerve cross, however, there was an increase in the frequency (from 6% in normal cats to 40% in operated cats) of synaptic input from lateral gastrocnemius muscle afferents onto the crossed peroneus moto-

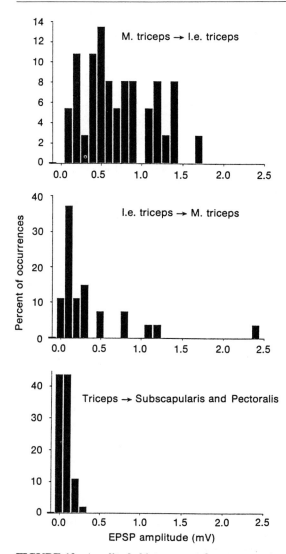

FIGURE 12. Amplitude histograms of monosynaptic triceps EPSPs in triceps and nontriceps (subscapularis and pectoralis) motoneurons in frogs whose DRG2 was removed at early larval stages. DRG3 sensory fibers innervating the triceps muscles make stronger projections to triceps than to nontriceps motoneurons, just as DRG2 triceps sensory fibers do in normal frogs.

neurons that now innervated the lateral gastrocnemius muscle. These are antagonistic muscles whose motoneuronal pools lie close together in the spinal cord and where excitatory connections are occasionally seen in normal cats. Unfortunately, this result was obtained for only one type of nerve cross and the data came from a single litter of kittens, so these studies are usually quoted as providing evidence *against* the idea of a peripheral influence on central connectivity (see, for example, Mark, 1969).

In a similar set of experiments, Mendell and Scott (1975) crossed synergistic muscle nerves in the hindlimbs of young kittens and used the technique of spike-triggered averaging to assess the central projection pattern of individual muscle afferents that had reinnervated foreign muscles. They found no measurable change in the probability of synaptic connections between crossed muscle afferents and uncrossed motoneurons, or between uncrossed muscle afferents and crossed motoneurons.

An important difference between these studies and our own is that the former were attempts to change connections that had already been made, whereas in our studies the peripheral rearrangements were made well *before* the time of initial synapse formation. DRGs were removed and/or transplanted before stage X, and monosynaptic connections between triceps sensory and motor neurons first begin to form at stage XVI or XVII (Frank and Westerfield, 1983). Recently, we have transposed muscle nerves in young postmetamorphic frogs (E. Frank and C. Smith, unpublished data), and, in accordance with the results of neonatal kittens, we found that triceps muscle afferents selectively innervate triceps rather than subscapularis motoneurons even though they have made functional contact with spindles in the subscapular muscle. Taken together, these studies suggest that peripheral influences may be operative only before or during the period when central connections are normally developing.

The effects of removing groups of sensory neurons in developing chick embryos have been studied by Jansen and his collaborators (Eide et al., 1982). Small lesions of the neural crest were made at spinal levels where sensory ganglia that innervate the shank would normally form. These lesions were made at stages 17 to 18 (Hamburger and Hamilton, 1951), before the crest cells had migrated out to form sensory ganglia. Shortly before hatching, the pattern of synaptic connections between muscle sensory and motor neurons was measured by stimulating various nerves in the shank and recording the resulting synaptic potentials in shank motoneurons. In contrast to the effects of DRG removal in frogs (Miner, 1956; Frank and Westerfield, 1982b), sensory ganglia in adjacent spinal segments did not take over the vacant peripheral targets normally supplied by the missing ganglia. Eide and coworkers (1982) suggest that this may result from a difference between neurons in lower vertebrates and neurons in chickens in their ability to sprout to innervate novel targets. Another contributing factor may be the method by which the sensory neurons were removed. In the frog, the removal of one sensory ganglion results in a large hyperplasia of the adjacent ganglia (Bibb, 1977, 1978; Davis and Constantine-Paton, 1983a,b); presumably these excess neurons are supplying much of the novel peripheral target. In contrast, Carr (1984) has found that partial ablation of the neural crest in chick embryos does not lead to an increase in the number of neurons in the adjacent DRGs. Strikingly, Carr finds that sensory ganglia two to three segments away *are* hyperplasic. Perhaps the lesions deplete *all* of the neuroblasts in one segment and *some* of the neuroblasts in adjacent segments. Neurons in these adjacent, partially depleted ganglia might then be incapable of supplying additional peripheral targets. It may be possible

to test this idea by removing sensory neurons in chick embryos *after* the formation of distinct ganglia, so that adjacent ganglia are not disturbed.

Despite the difference in the ability of adjacent ganglia to sprout peripherally, the central projections of muscle afferents in both chicks and frogs were strictly analogous. The few afferents in adjacent segments that did supply shank muscles in chicks also innervated the correct shank motoneurons, and there was no indication that cutaneous afferents or antagonistic muscle afferents ever established monosynaptic connections with motoneurons. Similarly in frogs, when DRG2 is removed slightly later in development (stages X to XII), relatively few DRG3 afferents sprout into the forelimb (P.C. Jackson and E. Frank, unpublished data). Just as in chicks, adjacent sensory ganglia are unable to respond to an expanded peripheral field. Under these circumstances, we find no evidence for monosynaptic connections between DRG3 sensory neurons and forelimb motoneurons, just as in normal frogs. Thus in both systems, there is a strict matching of peripheral and central targets.

Mechanism of Peripheral Specification

The fact that novel peripheral targets of sensory ganglion cells result in the formation of novel but functionally appropriate central connections implies that some aspect of the periphery can modulate the developmental fate of sensory neurons. How is this peripheral influence mediated? Two possible mechanisms are widely recognized as being important in the establishment of synaptic specificity. The chemoaffinity hypothesis, as proposed by Sperry (1963), postulates the existence of chemical labels on pre- and postsynaptic neurons such that appropriate pairs of cells can recognize one another chemically. An alternative idea is that patterned electrical (i.e., impulse) activity may selectively reinforce the synaptic connections between pairs of pre- and postsynaptic cells that have temporally correlated activity.

In situations where the pattern of neuronal connections is determined chemically before the period of synaptogenesis, synaptic connections could be specific from the outset. Synaptic patterns could develop in a highly stereotyped manner independent of functional consequences or correlated patterns of electrical activity. On the other hand, when patterned impulse activity plays a key role in establishing specificity, the specificity can evolve only *after* the connections begin to be made. Alterations in the coordinated activity of pre- and postsynaptic cells would lead to alterations in connectivity. We have studied the development of the stretch-reflex pathway to see whether these mechanisms are important in the establishment of specific sensory-motor synapses.

Normal Development of Sensory-Motor Synapses

The normal development of synaptic connections between triceps muscle sensory and motor neurons in tadpoles was studied by labeling triceps sensory and motor cells with HRP and making intracellular recordings from forelimb motoneurons at various stages of development (Frank and Westerfield, 1983; Jackson and Frank, 1987). Sensory and motor axons are present in the triceps brachii muscle nerves by stages XIII to XIV, and within the spinal cord the sensory fibers have already begun to arborize in the ventral neuropil (where sensory-motor synapses are located). However, the dendrites of triceps and other brachial motoneurons are rarely present within the ventral neuropil region until stage XVI, and only by stage XVII are they common. These anatomical observations imply that synaptic connections between triceps sensory

and motor cells can only begin to form at or after stages XVI to XVII.

These are just the stages when short-latency EPSPs can first be evoked in brachial motoneurons by electrical stimulation of triceps sensory axons. Monosynaptic triceps input, as defined by a central latency of 3 msec or less, was first seen at stage XVII, when metamorphic climax begins. By stage XIX, about four to seven days after monosynaptic triceps inputs first appeared, about 50% of triceps motoneurons had short-latency input; this figure rose to 86% by stage XXII, and was nearly 100% in juvenile frogs. Thus the great majority of monosynaptic triceps inputs to motoneurons are made at or after stage XVII, and these contacts must be functional very soon after they first form anatomically.

The connections of triceps sensory afferents with motoneurons were remarkably specific even at early stages in their formation. For example, at stage XVIII, just one stage after connections begin forming, EPSPs evoked by stimulation of triceps afferents were five to ten times larger in triceps than in subscapularis or pectoralis motoneurons. This ratio is nearly identical to that observed in normal adult frogs (Frank and Westerfield, 1982a). As shown in Figure 13, triceps sensory afferents discriminated appropriately among various kinds of motoneurons at every developmental stage tested. We thus found no evidence for any rearrangement of synaptic connections.

These results should not be interpreted as demonstrating that no errors are made during the period of synaptogenesis. This period probably lasts for at least one to two weeks in the frog; if errors were made but corrected quickly (within a day, for example) they would not be detected by our electrophysiological methods. Rather, the data suggest that once the initial pattern has been established, there is not a major rearrangement of connections such as occurs in several peripheral systems

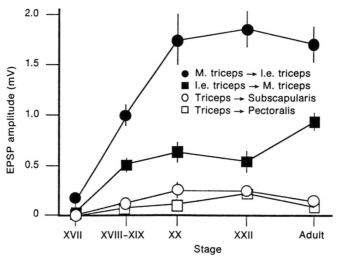

FIGURE 13. Development of triceps sensory projections to four types of brachial motoneurons. The first mono-synaptic connections are detectable by intracellular recording around stage XVII. The adult pattern of these connections is apparent from the outset; triceps sensory axons innervate triceps motoneurons but never provide much input to subscapularis or pectoralis motoneurons. (From Frank and Westerfield, 1983.)

(Purves and Lichtman, 1980) and in the mammalian visual system (LeVay and Stryker, 1979), where electrical activity is thought to be important in determining synaptic patterns.

Reduced Numbers of Sensory Afferents Still Make a Stereotyped Pattern of Projections

If competition between sensory fibers played a role in determining their target choices, one might expect the central and peripheral connections made by individual sensory afferents to be influenced by the presence or absence of other sensory inputs. Examples of such interactions among presynaptic fibers come from studies of partially de-afferented superior cervical ganglia (Murray and Thompson, 1957), optic tecta in hamsters (Rhoades and Chalupa, 1980) and goldfish (Schmidt et al., 1978), and the hippocampus (Lynch et al., 1978). A reduction in the number of sensory fibers could cause the remaining af-

ferents to form stronger than normal connections with spinal neurons or even establish inappropriate patterns of central or peripheral connections.

Because the numbers of sensory neurons innervating the forelimb were greatly reduced in experimental frogs when the forelimb was innervated by thoracic ganglia, we were able to test these possibilities by examining the projections these sensory afferents made. The strength of the connections with motoneurons was assessed by measuring the average amplitude of unitary EPSPs produced by triceps sensory afferents in triceps motoneurons. This amplitude was calculated by dividing the amplitude of the EPSP evoked by electrical stimulation of an entire muscle nerve (and hence *all* the afferents in that nerve) by the number of sensory afferents in the nerve. In 17 of 29 cases where the number of medial or internal-external triceps afferents was between 1 and 12, the unitary amplitudes were within the normal range (0 to 100 µV, average of 51 µV). Nor did the larger amplitudes correlate with fewer afferents. When the number of afferents was greater than 5 (n = 11), the unitary amplitude was 150 ± 107 µV (mean ± 1 SD), whereas when the number was less than 5 (n = 17), the unitary amplitude was only 100 ± 100 µV. Thus there is little evidence for a compensatory sprouting in this system.

Similarly, the ability of sensory afferents to discriminate among different kinds of motoneurons was independent of their number. Electrical stimulation of thoracic sensory neurons innervating triceps muscles evoked consistently larger monosynaptic EPSPs in triceps motoneurons than in nontriceps motoneurons (Figure 14), even when the number of afferents was small. Therefore the specificity does not appear to depend critically on interactions among a number of afferent neurons competing for a restricted postsynaptic space. These observations are consistent with the idea that once a muscle sensory neuron has made contact with a particular peripheral target, the number and detailed pattern of its central synaptic connections are rather rigidly determined.

Finally, the relative proportions of sensory axons innervating each of the three heads of the triceps muscle were also unaffected by the reduction in numbers of afferents. In normal frogs, the three heads are innervated by quite different numbers of sensory fibers (30–40 to medial triceps, ~15 to internal triceps and 0–2 to external triceps), yet the size of the muscles and the numbers of motoneurons supplying each head are comparable (Lichtman and Frank, 1984). If the amount of sensory innervation of each head were normally determined by competition among the afferents for a restricted number of target sites or a limited supply of trophic substance, one would expect that a reduction in this competition would result in a more equal innervation of the three heads. Instead, reduced numbers of sensory

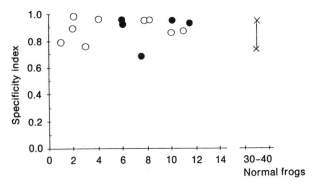

Number of triceps sensory fibers

FIGURE 14. Specificity of sensory-motor synapses plotted as a function of the number of medial triceps spindle afferents. Each point represents an animal in which the triceps muscle was innervated by DRG3 (filled circles, DRG2 removed) or transplanted DRGs 4 and 5 (open circles, DRGs 2 and 3 removed). The number of sensory fibers was estimated by recording from the cut peripheral end of the dorsal root while stimulating individual triceps nerve branches electrically at progressively increasing strengths (for details see Sah and Frank, 1984). Values for normal frogs are from Sah and Frank (1984). The specificity index is calculated as 1 − (triceps EPSP amplitude in nontriceps motoneurons)/(triceps EPSP amplitude in triceps motoneurons). The fact that the specificity indices are all greater than 0.67 indicates that medial triceps afferents projected more than three times as strongly to internal-external triceps motoneurons than to nontriceps (subscapularis and pectoralis) motoneurons in all 14 animals.

fibers maintained their differential pattern of distribution. As illustrated in Figure 15, the medial triceps branch usually contained substantially more sensory axons than the combined internal-external branches, as in normal frogs. This result suggests that the density of sensory innervation is controlled not by competition among the afferents but by some factor intrinsic to the muscle itself or the neural pathways that lead to it.

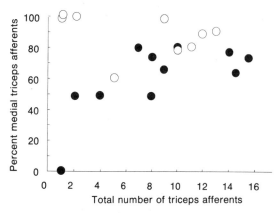

FIGURE 15. Proportion of triceps sensory afferents innervating the medial head of the triceps muscle in the two types of experimental frogs. The number of sensory axons was estimated as described in Figure 14. Open circles, DRG transplants; filled symbols, DRG2 removals. The shaded line shows the normal ratio of medial to total (medial, internal, and external) triceps afferents. Even when the muscles are innervated by thoracic sensory ganglia and relatively few axons, this ratio is approximately maintained.

Influence of Patterned Neural Activity on Synaptic Connections in the CNS

We have examined the effect of impulse activity on the formation of specific muscle sensory-motor connections by disrupting this activity during the period in development when the central synapses are forming (Frank and Jackson, 1986). In one series of experiments, the distal tendon of the medial triceps muscle (an elbow extensor) was cut and sutured to the tendon of an elbow flexor (the sternoradialis muscle). Because extensions of the elbow now stretched the medial triceps muscle rather than slackening it, the temporal pattern of impulse activity in stretch-sensitive sensory afferents in this muscle was dramatically altered. In a second series of experiments, stretch-evoked impulse activity in the sensory afferents was reduced by cutting the distal ten-

don of the medial triceps muscle. Tenotomy caused a drastic reduction in the discharge of muscle spindles by eliminating the ability of elbow flexion to stretch the medial triceps muscle.

After the tadpoles metamorphosed and the period of normal synaptogenesis was over, the central projections of the affected sensory fibers were assessed by intracellular recordings from forelimb motoneurons. As shown in Figure 16, the projections were normal in both sets of animals. Medial triceps sensory axons made strong direct connections with other triceps motoneurons but not with motoneurons innervating subscapular or pectoralis muscles. Apparently the abnormal pattern or level of impulse activity in these sensory neurons did not impair their ability to discriminate among different types of motoneurons.

A possible difficulty with the interpretation of the crossed-tendon experiments is that the operation was only technically feasible at stage XX, which is a few stages *after* sensory-motor synapses begin to form (stage XVII). Sensory neurons that had already formed connections might no longer be sensitive to the influence of patterned electrical activity. However, even at stage XXII, not all triceps motoneurons have received triceps input (see earlier discussion). If activity does play a role in specifying these connections, sensory neurons that had not yet made connections should have been influenced by the abnormal pattern of activity. Moreover, in other systems where neural activity is important in determining connectivity, the sensitive period usually extends beyond the time of initial synapse formation (Hubel and Wiesel, 1965; Hubel et al., 1977; LeVay et al., 1977; Dubin et al., 1986; Stryker and Harris, 1986). The fact that the connections were just as specific as those in normal frogs suggests that the establishment of normal connections is not critically dependent on normal patterns of

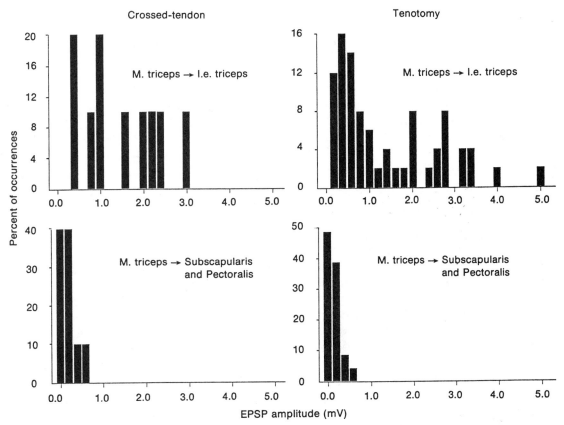

FIGURE 16. **Effects of abnormal neural activity on the pattern of muscle sensory-motor connections illustrated with amplitude histograms of medial triceps EPSPs in synergistic (internal-external triceps) and unrelated (subscapularis and pectoralis) motoneurons. Whether the pattern of sensory impulses in medial triceps afferents was modified by moving the insertion point of the distal tendon (crossed-tendon experiments) or by removing the tension in the muscle (tenotomy experiments), medial triceps afferents still discriminated normally between triceps and nontriceps motoneurons. (After Frank and Jackson, 1986.)**

electrical activity. Furthermore, in the tenotomy experiments the level of activity in triceps muscle afferents was probably reduced throughout the period of synaptogenesis, because the tendons were cut at stage XIV, well before sensory-motor synapses begin to form. Although we did not monitor the sensitivity of medial triceps sensory afferents to stretch *during* the period of synaptogenesis, at the time of the final experiment we always checked that the normal range of elbow movements did not elicit any detectable impulse activity in medial triceps sensory afferents. The tenotomy experiments are somewhat analogous to the monocular closure experiments in the visual system of kittens (Hubel and Wiesel, 1965), in which the projections from the deprived eye were much weaker than normal. In contrast, in the tenotomy experiments the pattern of connections made by silenced muscle afferents is indistinguishable from normal.

Summary

Specification of Neurons By Their Inputs

A major conclusion of these studies is that the peripheral targets of developing sensory neurons have an important influence on the choice of their targets within the central nervous system. Although it is possible that this effect is mediated simply by the selective survival of predetermined sensory cells that find their correct peripheral targets, the available evidence points instead to an instructive effect of these targets.

This specification of muscle sensory neurons by their peripheral targets could be mediated chemically. Different limb muscles, or the pathways that lead to these muscles, might have distinctive chemical labels, for example, in surface membranes or surrounding extracellular matrix. These macromolecules could serve as the means by which muscles are selectively innervated by distinct populations of motoneurons. Sensory neurons innervating a particular muscle could interact with these same labels, perhaps transporting them in a retrograde manner to the cell body, where they could regulate the expression of a particular set of genes. The products of these genes would then be transported to the terminals of the sensory neurons within the spinal cord, where they would serve as recognition molecules for particular classes of motoneurons. To account for the observed projection strengths to different types of motoneurons, one could imagine that the affinity of a "labeled" sensory axon for motoneurons is graded. The chemical affinity could be highest for motoneurons projecting to its own muscle (homonymous connections), intermediate for motoneurons projecting to "similar" muscles (synergistic connections), and very low for antagonistic or unrelated motoneurons. An essential feature of any proposed mechanism is that sensory axons acquire chemical information from the periphery that is somehow transmitted to their axon terminals in the spinal cord, permitting them to recognize appropriate synaptic partners. Individual sensory neurons could thereby be capable of innervating any one of a variety of peripheral targets yet could still make the central connections appropriate for the particular target they happened to innervate.

The specification of neuronal characteristics by their inputs may be an important principle in the development of many parts of the nervous system. In the stretch reflex pathway, the peripheral targets of muscle sensory neurons are muscles, which provide their functional input, and these "inputs" specify the particular types of motoneurons the sensory afferents are to innervate in the spinal cord. A more common situation may be one where the inputs are neurons, for example, the projection of neurons in the lateral geniculate nucleus onto cells in layer IV of the mammalian visual cortex. The geniculate inputs are strictly monocular, and in the adult the first-order cortical cells are also primarily monocular. However, these layer IV cells initially receive binocular input; during development they come to be dominated by input from one or the other eye, depending on how the animal is raised (Hubel and Wiesel, 1970; LeVay et al., 1977). This suggests that these cells are not committed to being "right-eye" or "left-eye" neurons at the time they receive their initial inputs from the geniculate. Instead, competitive interactions among the inputs to each cell apparently determine which eye becomes the dominant input and may influence the connections the cortical cells will ultimately form.

Whenever presynaptic inputs *impose* a functional identity on their postsynaptic targets, there is no need for the target neurons to be chemically distinct from one another. And if these target neurons are *not* chemically distinct, a chemoaffinity mechanism for synaptic specificity is simply not applicable. In contrast, muscle sensory neurons cannot impose a functional identity on the motoneurons they innervate. Because it has already innervated a particular muscle, a motoneuron's function is already determined by the time it receives its sensory inputs. Neurons belonging to different classes (different cortical layers or different synaptic chemistries, for example) or projecting to different targets (as for motoneurons) may often be chemically distinct from one another, and in such cases selective synapse formation by their presynaptic inputs may depend on a chemical recognition between appropriate pre- and postsynaptic partners.

Lability versus Rigidity in Developing Synaptic Pathways

A second observation that can be made from the studies described here is that once a sensory neuron has innervated a particular muscle, its projections to neurons within the CNS appear to be rigidly determined. The specificity for particular kinds of motoneurons is present from the onset of synaptogenesis; there is no major rearrangement of the synaptic pattern during development. Neither the number of sensory afferents supplying a muscle nor the pattern or level of impulse activity in those afferents has an appreciable effect on the connections made with motoneurons in the spinal cord.

Synaptic connections in other neural pathways can be quite labile as they develop, however. In the developing visual and auditory pathways, there are several examples in which the initial pattern of connections is rearranged or refined, and competitive interactions are important in this process (Boss and Schmidt, 1984; Hubel and Wiesel, 1965; LeVay et al., 1977; Sanes and Constantine-Paton, 1985; Schmidt and Eisley, 1985; Stryker and Harris, 1986; Knudsen et al., 1984; Knudsen and Knudsen, 1985). Why might synaptic

lability be present in some systems but not in others?

One situation in which synaptic refinements may be important is when the number of possible synaptic targets makes it impractical to specify the targets of individual presynaptic neurons in advance and with the necessary precision. The pattern of connections required to account for the visual acuity of certain mammals and birds demands that individual retinal ganglion cells recognize the correct one of perhaps millions of potential postsynaptic targets. This level of matching may be impractical through use of chemical mechanisms alone. Although the basic retinotectal map is almost certainly established chemically (Sperry, 1963; Harris, 1980), patterned neural activity is apparently used to sharpen and refine the initial map (Schmidt and Edwards, 1983). Refining initial synaptic patterns that are approximately correct by activity-dependent interactions among neighboring neurons could be a more efficient way of producing fine-grained maps than specifying each neuron chemically.

Rearrangements of connections may also be important when two or more topographic maps must be superimposed on one array of postsynaptic neurons. For example, left-eye and right-eye representations of the visual world must be accurately superimposed in the visual cortex, where most neurons are sensitive to inputs from both eyes. Another example occurs in the tecta of owls, where there is a precise superposition of the visual and auditory worlds (Knudsen and Konishi, 1979, Knudsen, Chapter 20). Even if each of the millions of cortical or tectal cells were uniquely identifiable in these systems, slight differences in the direction of gaze of the two eyes, or in the sensitivity of the two ears to sound, would produce serious misalignments of the two maps. The formation of topographic maps that were approximately correct but slightly diffuse, coupled with a mechanism for refining these maps by selective reinforcement of correct connections, however, would provide a method by which small misalignments could be corrected. Although there is no direct evidence demonstrating that impulse activity acts in these systems to correct such errors, Knudsen's experiments in owls (Chapter 20) clearly demonstrate that rather large perturbations to either the visual or the auditory maps *can* be corrected during development.

In summary, a comparison of the mechanisms used to establish synaptic specificity in different systems shows that they are matched to the particular requirements of the system.

Muscle sensory and motor neurons innervate their peripheral targets *before* they make connections with each other in the spinal cord, so their appropriate central connections have already been uniquely determined in advance. In such a system, synaptic patterns can be rigidly prespecified, and mechanisms for the refinement of these connections are apparently unnecessary. In contrast, when the correct pattern of presynaptic inputs cannot be predicted in advance, a developmental mechanism allowing for the refinement or rearrangement of connections is provided.

V

PLASTIC AND MOSAIC DIFFERENTIATION IN THE NERVOUS SYSTEM

RICHARD I. HUME

An important aspect of development is the creation of diverse cell types from a uniform pool of precursor cells. In fact, many scientists would argue that the central goal of developmental biology is to understand how a single cell, the fertilized egg, gives rise to the variety of cell types that characterize an embryo. The logical possibilities for how cell differentiation might occur are no different for the nervous system than for any other tissue (see Slack, 1983); however, the magnitude of the task is most evident in neural tissue. Classic and modern neuroanatomical techniques allow the use of morphological criteria such as the size and position of the cell body, the shape of the dendritic tree, and the site of axonal projections to describe hundreds of neuronal types in the mammalian brain. Biochemical markers such as neurotransmitters, cell surface receptors, and metabolic enzymes provide additional cell types. There is no reason to believe that our current ability to catalog neurons comes close to defining the amount of cellular diversity in vertebrate brains. Although the nervous systems of higher invertebrates (annelids, molluscs, arthropods) often have only a few thousand cells, the prospect of defining the number of cell types

203

in their nervous systems is not necessarily any less daunting, for in these organisms, many (perhaps most) of the neurons perform unique functions, and therefore are presumably individually specified during development. Only in the nervous systems of the simplest animals (such as the nematode *Caenorhabditis elegans;* see Sulston, 1983) is the number of cell types in the nervous system small enough to allow a complete catalog. Given the great complexity of the nervous system, an issue that must be faced is, Why study cell differentiation in the nervous system rather than in some simpler part of the developing animal?

One answer is simply that we wish to understand the nervous system, so whether or not it is the simplest system for studying differentiation is irrelevant. However, as the chapters in this section make clear, there are in fact some special advantages to studying differentiation in a complex tissue. The typical approach of an experimental embryologist is to alter the number or position of cells in an early embryo, and then to observe the way in which development occurs. The resolution of such experiments is limited by the number of markers available for differentiated cell types. The large number of cell types that occur in the nervous system can often be used to define the results of a manipulation with great precision. Thus the complexity of the nervous system can provide substantial advantages in interpreting experimental manipulations.

Experiments on cell differentiation typically follow a series of logical steps. The initial step is usually to establish the pattern of normal development. In the embryological literature this is referred to as constructing a fate map. Attempts to construct a fate map can produce one of two extreme results, or an intermediate between them. At one extreme, one can predict the outcome of every cell division with a high degree of certainty. An animal that would fall into this category is the nematode *C. elegans.* To the extent that an animal has a predictable pattern of cleavages Weisblat (Chapter 14) refers to it as having *determinate* differentiation. An example of the other extreme, which Weisblat refers to as *indeterminate,* is found in early mammalian embryos such as the mouse. None of the cells of an early mouse embryo has a defined fate, since any cell may give rise to either embryo or extraembryonic membranes (Tarkowski, 1959). In most animals it is possible to find at least some stages of development for which a fate map can be drawn, since all cells must eventually differentiate, and the final divisions of most developmental sequences are determinate. However, different animals, and even different stages within the same animal can vary in the extent to which they are determinate. For instance, Weisblat describes an example in which a small degree of indeterminacy is found in an otherwise highly predictable sequence: two sibling cells of one generation will always give rise to the same progeny, but which cell is the "father" and which the "uncle" is unpredictable.

A point that cannot be emphasized too much is that the existence of a determinate pattern of embryonic development does not provide information as to when, and by what mechanism, specification of cell fate has occurred. Such information can only be gained by placing cells in novel environments. However, the absence of a predictable fate map at a particular stage of development is usually taken to mean that commitment has not yet occurred (Slack, 1983). It is the goal of the experimental embryologist to discover whether at any particular time the fate that a cell would have achieved during normal development was its only possible fate (this case would be referred to as mosaic development), or whether if placed in a different environment it might have taken on some other fate (plastic development). Given that a particular developmental process has been shown to be plastic or mosaic, it is usually next asked whether the signal that triggers differentiation is derived from within the precursor cells, or received from an external source. Both cell intrinsic and cell interactive mechanisms are capable of producing a highly precise, repeatable pattern during development, and many examples of each are known. For example, the anterior–posterior axis of developing frog embryos is established by the distribution of material within the cytoplasm of the egg (that is, by an intrinsic mechanism), while the dorsal–ventral axis of the same embryo is established by the point of sperm entry (an extrinsic trigger).

The methods used in constructing a fate map, and in testing the amount of plasticity in a developing population of cells, vary with the type of embryo. In very simple animals with clear tissues, the entire fate map can be established by direct microscopic observation. By patient observation the entire lineage of *C. elegans* was described (Sulston, 1983), as was much of the early development of ascidians (Conklin, 1905). In more complex, or opaque, embryos, other methods must be used. The chapters in this section use three quite different methods.

Weisblat pioneered the technically elegant method of injecting intracellular marker dyes into single cells of developing embryos. Since the markers are too big to pass through gap junctions, they can only appear in the progeny of the injected cells. Weisblat initially used the marker enzyme horseradish peroxidase, for which sensitive histochemical detection methods are available, but more recently he and his colleagues have used fluorescent dyes, which allow observations to be made in living embryos. The great advantage of labeling cells in living embryos is that researchers are then able to ablate individual cells by enzymatic injection or with a laser microbeam. This method allows them to directly test the importance of cell–cell interactions between particular cells during neural development.

Weisblat and his colleagues have concentrated on the development of the nervous system of leeches. Leeches offer the advantage that the number of cells in each segmental ganglion is quite small, and that many of the neurons of the segmental ganglia are individually identified. The principal result they report is that although much of leech neurogenesis is determinate and mosaic, interesting cases of plasticity can also occur. Furthermore, the decisions selecting between possible fates appear to be made in a hierarchical manner. For example, the OP teloblasts are precursor cells that give rise to many neurons, as well as some other cell types. If only one of the OP teloblasts is present it always makes the P set of progeny and the O progeny are absent, while if two OP teloblasts are present, then both sets of progeny are made. In the concluding section of Chapter 14, Weisblat argues that cell death and quiescence should be considered as differentiated cell fates. Using this simple idea he then suggests that much of the development of regulative vertebrate embryos might be understood by the same set of hierarchical rules that he finds can explain the development of the leech nervous system.

The intracellular injection technique is well suited for use in embryos such as the leech and frog in which development partitions a preexisting mass into smaller and smaller compartments of cytoplasm. However, in embryos in which substantial growth occurs (whether by utilizing yolk, as in birds, or material obtained from the mother, as in mammals) this type of label will quickly be diluted to undetectability. Thus Weston (Chapter 15) and Herrup and Vogel (Chapter 16) use labels that are regenerated by the animal with each cell division.

Weston describes cell differentiation among neural crest cells. Our knowledge of the fate map of neural crest cells is largely derived from studies using chick/quail chimeras. In this technique, which was pioneered by Le Douarin (1982), a piece of neural crest from a quail embryo is transplanted into a chick host. Quail cells possess a distinct nuclear staining pattern that allows them to be unequivocally identified, but the transplanted cells develop in their host in the same manner as host cells. By making transplants at various places and times it is possible to determine the normal and potential progeny of each region of the neural crest. All investigators in this field agree that the crest cell population of each region is capable of generating a diverse array of cell types, and that the particular cell type that appears at a particular site depends on the environment into which crest cells migrate. The critical issue that Weston considers is whether the crest cells entering the target are pluripotent stem cells, each capable of giving rise to all types of progeny, or whether the environment selects from a set of already committed cells. The former idea has held sway in this field for a number of years, but Weston presents evidence that there is substantial differentiation of crest cells before they migrate into their targets, and that selection is likely to be at least as important as instruction in bringing about crest cell diversity. In the terminology defined above, the differentiation of neural crest cells would be more mosaic (or less plastic) than had previously been thought to be the case.

Analyzing the development of mammals presents special difficulties. Fertilization and development take place in an internal environment that is relatively inaccessible for experimental embryology. Furthermore, as noted above, the early development of the mouse (and presumably other mammals) is indeterminate, with even the initial commitment steps occurring only after the fertilized egg has divided many times. In Chapter 16 Herrup and Vogel describe one way that developmental biologists have approached early mammalian development. They make chimeric mice by combining the cells of two individual mice at the eight-cell stage of development. These cells intermix and give rise to a single mouse when reimplanted into a foster mother. Because commitment had not yet occurred, the adult mouse can consist of any proportion (from 0 to 100%) of cells of each genotype (the rest of the cells form extraembryonic structures), but typically the animals contain some cells of each genotype. This technique does not produce a true fate map, since you do not know where in the embryo the cells of each genotype were originally located. However, if the cells of each genotype carry detectable markers this technique can be used to determine cell lineage relationships, since all cells derived from a single precursor cell must be of the same genotype. Furthermore, this technique has the advantage that it can produce mixtures of cells within the chimeric embryos that would be impossible to create by surgical means.

Herrup and Vogel studied the development of the cerebellum, because a number of genetic mutants that have their primary effect on this tissue were known. Their initial conclusion from studying chimeric mice carrying these defects is that there is no unique precursor cell to any class of mouse neuron. This idea is in accord with traditional ideas of neuroembryology. However, by analyzing quantitative data on the number of Purkinje cells present in the chimeras, they argue that all Purkinje cells are derived from a very small number of precursor cells (8–12 per side), that commitment of these cells occurs at a very early stage of neural development, and that each precursor gives rise to a fixed number of Purkinje cells, presumably because there is a regular pattern of cell divisions. This idea is an example of relatively mosaic development in the early stages of vertebrate development, and is rather unexpected. Herrup and Vogel then present arguments, again based on quantitative data from chimeric embryos, that the number of cells of a second type of cerebellar neuron—the granule cell—is regulated very differently from Purkinje cells. Rather than being determined by very early commitment and a fixed lineage, the number of these cells is determined by cell—cell interactions that occur quite late in development.

Taken as a group, the chapters in this section challenge the notion that plastic development is a characteristic of vertebrate development, and mosaic development of invertebrate development. Rather they argue that both patterns are important developmental strategies that are likely to be used in many animals. It is a challenge for the developmental neurobiologist to discover which scheme an animal is using at each stage of development.

CHAPTER FOURTEEN

Equivalence Groups and Regulative Development

DAVID A. WEISBLAT

Introduction

The properties of particular cell types, such as nerve and muscle, are extensively conserved throughout the animal kingdom. Thus, it seems likely that developmental mechanisms for generating these cell types arose before the current groups of animals diverged in evolution and that these mechanisms might also have been conserved during evolution. If so, it should be possible to elucidate developmental mechanisms in simpler organisms and apply that knowledge to studying more complex organisms that are less tractable experimentally. Two hurdles face any developmental biologists who seek to follow this strategy. First, we must establish the extent to which apparently similar phenomena really do operate by unitary mechanisms, as opposed to being superficially similar manifestations of fundamentally different mechanisms. And conversely, we must be prepared to see how apparently distinct phenomena might in fact be homologous, having arisen by the modification of a basic developmental theme in evolution.

Toward these ends, this chapter summarizes work on phenomena in the development of the leech, wherein equally pluripotent cells assume distinct developmental fates as the result of hierarchical interactions. Such groups of interacting, pluripotent cells have been defined as "equivalence groups" in the nematode (Kimble et al., 1979). I also attempt to show connections between equivalence groups in the nematode and the leech and regulation, another type of developmental plasticity that is usually held to be qualitatively different and is commonly associated with the embryogenesis of higher animals.

Leeches are attractive for cell lineage studies because their nervous systems are relatively well characterized (Muller et al., 1981) and thus constitute well-defined developmental end points. Clearly, the better we know the end point of development, the more precisely we can phrase experimental questions and assess experimental results. Another attraction of leeches is that certain species produce eggs that develop normally in simple media and via relatively large, identifiable cells (Whitman, 1978). In addition to these intrinsic merits, leeches, as annelids, occupy the phylogenetic middle ground between nematodes and arthropods. Therefore they offer possibilities for comparative studies that can be wide-ranging without requiring the leaps of imagination and credulity called for by direct comparisons of, say, *Drosophila* and *Homo*.

Overview of Leech Development

Eggs of the glossiphoniid leech *Helobdella triserialis* are fertilized internally and deposited in cocoons on the venter of the hermaphroditic parent. Young leeches remain with the parent until at least their first feeding. (*Helobdella* feeds on freshwater snails.) But isolated embryos can develop normally, because the yolk-filled, 400-μm-diameter egg contains all the nutrients required to complete development.

The overall scheme of development, including the stereotyped early cleavage pattern that gives rise to identifiable blastomeres, was described by Whitman (1878, 1887, 1892). In the past decade, this description has been refined (Fernandez, 1980; Stent et al., 1982; Weisblat and Blair, 1984) and is summarized in Figure 1. The development of glossiphoniid

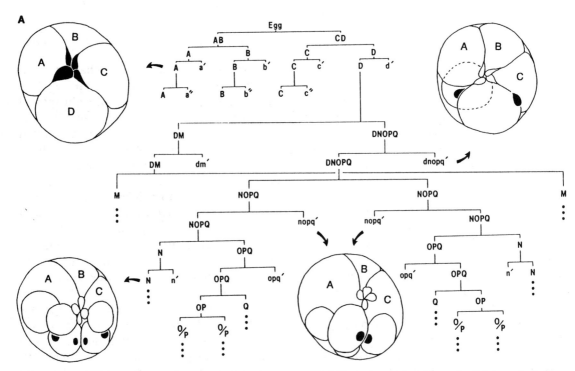

FIGURE 1. Summary of glossiphoniid leech development. (A) A lineage tree showing early divisions, including the generation of micromeres; four embryos are depicted as well, showing the positions of the blastomeres at four different stages (indicated by arrows) and the most recently born micromeres for each stage (in black). Stage 1 begins when the egg is laid and ends at first cleavage. Stages 2–6 begin with the two-cell embryo (cells AB and CD) and end with the cleavage of cells OP to form the O/P cells, last of the five bilateral pairs of teloblasts. Stage 7 includes the period during which most of the blast cells are produced, including the formation of left and right germinal bands, and ends when the bands start coalescing along the ventral midline to form the germinal plate. **(B,** *left*) Diagram showing the relationship of the teloblasts and the blast cell bandlets that coalesce to form the

leeches has been divided into 11 stages. Prior to the first cleavage (stage 1), cytoplasmic reorganizations form domains of yolk-deficient cytoplasm, or *teloplasm*. Teloplasm is segregated into particular cells during cleavage and is assumed to somehow govern the developmental potential of the recipients. Early divisions (stages 2 to 6) generate three classes of cells, *macromeres, teloblasts,* and *micromeres*. Macromeres, of which there are ultimately three (A, B, and C), contain most of the yolk and are incorporated into the gut by late embryogenesis. Five bilateral pairs of teloblasts arise by further cleavage of a fourth

macromere, D, the chief inheritor of the teloplasm. There is one bilateral pair of mesodermal precursors, the M teloblasts, and four bilateral pairs of ectodermal precursors, the N, O/P, O/P, and Q teloblasts. Micromeres, of which there are at least 15, arise at various points during cleavage, and cluster about the animal pole of the embryo.

Teloblasts are embryonic stem cells par excellence. Each one undergoes a series of several dozen highly unequal divisions (stages 6 to 8) that generate a coherent column or

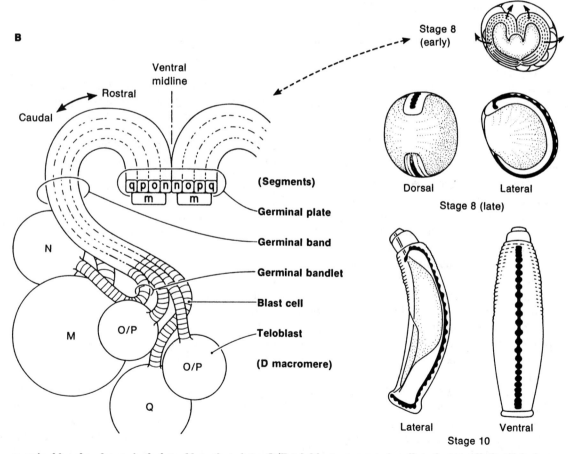

germinal band and germinal plate. **Note that sister O/P teloblasts generate bandlets that are distinguished as o and p on the basis of their relative positions in the germinal band. (*Right*) Representations of embryos at stages 8 and 10 of development. The definitive progeny of teloblasts injected with lineage tracers are identified in stage-10 embryos, which have almost the complete complement of adult structures. (From Ho and Weisblat, 1987.)**

bandlet of *primary blast cells,* at the rate of about one per hour. Within the bandlet, older cells lie further from the parent teloblast than do their younger siblings. Thus, a caudal-to-rostral examination of the blast cells and their clones in a bandlet reveals the developmental history of individual blast cells. The two O/P teloblasts on each side of the embryo were regarded as distinct O and P teloblasts until rather recently, and therein hangs the tale with which much of this account will be concerned. For now, suffice it to say that these are sister teloblasts formed by the cleavage of a precursor cell named OP. Curiously, the OP cells themselves undergo several highly unequal, blast cell–forming divisions before cleaving to form the two O/P teloblasts (Fernandez, 1980).

The distal ends of all the bandlets come to be attached, five to a side, at the animal pole, site of the future head of the leech (Fernandez, 1980). On each side, the distal portions of the five bandlets lie on the surface of the embryo in a close-packed parallel array called a *germinal band*. In each band the positions of the bandlets are stereotyped; the ectodermal bandlets lie superficially with the q bandlets closest to the animal pole, the n bandlets furthest from it, and the two bandlets derived from the O/P teloblasts in between. At this point, the bandlet next to n is designated o and the one next to q is designated p (see Figure 1). The mesodermal, or m, bandlets lie just beneath the others. The bandlets and the region of the embryo between them is covered by an epithelium derived from micromeres (Figure 2) (Weisblat et al., 1984; Ho and Weisblat, 1987).

As the teloblasts continue to produce blast cells, the bandlets and germinal bands get longer and move into the vegetal hemisphere, still at the surface of the embryo (stage 7). The micromere-derived epithelium expands simultaneously, providing a provisional epidermis for the embryo. These morphogenetic

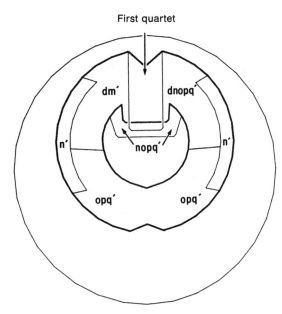

FIGURE 2. Stylized representation of an embryo at early stage 8 (see Figure 1), showing a map of the provisional epithelium over the germinal bands according to micromere of origin. Thick lines represent the outline of the germinal bands at this stage. Teloblasts and macromeres are not shown. Thin lines within the germinal bands delineate the idealized boundaries of the domains of micromere progeny. Fate-determining interactions are occurring in roughly that part of each germinal band covered by the progeny of opq′ and the caudal progeny of n′. (From Ho and Weisblat, 1987.)

movements culminate as the left and right germinal bands meet one another (stage 8), coalescing like the two halves of a zipper, from future head to future tail along the ventral midline of the embryo. The coalesced germinal bands constitute a structure called the *germinal plate*. It is from this structure that the segmental tissues of the leech arise. As the cells in the germinal plate proliferate (stages 9 to 10), it uncurls, and the roughly spherical embryo takes a shape more like a cigar. The lateral edges of the germinal plate expand around the surface of the embryo and eventually meet at the dorsal midline, closing the body tube. This process also proceeds from anterior to posterior. By this time in development, many of the organ systems, including the nervous system, closely resemble the adult form, especially in the anterior segments of the animal.

Stereotypy and Commitment

Individual cells in the germinal bands are small and numerous; moreover, the primary blast cell and its definitive progeny are separated by several days, many cell divisions, and hundreds of microns. Thus it was impossible to follow the fates of individual cells in the leech until it was shown that tracer substances could be microinjected into blastomeres of living embryos, marking the injected cell and its descendant clone without appreciably disrupting normal development (Weisblat et al., 1978, 1980a). The basic technique has been improved by the development of new tracers (Gimlich and Braun, 1985; Ho and Weisblat, 1987) adapted for ablation as well as lineage tracing (Blair, 1982; Shankland, 1984) and combined with techniques from physiology (Kramer and Weisblat, 1985), histochemistry (Blair, 1983; Stuart et al., 1987), and autoradiography (S.T. Bissen and D.A. Weisblat, in preparation). As a result, it has been possible to analyze leech development with the same cellular precision as in the nematode *Caenorhabditis elegans*. And as in the nematode (Sulston et al., 1983; Sulston and Horvitz, 1977), the dominant impression is one of great stereotypy, especially when confronted with the segmentally iterated sets of labeled cells obtained when individual teloblasts are injected with lineage tracers.

Corresponding to the five bandlets from each side of the embryo, five distinct patterns of labeled progeny are observed. The segmental patterns derived from the four ectodermal bandlets are shown schematically in Figure 3. Each of the five distribution patterns (M, N, O,

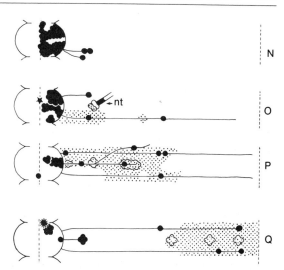

FIGURE 3. Ectodermal kinship groups N, O, P, and Q. In each panel, one segment is shown. Anterior is up, and the representation is that of an animal that has been opened along the dorsal midline and laid out flat. One segmental ganglion (and parts of two adjacent ones) are indicated by the large curved outline. The ventral midline is indicated by a dashed line along the length of the ganglion; the dorsal midline is indicated by a dashed line at the right edge of the figure. Stippled areas indicate domains of labeled epidermis, whose shapes have been idealized for this representation. Cell bodies of neurons and glia are shown in solid black, as are afferent axons of peripheral neurons. Lobed shapes within the stippled areas indicate epidermal specializations called cell florets. Cells in the O kinship group that contribute to the distal end of the nephridial duct are labeled nt.

P, and Q) is ipsilateral to the injected teloblast, mirror-symmetric to the distribution of cells derived from the corresponding bandlet on the other side, and reproducible from segment to segment, from individual to individual, and also between different leech species (Weisblat et al., 1980b, 1984; Kramer and Weisblat, 1985; Torrence and Stuart, 1986). The stereotyped group of cells in each segment descended from one teloblast are referred to as a *kinship group*. (Kinship groups are not clones because they do not include *all* the descendants of one teloblast or even one blast cell [Weisblat and Shankland, 1985].) Within both the central and the peripheral nervous systems, individual pattern elements correspond to neuron clusters or individually identifiable cells that invariably arise from a particular kinship group in normal development (Kramer and Weisblat, 1985; Weisblat and Shankland, 1985). In contrast to classical notions about cytoplasmic determinants, neurons arise from all five kinship groups. Even the cells constituting the squamous epidermis of the stage 10 embryo fall into segmentally repeating patterns characteristic of the bandlet of origin.

Like the cleavages leading from the egg to the teloblasts, the mitoses leading from the blast cells to their definitive progeny are also stereotyped, to the extent that the blast cells can be distinguished not only by their position in the germinal band, but also by the timing and spindle orientation of their initial mitoses (Zackson, 1984). For example, blast cells in the o bandlet divide unequally, about 20 hours after they are born and with the spindle roughly parallel to the long axis of the bandlet, so that pairs of larger anterior and smaller posterior sister cells can be distinguished. Blast cells in the p bandlets divide at about the same time and with the same spindle orientation, but equally, so that anterior and posterior sister cells are of the same size.

In normal development there is little variation from these patterns. Knowing the identity of a given cell in the leech embryo, we can predict its fate with relative certitude; thus, we say that leech development is quite *determinate*. As used here, the terms determinate and indeterminate refer only to the predictability, or *stereotypy,* of cell fates in normal development (Stent, 1985). Leeches, nematodes, and ascidians are quite determinate; hydra, slime molds, and the early embryos of vertebrates and arthropods are relatively indeterminate. From these examples, it is apparent that, in ac-

cord with the ideas set forth in the introduction to this chapter, determinacy and indeterminacy are not associated with any one part of the phylogenetic tree.

How readily will a cell deviate from its normal fate in response to some experimental challenge? Any such challenge tests the degree of *commitment* of the cell to its normal fate. Cells that adopt new fates in response to an experimental challenge are termed *plastic;* cells that resist experimentally induced fate changes are called *mosaic.* Even if a cell's fate is unaffected by one perturbation, there is always the chance that it would be affected by a different one, especially as the palette of possible interventions is expanded from cell ablation, explantation, and transplantation to include removal or introduction of cytoplasm or nuclei and transformation by viral or other agents. Thus, we can only speak of a cell's commitment to its fate *with respect to a particular experimental perturbation* (Stent and Weisblat, 1985). This definition of commitment intentionally combines the three types of commitment (specification, determination, and potency) proposed by Slack (1983) because the fine distinctions drawn between them are meaningless in the face of experimental ambiguities.

Notice that the concepts of stereotypy (measured on the determinate-indeterminate axis) and commitment (measured on the mosaic-plastic axis) rest on the fates of individual cells in normal or experimentally perturbed development, respectively. Since different cells within one embryo often exhibit different properties, it is imprecise to refer to the development of a whole embryo as determinate or indeterminate, or plastic or mosaic, unless we take some weighted average of all the cells in the embryo.

In the leech embryo, the overall impression is indeed one of great determinacy, but there are clear instances of indeterminacy as well. For instance, in addition to the roughly 200 bilateral pairs of neurons in each mature segmental ganglion, there are some neurons that are unpaired, including a serotonin-containing posterior medial (PM) neuron. Although the genesis of PM neurons is determinate to the extent that they invariably arise from N teloblasts, the PM neuron in any one ganglion arises with equal probability from the left or the right N teloblast (Weisblat and Kristan, 1985; Stuart et al., 1987).

Earlier in development, the O/P teloblasts and their progeny provide another example of indeterminacy (Weisblat and Blair, 1984). The two sister O/P teloblasts on either side of the embryo invariably give rise to ipsilateral O and P progeny distribution patterns in normal development. But among these two cells, the one lying closest to the N teloblast generates the O pattern 70% of the time and the P pattern 30% of the time. Its sister O/P teloblast, lying further from N, makes the complementary P pattern in 70% of the embryos and the O pattern in the remaining 30%. Further analysis showed at just what level this indeterminacy operates. Usually (70% of the time) the o bandlet (i.e., the O/P-derived bandlet next to the n bandlet) arises from the O/P teloblast closest to N and gives rise to the O pattern. In the remaining embryos, the bandlets of the two O/P teloblasts cross one another at the point where they enter the germinal band; then the o bandlet arises from the O/P teloblast lying further from N. It still gives rise to the O progeny distribution pattern, however. Thus, the fate of the O/P blast cells is determined by their relative positions in the germinal band, not by the position of their teloblast of origin (Figure 4A and Figure 5, row A).

Similarly, in ablation experiments carried out on the leech embryo, an overall impression of mosaicism is achieved, but with clear examples of plasticity. For example, if an N, OP, or Q blastomere is ablated, almost all the identifiable central and peripheral neurons normally arising from it are absent. Known

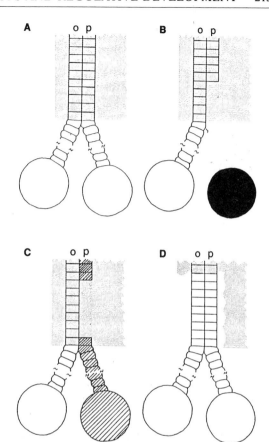

FIGURE 4. Manipulations affecting cell fates in the lefthand O-P equivalence group of *Helobdella triserialis*. In each panel only the O/P teloblasts (circles), their bandlets (o and p), and the overlying provisional epithelium (stippling) are shown. Anterior is up. (A) Normal development, in which all components are left intact. (B) Teloblast ablation, in which the O/P teloblast generating the p bandlet has been killed by injection of DNase after it had already made some blast cells. Anterior germinal band is normal; posterior germinal band lacks the p bandlet. (C) Commitment experiment, in which the O/P teloblast generating the p bandlet was injected with fluoresceinated dextran (FDX); some of its labeled progeny have been photoablated after entering the germinal band. (D) Epithelial interactions, in which n' and opq' micromeres were injected with FDX, and the epithelial domain derived from these cells has been photoablated.

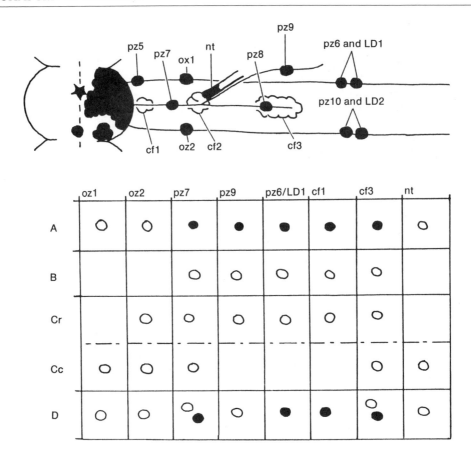

exceptions include just those unpaired cells (e.g., the N-derived PM neurons) whose origins were somewhat indeterminate in normal development. When one N teloblast is ablated, PM neurons are still present in every ganglion and they all now arise from the surviving N teloblast. It turns out that two equivalent cells, each capable of becoming a PM neuron, arise from the two N kinship groups in each ganglion; one survives and the other dies (Weisblat and Kristan, 1985; Stuart et al., 1987). Similarly, when either one of the O/P teloblasts on a side is ablated (Figure 4B and Figure 5, row B), the ipsilateral survivor's blast cells generate the P pattern of central and peripheral neurons, and the O pattern is missing in the affected part of the embryo (Weis-

blat and Blair, 1984). In both these examples, a finding of plasticity was presaged by a finding of indeterminacy. This is not always the case, however. Within the epidermis, OP- and Q-derived domains of epidermis appear highly determinate, but any ablation removing epidermal progenitors results in a rearrangement of surviving epidermal cells to cover the territory that would otherwise be left uncovered (Blair and Weisblat, 1984).

Equivalence Groups in the Leech

The two sets of developmentally equivalent and pluripotent cells described above, the precursors of the unpaired PM neuron in the N kinship groups and the pairs of O/P-derived

◄ FIGURE 5. Fates of diagnostic cells in the O-P equivalence group after experimental manipulation. The top panel represents the ventral part of one segment, showing the combined elements of the O and P kinship groups (compare with Figure 3; squamous epidermis is not shown). LD2 and oz(1–2) denote peripheral neurons in the O kinship group. LD1 and pz(5–10) denote peripheral neurons in the P kinship group. Cell florets (cf) 1 and 3 derive exclusively from the P kinship group. Cell floret 2 contains both O and P progeny. Nephridial tubule cells (nt) derive from O. The bottom panel is a matrix showing the disposition of cells diagnostic of the O and P kinship groups under various experimental paradigms. The letter on each row refers to the corresponding paradigm illustrated in Figure 4. Open circles indicate cells derived from the o bandlet; closed circles indicate cells derived from the p bandlet; and absence of a circle indicates that the cell is not formed. Cc and Cr depict the caudal and rostral classes of hybrid patterns observed in the commitment experiment. Row D shows a typical result after epithelial ablation. (Considerable variability is observed in the patterns obtained in this experiment; supernumerary neurons homologous to pz8 are also common after epithelial ablation.)

blast cells lying side by side in the germinal band, constitute *equivalence groups,* as defined in the introduction to this chapter. Each group consists of cells that normally follow different fates, but which have been shown to be developmentally equivalent, assuming different fates on the basis of hierarchical interaction with their environment. There are also differences between these two equivalence groups. The PM neuron equivalence group involves two postmitotic neurons and has serotonergy as the "preferred," or primary, fate and death as the secondary fate. But in the O-P equivalence group, both cells survive and both undergo several more rounds of division, ultimately generating a variety of epidermal, neuronal, and glial progeny. In the O-P equivalence group, the P fate is primary and the O fate is secondary.

The O-P Equivalence Group

Commitment in the O-P Equivalence Group

The earliest difference between equipotent O/P-derived blast cells that allows us to predict their fate unambiguously in normal development occurs when they assume different positions in adjacent o and p bandlets. A second difference appears at their first mitoses, during which p blast cells divide equally and o blast cells divide unequally, as described in an earlier section of this chapter. When the p bandlet is absent after ablation of an O/P teloblast, nominal o blast cells first indicate their change of fate by dividing symmetrically (Zackson, 1984). Thus, it was expected that either of these differences, especially the mitotic asymmetry, with its attendant cytoskeletal differences, would either cause or result from intracellular events that would make it difficult for the cell to change fate, i.e., that would "commit" it to one fate or the other. It is hard to devise a serious test for the commitment of the p blast cells; because they occupy the dominant place in the hierarchy, they are unlikely to change fate in response to ablation of other cells. But the capacity of nominal o blast cells to change fate in the absence of the p bandlet presents an excellent opportunity for examining the phenomenon of commitment as it pertains to individual cells in the intact embryo (Shankland and Weisblat, 1984).

Thanks to the temporal gradient in the bandlets, it has been possible to screen o blast cells and their clones at stages ranging from premitotic cells that had just entered the germinal band to clones containing several cells and many hours' experience in the germinal band, simply by killing the p bandlet in the germinal band of a stage 7 or 8 embryo. The technique that made this experiment feasible was that of selectively photolesioning cells containing fluoresceinated dextran (FDX) as a lineage tracer (Figure 4C) (Shankland, 1984). Thus, hierarchically dominant p bandlet cells were killed and the plasticity of the adjacent o blast cells and their clones was monitored by having that bandlet labeled with horseradish peroxidase (HRP).

Four different patterns of definitive progeny resulted, depending on the age, and thus the position within the germinal band, of the nominal o blast cells and their clones at the time the p bandlet was ablated. The youngest, most caudal cells in the zone of the ablation seemed completely plastic and generated the same set of P progeny as when the teloblast was ablated (see Figure 5, row B). On the other hand, the oldest, most rostral cells seemed completely committed to the O fate, giving the same set of progeny as in the normal embryo (Figure 5, row A). But cells in the sector between these two extremes generated hybrid patterns containing cells in both the O and the P kinship groups. The younger, caudal part of this sector gave clones of definitive progeny in which the P pattern was almost normal, with just a few cells normally arising from the O kinship group (Figure 5, row Cc). The older, rostral part of this sector produced clones of definitive progeny containing relatively more cells of the O kinship group and fewer cells of the P kinship group (Figure 5, row Cr).

The fact that four distinct patterns of cells (O-like, P-like, and two hybrids) are observed in this experiment indicates that progeny of the o blast cells become committed to the O fate in a stepwise manner, not all at once. Moreover, further analysis showed that the three transitions between these four states of commitment correspond to the second, third, and sixth mitoses in the nascent o blast cell clones (Shankland, 1987). It appears that each of these three divisions gives rise to two cells, one of which is mosaic, so that when the adjacent p bandlet is ablated, the cell is committed to making a certain subset of the normal O kinship group. The other cell is still plastic; that is, it is able to change its fate and make a subset of P kinship group cells. With each mitosis of a plastic cell, the cumulative set of O

kinship group cells to which the nascent o clone is committed grows and the potential for generating P kinship group cells declines. At first glance, the unequal first mitosis of an o blast cell seems not to generate a mosaic cell. The larger anterior cell seems capable of generating all identified elements of the P pattern, and no remnants of the O pattern are observed in these segments. But, in fact, this cell also seems capable of generating the entire set of O pattern elements (Shankland, 1987). There are no known progeny for the posterior cell, and it is believed to undergo programmed death (R.K. Ho and M. Shankland, unpublished data). Thus, the first mitosis of the o blast cell also generates one cell that is mosaic and one that is plastic.

Epithelial Interactions in the O-P Equivalence Group

By what mechanism(s) is the hierarchy between the two bandlets in the O-P equivalence group established? Ablation studies suggest that neither the n, the q, nor the m bandlet is required for the initial differences in mitotic symmetry between o and p blast cells to occur (Zackson, 1984). Another possible source of information affecting interactions between o and p blast cells is the epithelium that lies over the germinal bands. To determine whether the epithelium is involved in this process, it was necessary to perturb it selectively and observe the fates of the underlying blast cells (Ho and Weisblat, 1987).

For this purpose, the contributions of various micromeres arising during stages 4 to 6 to the epithelium of the embryo at stages 7 and 8 were mapped. It was found that progeny of the micromeres n' and opq' lie over the critical region where the fate-determining interactions occur (see Figure 2). The next experiment was to make this region of epithelium photosensitive by injecting those micromeres with FDX and to inject an O/P teloblast with a different tracer so that the fates of the o blast cells could be followed. For this purpose, a new tracer biotinylated fixable dextran (BFD) was devised that is even less sensitive to photo-degradation than HRP. The labeled epithelium was photolesioned when the injected embryos had developed to stage 7 or 8 (see Figure 4D). Some embryos were fixed a few hours later to observe the initial mitoses of blast cells under the lesion; others were allowed to develop to stage 10, so that the distribution patterns of the definitive progeny could be ascertained.

In the first group, it was found that both o and p blast cells under the photolesioned epithelium divide equally, like normal p blast cells. In stage-10 embryos, while hybrid patterns containing both O and P pattern elements were again found (Figure 5, row D), they differed significantly from patterns seen in the commitment experiment. In affected segments the O pattern was largely or entirely complete, as judged by the presence of those identifiable cells normally seen only in the O kinship group. However, these segments also contain supernumerary cells, most of which show close positional and morphological homologies with cells of the P kinship group. We know that these extraordinary cells are not *replacing* their normal p bandlet–derived homologs, for two reasons. First, similar embryos in which the p bandlet was labeled with lineage tracer contain the normal complement of P kinship group cells. And second, unlabeled, supernumerary cells in those embryos can be seen next to labeled cells of the normal P kinship group. These unlabeled cells are presumably derived from the affected o bandlet.

Two possible explanations exist for the fact that o blast cells in the photolesioned embryos did not change fate completely and generate pure P kinship groups. One is that the wounded epithelium may heal before the nominal o blast cells become committed to the P fate. Another is that the epithelium may not be the

only factor determining cell fates in the O-P equivalence group.

The results summarized above suggest that the epithelium is required for the normal interactions that cause developmentally equivalent o and p blast cells to follow different fates. This experiment does not distinguish between permissive and instructive effects of the epithelium, however. For instance, the results could be explained by models in which the epithelium acts either by restricting the diffusion of some substance, by restricting inductive contacts between blast cells and other cells, by generating and shaping electric currents and fields, or by direct inductive contact on the blast cells.

Equivalence Groups and Regulative Development

The equivalence group might be envisioned as a group of cells with a peculiar sort of plasticity unique to simple organisms that are otherwise strictly mosaic. But the operational principles of equivalence groups can also be applied to more complex systems, providing at least a theoretical bridge to other types of developmental plasticity collectively known as regulation. Regulation is perhaps the most troublesome word in embryology. A minimal definition of regulation that automatically embraces equivalence groups is "the generation of structures from alternative progenitors after the destruction of their normal progenitors" (Sulston and White, 1980). For our purposes, the definition will be modified to include the provision, implicit in most discussions of regulation, that the generation of structures from alternative progenitors occurs *without sacrificing other structures*. In particular, the highest form of regulation, in which all the usual structures arise after ablation and *contain their normal numbers of cells*, will be referred to as numerical regulation.

To make a connection between equivalence groups and regulative development, we need only to take into account two additional notions that are already well established. The first, as noted earlier, is that death can be a bonafide, developmentally important cell fate (Hamburger et al., 1981; Horvitz et al., 1982). The second is that single progenitor cells and the clones to which they give rise in simpler organisms may be replaced in more complex organisms by groups of progenitor cells and the polyclones to which they give rise (Crick and Lawrence, 1975).

In Figure 6 the lineal origins of hypothetical cells D, E, and F are presented. In this scheme, members of an equivalence group in each generation share a common designation (A, B, or C). Cell death occurs as a fate in all three equivalence groups, occupying the lowest priority fate. In the example constructed here, any one of the founder cells is capable of generating the entire complement of definitive progeny; even after the first mitotic cycle, half of the cells could be removed without affecting the final pattern either qualitatively or quantitatively. Thus, this group of cells would be defined by such ablation experiments as exhibiting complete numerical regulation. Obviously, the higher the proportion of normal cell death in the system, the more robust it should be in terms of regulation after partial ablation. Other aspects of this model are listed briefly here:

1. To generate larger populations of cells, additional rounds of proliferative divisions could be inserted at any point in this scheme without changing its fundamental properties. Thus, it is conceivable that the only difference between the equivalence group in simpler embryos and regulation in more complex ones is that in the latter there are too many cells to keep track of individually.

2. This sort of scheme could account for regenerative phenomena and for regulative phenomena in the absence of cell

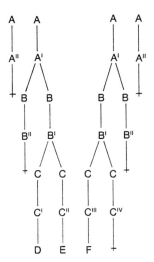

FIGURE 6. **Extension of the equivalence group formalism to account for regulative phenomena in development. A hypothetical lineage tree leading to the formation of definitive progeny D, E, and F. The hierarchical assumption of nonequivalent fates (primary, secondary, and so on) is indicated by superscripts. Cell death (the lowest priority fate in each equivalence group) is indicated by crossed lines. Note that any manipulation leaving even one A cell intact would have no effect on the number or type of definitive progeny.**

death if "quiescence" is included as a possible cell fate. Like death, quiescence would occupy a low rank in the hierarchy of possible fates.

3. From this model, one would predict that ablations would have different effects from generation to generation and even within cell generations if carried out before and after the assumption of distinct fates within an equivalence group (B, B', and B″ in Figure 6, for example). This effect would become less noticeable with asynchrony in the cell divisions. Such asynchrony at first seems antithetical to the notion of the equivalence group, but in fact it is not. As long as the determinative signals are constant, the equivalence groups can be dynamic, enduring pop-

ulations of whatever cells are in state A, B, or C at the time of interest.

4. In the example given, deleting more than 25% of the cells after the second round of division would result in the loss of specific cells (or cell populations) from the embryo, beginning with F. This effect, which is in accord with the fact that this model for regulative development is based on the equivalence group formalism, could be used not only to explain effects of ablations, but also some types of phylogenetic change. For example, different groups of plethodontid salamanders have evolved species that are greatly reduced in overall size (Wake, 1966). These species also vary dramatically in genome size, with corresponding variations in average cell volume (Sessions, 1984). Dwarf species with large genomes have less well-developed skeletons than dwarf species with small genomes and lack certain bones altogether (Wake, 1966; Sessions, 1984; Hanken, 1984). These observations might be explained as consequences of the fact that the combination of dwarfism and large genomes would greatly reduce the number of cells that can be contained in any given volume of the embryo. If some particular volume element serves to define an equivalence group, the combination of extreme dwarfism and large cell volume could reduce the number of cells in the equivalence group to the point at which low-priority lineages are lost, just as if an ablation had been carried out that exceeded the regulative capacity of the affected equivalence group.

Equivalence Groups in Other Organisms

In the preceding section, it was suggested that the formal principles of the equivalence group can be expanded to explain a variety of regulative phenomena in development. But in any attempt to extend comparisons of different equivalence groups from the phenomenological to the mechanistic level, we must be wary of assuming that two things are the same merely because they are given the same name. To what extent do cells within different equivalence groups interact by common mechanisms? No definitive answer to this question is yet possible, but several comparisons can be made.

The concept of the equivalence group was first defined from examples of plasticity uncovered in the course of lineage and ablation studies in the nematode *Caenorhabditis elegans* (Kimble et al., 1979; Sulston and White, 1980). The best-studied example in the nematode is in the vulval equivalence group (Kimble, 1981; Sternberg and Horvitz, 1986; Greenwald et al., 1983), in which each of six equipotent cells assumes one of three distinct fates. In this case, the hierarchy of cell fates seems to be established on the basis of distance between members of the equivalence group and an inducer, named the anchor cell, that is not in the equivalence group (Figure 7). It is suggested that the equivalent cells respond to a gradient of a substance generated by the anchor cell. All members of the equivalence group adopted the lowest-priority (tertiary) fate when the anchor cell was killed by ultraviolet microbeam. Comparing the O-P equivalence group in the leech and the vulval equivalence group in the nematode is not easy. Is the leech epithelium acting instructively, like the nematode anchor cell? If so, it must be by some means other than a chemical gradient, since both o and p bandlets contact the epithelium (see Figure 7). Moreover, when the epithelium is lesioned in the leech embryo, both bandlets tend toward the highest-priority (primary) fate. If the epithelium is acting permissively instead, then the main fate-determining interaction may turn out to take place between cells in the equivalence group themselves, again in contrast to the results in the nematode.

Other equivalence groups have been dis-

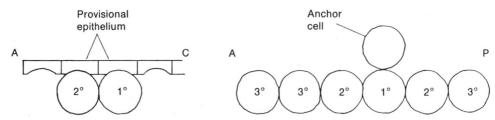

FIGURE 7. Schematic comparison of the O-P equivalence group in *Helobdella* (left) and the vulval equivalence group in *Caenorhabditis* (right). Cells in each equivalence group are labeled according to the usual fate associated with that position. The only other cells shown are those that are known to affect cell fates within the equivalence group. The polarity indicated for the *Helobdella* embryo is that defined by moving towards the center of the epithelium, or micromere cap (C, capward), or towards its edge (A, anticapward). For *Caenorhabditis*, the anterior-posterior (A-P) axis is indicated.

covered in insect neurogenesis. In one case, sister cells generated by the mitosis of a particular ganglion mother cell are initially equipotent, but then follow two distinct fates (Kuwada and Goodman, 1985). Earlier in embryogenesis, cells in the neuroepithelium appear competent to become any of various neuroblasts or epithelial support cells (Doe and Goodman, 1985a,b). It is interesting that insects have both classical equivalence groups, involving small numbers of identified cells, and what would pass for regulative development in the epidermal compartments and in the generation of neuroblasts from neuroepithelium. This is further circumstantial evidence for the hypothesis that the two phenomena may be related. Moreover, genetic and molecular analyses of nematode and insect suggest that meaningful comparisons might be made at the molecular level as well. The evidence consists of findings that genes affecting cell fates in nematode equivalence groups (lin 12) and in the neuroepithelium of *Drosophila* (notch) both code for proteins containing domains homologous to epidermal growth factor (Akam, 1986). It would be interesting to know if a homologous gene exists in the leech and, if so, whether its expression is linked to the cellular interactions in the O-P (or PM neuron) equivalence groups as well.

The discovery of these genes has generated excitement, but a molecular explanation of equivalence groups is not yet in hand. In addition to lin-12, other genes have been discovered that, when mutated, affect equivalence groups in *Caenorhabditis* (Fixen et al., 1985). Not all these genes affect all equivalence groups, which suggests the possibility that even within this one species, the equivalence groups are a family of related, yet distinct phenomena. And even when all the genes are known for all the equivalence groups, the hard task of understanding how they affect cell properties so as to generate the observed behaviors will still remain.

Summary

The results presented in this chapter show once again that the relative simplicity of invertebrates such as the leech allow precise cellular analyses of developmental processes. Previous work on the leech has dealt mainly with fate-mapping the progeny of the teloblasts, which generate the definitive segmental tissues (Blair, 1983; Weisblat et al., 1984; Kramer and Weisblat, 1985; Weisblat and Shankland, 1985; Zackson, 1982, 1984), and on the interactions between those progeny (Blair, 1982; Blair and Weisblat, 1982, 1984; Weisblat and Blair, 1984; Shankland, 1984; Shankland and Weisblat, 1984). The primary conclusion of the new work summarized here (Ho and Weisblat, 1987) is that a provisional epithelium derived from the micromeres is necessary for the interactions that normally cause equipotent cells to follow different fates.

The terminology used here for basic concepts of cell fate in development differs somewhat from standard usage, being a further extension of recent attempts to clarify muddy waters (Slack, 1983; Stent and Weisblat, 1985; Stent, 1985). In essence, there is little point in attempting to generalize fine conceptual distinctions between various developmental phenomena while we remain so ignorant of basic developmental mechanisms and how they relate to the phenomena we observe. Two basic ideas distinguished here are that (1) cells have fates in normal development that are predictable to varying degrees (stereotypy), and (2) the normal fate of a cell may change in response to any of various perturbations of normal development (commitment). Stereotypy can be thought of as one embryonic dimension, with determinacy and indeterminacy at opposite ends of the axis. Commitment corresponds to another dimension, defined by a mosaic-plastic axis. In this usage, determinacy refers only to the predictability of normal fate and is quite distinct from mosaicism (Stent, 1985). Stereotypy and commitment are most easily assessed in embryos where cells are individually identifiable, and commitment is most easily measured in systems that are also highly determinate. But in principle, stochastic approaches are possible where cells are not highly determinate and even where cells are not uniquely identifiable. Ipsilateral O/P teloblasts in the leech constitute an example of cells that are neither purely determinate nor uniquely identifiable, since they are lineally equivalent sister blastomeres and also somewhat variable in their relative positions. It could be argued that this is a trivial example, involving ambiguity between only two cells. But to do so would miss a basic contention of this chapter, namely, that basic mechanisms may well be conserved between simple and complex organisms and that the apparent differences are consequences of large differences in the size of cell populations on which the mechanisms operate.

Development presents interesting problems at many levels of organization, from subcellular to behavioral; a correspondingly broad range of techniques must be used to reach a satisfactory explanation of the process as a whole. Phenomena observed at higher levels of organization, for example, cell movements during morphogenesis, are certainly rooted in the cell and the molecular biology of the system. But we can no more *predict* one from the other than we can predict the existence of DNA from quantum mechanics. (The question of whether or not such predictions are even *theoretically* possible is intriguing but irrelevant to the present discussion.) Thus, we should continue to analyze development at all levels and to integrate the results of our efforts.

Another lesson to be learned here is that ideas arising from studying the minimal plasticity of the equivalence groups in the nematode and leech might be applicable to the phenomena of regulation in higher invertebrates and vertebrates. The success of such applications depends on identifying phenomena that are meaningfully similar, and this in turn depends on the level of analysis undertaken. As shown here, similarity can be sought and found at the operational, cellular, and molecular levels. But equivalence at one level does not necessarily entail equivalence at the others. Which level of analysis is most meaningful? The answer to this question lies in the mind of the questioner; to debate it needlessly is to fall into the same trap as those who debate bitterly as to the true unit of natural selection.

Identification and Fate of Neural Crest Cell Subpopulations in Early Embryonic Development

JAMES A. WESTON, KRISTINE S. VOGEL,
AND MICHAEL F. MARUSICH

Introduction

Analyses of the mechanisms that establish cellular diversity and specific connections within metazoan nervous systems address two fundamental developmental issues: (1) how phenotypic differences arise and become stabilized within cell lineages, and (2) what role is played by environmental cues that are normally encountered by cells that express these differences. Since developing multicellular organisms are complex and dynamic systems, those who try to answer these questions must contend with a biological "uncertainty principle." According to this principle, when a cell is apparently undifferentiated, one cannot predict precisely what phenotype it will express, whereas if it has already differentiated, one cannot easily reconstruct what its antecedent could have become under other circumstances. Because of this dilemma, the genetic and molecular mechanisms involved in the choice and expression of specific cellular phenotypes have frequently defied analysis.

To deal productively with these issues and to allow coordinated experimental approaches at several levels of biological organization, we must identify systems where analysis of developmental abilities of individual cells can be combined with an assessment of the extent of stable, propagable restrictions in these abilities. We must then be able to establish when such restrictions are imposed on a cell or its progeny, so that the role of environmental cues that might elicit such developmental changes can be assessed.

In order to analyze the process of sequential developmental restrictions during normal development, moreover, several experimental criteria must be satisfied. First, because it is clear that we cannot sufficiently define a cell's phenotype solely in terms of expression of individual traits, it is essential that a variety of sensitive cell type–specific markers be systematically applied on individual embryonic cells. Second, the stem cell populations, and *their* developmental antecedents, that give rise to specific cell types should be known and accessible to direct observation and experimental manipulation. Finally, we must be able to perturb the macromolecular components of the environments normally encountered by individual identified cells, in order to analyze the role of these molecules in determining the time and stability of commitment to specific developmental pathways.

The Differentiation of Neural Crest Cells

The neural crest, a transient stem cell population present in vertebrate embryos, satisfies some of these requirements. It consists of a recognizable population of cells that disperses widely into different embryonic environments, proliferates rapidly, and gives rise to a variety of cell types. These include the pigment cells of the integument and iris, neurons and glial cells of the peripheral (sensory, autonomic, and enteric) nervous system, and neurosecretory tissues in the adrenal, heart, lungs, and pharyngeal structures (e.g., the carotid body and the C cells of the thyroid). In addition, so-called ectomesenchyme from the cranial neural folds ultimately differentiates as skeletal and connective tissue of the head and face (Le Douarin, 1982; Noden, 1983, 1984; Weston, 1982; Weston et al., 1984).

The neural crest precursor population can be isolated at early developmental stages, and this has allowed the developmental abilities of such populations to be analyzed both in vivo and in vitro. These analyses exploit the variety of useful markers for crest cell phenotypes, including morphology, cytochemical and immunological traits, and physiology (Weston, 1982, 1983). Likewise, the kinds and amounts of interstitial matrix macromolecules produced and encountered by dispersing crest

cells have now been examined in some detail (Weston et al., 1978; Glimelius and Pintar, 1981; Rovasio et al., 1983; Turley et al., 1985; Runyan et al., 1986; Rogers et al., 1986). The effects of the matrix components encountered by crest cells on their morphogenetic and differentiative behavior have also been investigated (Sieber-Blum et al., 1981; Loring et al., 1982; Derby, 1982; Rovasio et al., 1983; Tucker and Erickson, 1984; Perris and Lofberg, 1986).

As a consequence of these and other analyses over the last two decades, it has become widely accepted that environmental cues encountered during dispersal and localization influence crest cells' developmental fates. Thus, heterotopic transplantation of neural crest cells has clearly demonstrated that crest cell populations from every axial level maintain a wide range of developmental potentialities (Le Douarin, 1982; Noden, 1984). However, since these experiments have been performed with cell populations rather than single cells, inferences about the developmental abilities of *individual* cells, and their responses to environmental cues, are difficult or impossible to make. In fact, it has proved extraordinarily difficult to determine with certainty the extent to which environmental cues (1) induce homogeneous, pluripotent neural crest cell populations to express distinct phenotypes, or (2) differentially exert their effects on developmentally distinct subpopulations that already exist in the early embryo.

In spite of some assertions to the contrary (Weston, 1970; Le Douarin et al., 1978; Le Douarin, 1982; Anderson and Axel, 1985), the role of environmental cues on the development of crest derivatives remains poorly understood. In contrast to the conventional hypothesis invoking crest cell pluripotentiality, the environment might differentially affect the survival, proliferation, or maturation of cells whose developmental repertoire was at least partially restricted. *The implicit, but cru-*

cial, distinction between the two alternatives is how early in development the crest cells undergo restrictions of their developmental abilities. Thus, in order to understand the role of the environment in phenotypic diversification, it is essential to ascertain when, and under what conditions, developmental restrictions arise.

In cases where the developmental abilities of single crest-derived cells have been analyzed, either by in vitro cloning or by using cell type–specific markers for crest derivatives in vivo or in vitro, the results have demonstrated the existence of some crest cells that are at least bipotent. Thus, clones derived from the early neural crest outgrowth of explanted avian embryonic neural tubes contain cells that express an adrenergic phenotype as well as cells that express melanogenic traits (Cohen and Konigsberg, 1975; Sieber-Blum and Cohen, 1980). Similarly, secondary cultures of neural crest cells occasionally contain single cells that express both melanogenic and adrenergic traits (Loring et al., 1982; Weston, 1986). It is not known if either the founder cells of mixed clones or the dual-phenotype crest cells are also capable of generating other crest derivatives.

Heterogeneity of Developmental Potential Among Crest–Derived Cells

Partially Restricted Cells in Migrating Crest Populations

Although the early neural crest contains some cells that are developmentally labile, there is also evidence that some developmental restrictions may already exist within the early migrating crest cell population. For example, although cloning experiments (Sieber-Blum and Cohen, 1980; Kahn and Sieber-Blum, 1983) demonstrate that some progenitor crest cells are at least bipotent, other clonal progenitors produce colonies expressing a single cellular phenotype. Likewise, hetero-

chronic grafts of crest-derived structures (e.g., sensory and autonomic ganglia, and branchial arches) into the crest migratory spaces of younger host embryos indicate that although some cells are able to migrate, localize, and differentiate appropriately, other crest-derived cells in the grafted tissues are unable to do so (see discussion later in chapter; Erickson et al., 1980; Le Lievre et al., 1980; Ciment and Weston, 1983).

The presence of phenotypically distinct neural crest subpopulations at early migratory stages has now been directly demonstrated by the use of monoclonal antibodies (Barald, 1982; Ciment and Weston, 1982, 1985; Payette et al., 1984; Vincent and Thiery, 1984; Barbu et al., 1986). Some of these studies indicate that developmentally distinct populations exist even before they can be detected by sensitive immunological criteria. Thus, although local environmental cues encountered at the earliest stages of crest cell dispersal may influence developmental fates, the possibility also remains that intrinsic cellular mechanisms generate subpopulations within the premigratory crest upon which environmental factors act (Girdlestone and Weston, 1985; Weston, 1986). It should be emphasized, however, that although early *phenotypic* diversity is clear, the *developmental* abilities of most of these subpopulations and the mechanisms responsible for their generation still remain to be elucidated.

Developmentally Restricted Cell Populations Segregate Sequentially

We have suggested that neural crest cells undergo phenotypic diversification by a *sequence* of developmental restrictions in the crest lineage (Weston, 1981; 1982) and have proposed that such restrictions produce subpopulations of cells with different developmental abilities. Within each partially restricted, developmentally intermediate subpopulation, local environmental stimuli could presumably elicit expression of one or more of the phenotypes that remain in the subpopulation's repertoire. Conversely, in the absence of appropriate local cues, some intermediate cell types might fail to thrive. If such cells survive, however, parts of their repertoire might remain latent or be further segregated by subsequent developmental events into more highly restricted intermediate populations. Figure 1 diagrams one possible scheme of sequential segregation of developmental potential within a stem cell lineage.

Although we need additional experiments to establish the existence, location, and properties of such partially restricted cells, some

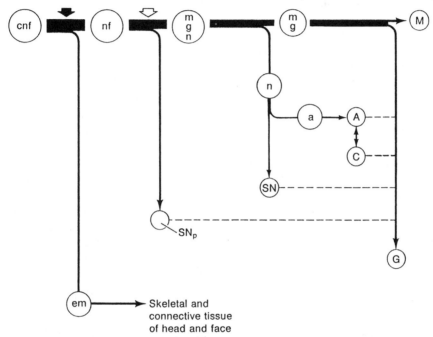

FIGURE 1. Segregation of developmentally restricted subpopulations from the neural crest lineage. The diagram, which reconciles many published results (summarized in Weston, 1982; 1986; Weston et al., 1984), suggests one possible sequence of developmental restrictions resulting in various intermediate subpopulations. The circles represent subpopulation classes that may be present at various times and embryonic locations. Lower case letters within the circles indicate the developmental capabilities remaining in each population. The time that certain restrictions occur is suggested on the horizontal axis relative to the onset of migration from the cranial neural folds (filled arrowhead), and then the trunk crest (open arrowhead). Some ectomesenchymal cells, for example, may be specified before cell dispersal begins in the head. Likewise, some sensory neurons may be precociously determined before the onset of cell migration from the neural folds. cnf, cranial neural folds; nf, trunk neural crest and cranial neural crest after ectomesenchymal (em) cells have departed; a, adrenergic neuroblasts (FIF-positive and SIF cells); A, adrenergic autonomic neurons; C, cholinergic autonomic neurons; g, supportive cell precursors; G, differentiated glial or satellite cells; m, melanoblasts; M, melanocytes; n, neurogenic precursors (neuroblasts); SN_p, early differentiating (primary) sensory neurons; SN, late differentiating neurons of the senory ganglia. Dashed lines denote putative interactions between neurons and supportive cell precursors (see Holton and Weston, 1982). Small arrowheads indicate terminal differentiations or modulations (Modified from Weston, 1981, 1982).

crest-derived populations with partial developmental restrictions have been demonstrated. For example, when grafted into suitable embryonic locations, the cranial crest can produce every known neural crest derivative, whereas early restrictions seem to be imposed on the trunk neural crest that prevent it from producing skeletal and connective tissue derivatives in appropriate embryonic environments (Noden, 1980; Weston, 1982; but see Nakamura and Ayer-Le Lievre, 1982). Likewise, restrictions that limit the ability of trunk crest cells to produce sensory neurons seem to be imposed prior to, or immediately after, the onset of crest cell migration (Erickson et al., 1980; Ziller et al., 1983; Ayer-Le Lievre and Le Douarin, 1982; Girdlestone and Weston, 1985; Ciment and Weston, 1985). In addition, nonneuronal cells of nascent sensory and sympathetic ganglia can give rise to melanocytes when their association with neurons is disrupted (Nichols and Weston, 1977; Nichols et al., 1977). Although this ability is stably repressed in the non-neuronal ganglionic cells of older embryos, experimental treatment with the cocarcinogen TPA partially restores it (Ciment et al., 1986a).

Finally, results of heterochronic and heterotopic grafting of crest-derived branchial arch mesenchyme suggest that these cells undergo progressive partial restrictions during development (Ciment and Weston, 1983, 1985). The crest-derived cells that populate these structures contain a neurofilament-associated protein, NAPA-73 (Ciment et al., 1986b), recognized by the monoclonal antibody E/C8 (Ciment and Weston, 1982). These cells normally produce connective tissue of the head and face, as well as enteric neuron precursors that enter the contiguous primordial gut (Allan and Newgreen, 1980; Le Douarin and Teillet, 1974; Le Lievre and Le Douarin, 1975; Smith-Thomas et al., 1986). When cells from posterior (third and fourth) branchial arches of four-day avian embryos are grafted heterotopically to trunk crest migratory spaces, they can give rise to neuronal and non-neuronal neural crest derivatives (e.g., glandular and connective tissues; Ciment and Weston, 1983, 1985). In contrast, under comparable environmental conditions both in vivo and in vitro, these four-day branchial arch cells lack the melanogenic ability that their rhombencephalic neural crest cell antecedents possessed (Ciment and Weston, 1983).

The Role of the Environment in Neural Crest Cell Differentiation

Environmental cues can affect the phenotype of crest-derived cells by modulating the traits expressed by members of an established class of cells, for example, the choice of transmitter by autonomic neurons (Black et al., 1984; Furshpan et al., 1976; Patterson, 1978; Potter et al., 1986; Coulombe and Bronner-Fraser, 1986) or the proportions of neuroactive substances contained by such cells (Hokfelt et al., 1984; Doupe et al., 1985; Hayashi et al., 1985). Environmental cues may also elicit more fundamental developmental decisions leading to the appearance of new classes of embryonic cells. Such decisions might, for example, involve the segregation of neuronal and glial cell lineages, or the sequential segregation of distinct neuronal or neurosecretory subclasses within a larger neuronal class.

Neuronal Precursors in Crest Cell Cultures

Recently, we have used the monoclonal antibody A2B5 (Eisenbarth et al., 1979) to demonstrate that cells with a neuronal phenotype arise early in cultures of neural crest cells (Girdlestone and Weston, 1985). When embryonic quail neural tubes are cultured on nonadhesive substrata, neural crest cells separate from the neural epithelium and form clusters (Loring et al., 1981; Glimelius and

Weston, 1981). After various intervals of primary culture, these crest cell clusters can be detached from neural tube and subcultured. When crest cell clusters are isolated from 20 to 24-hour explants of neural tube, about 1% of the crest cells already exhibit neuron-specific A2B5 immunoreactivity (Figure 2A,B). Additional A2B5$^+$ cells appear during subsequent days of secondary culture (Figure 2C,D). Ultimately, the proportion of A2B5$^+$ cells reaches a maximum and then declines after the fourth day of secondary culture (Figure 3, open circles).

Thymidine-incorporation studies indicate that A2B5$^+$ cells in four-day crest cultures are postmitotic, but that virtually all such cells could incorporate label early in the culture period. During subsequent days of culture, the labeling index for cells that revealed A2B5 immunoreactivity at the end of the four-day culture period dropped progressively (Figure 4, hatched bars), whereas the cell population as a whole continued to incorporate label throughout the culture period (Figure 4, open bars). We conclude that the specific A2B5$^+$ subpopulation that appeared during this interval

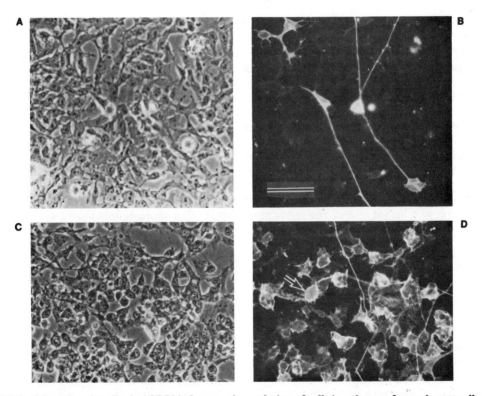

FIGURE 2. Monoclonal antibody A2B5 binds to a subpopulation of cells in cultures of neural crest cell clusters. Neural crest clusters isolated from 24-hour explants of neural tubes and subcultured for one (A, B) or two (C, D) days before immunolabeling. Immunoreactive cells in one-day cultures are usually unipolar or bipolar cells. Such cells are present in older cultures (arrow), but immunoreactive cells that appear after the first day of culture have a different (flatter, stellate) morphology. Phase-contrast (A, C; epifluorescence (B, D). Bar = 50 μm. (From Girdlestone and Weston, 1985.)

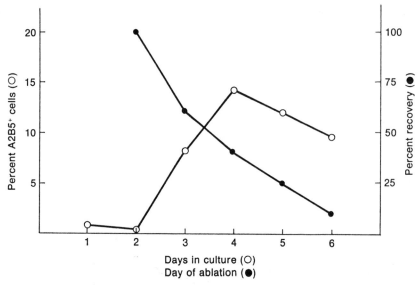

FIGURE 3. Proportion of A2B5⁺ cells in secondary cultures of 24-hour neural crest cell clusters. Open circles indicate the changes in the estimated proportions of immunoreactive cells throughout the culture period. The decrease observed during first day of subculture of one-day clusters is probably due to proliferation of A2B5⁻ cells. The decrease in the proportion of A2B5⁺ cells after the fourth day of culture is believed to result when differentiation of new A2B5⁺ cells declines while other crest-derived cells in the culture continue to proliferate. Filled circles indicate the remaining neurogenic precursors inferred from the proportions of immunoreactive cells present one day after complement-mediated immunoablation. Quail neural crest cell cultures were treated with antibody A2B5 and complement at various times of secondary culture. No A2B5⁺ cells were observed in complement-treated crest cultures immediately after treatment. Reappearance of A2B5⁺ cells was assessed 24 hours after ablation and is graphed as a percent of the proportion of A2B5⁺ cells present in parallel control (untreated) crest cell cultures. These data, represented as "percent recovery," are believed to reflect the progressive depletion, due to differentiation, of a limited pool of neurogenic precursors. Each point represents pooled data from 4 to 6 cultures; 500 to 1000 cells were counted in each culture.

arose from precursors that divided at least once in culture (Girdlestone and Weston, 1985).

Since the A2B5⁺ cells are postmitotic, the absolute increase in their number might be due to *recruitment* of cells with A2B5 (neuronal) immunoreactivity from a developmentally labile population of A2B5⁻ cells. Alternatively, A2B5⁺ cells might arise from the maturation of a specific A2B5⁻ lineage antecedent with limited proliferative potential. Although other mechanisms are possible, the former hypothesis suggests that the ability of cultures to produce A2B5-immunoreactive

cells would persist, whereas the latter predicts a limited and progressively diminishing source of precursor cells.

To test these predictions, A2B5-immunoreactive cells were selectively ablated by the addition of A2B5 antibody and complement (Vogel and Weston, 1986), and the ability of the culture to produce new A2B5⁺ cells was assessed. When two- or three-day subcultures of 24-hour clusters were immunoablated, new A2B5⁺ cells reappeared after one to two days of additional culture. In contrast, the ability to generate new A2B5⁺ cells in progressively older cultures is dramatically and consistently reduced, even after extended periods of culture (Figure 3, filled circles).

We conclude that the early subcultures contain some A2B5⁻ cells able to develop A2B5 immunoreactivity and that this ability is not adversely affected by treatment with antibody and complement. A parsimonious way to reconcile both the labeling and the ablation studies is to postulate that a limited, *specific* subpopulation of neurogenic precursors arises in these crest cultures and that the pool of precursors is depleted as they mature into postmitotic A2B5⁺ neurons. The observed decline in the proportion of postmitotic A2B5⁺ cells in older secondary cultures is consistent with the idea that this precursor pool depletes while

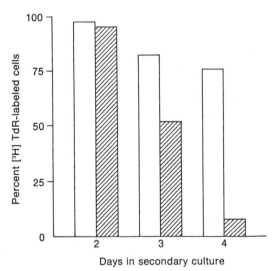

FIGURE 4. Incorporation of [³H]TdR by all cells (open bars) and by precursors of A2B5⁺ cells (hatched bars) in secondary cultures of neural crest cell clusters. Cultures of one-day crest clusters were given 24-hour pulses of [³H]TdR on the second, third, or fourth day after subculture. All cultures were prepared for immunocytochemistry and autoradiography at the end of the fourth day. In older cultures, thymidine incorporation by the precursors of A2B5⁺ cells is dramatically reduced compared to that by the total cell population. (From Girdlestone and Weston, 1985.)

other apparently undifferentiated crest-derived cells in the culture continue to proliferate. Alternatively, the decreased ability of cultures to produce new A2B5⁺ cells may reflect a decrease in the probability that pluripotent, undifferentiated cells in the culture can be recruited onto this developmental pathway. This hypothetical alternative might best be tested by culture experiments based on the Luria and Delbruck (1943) fluctuation test, whereas the most compelling evidence for the existence of a limited neurogenic precursor subpopulation will be provided if suitable independent markers for such precursors can be devised. Additionally, the analysis of whether other environmental conditions in vitro or in vivo can augment the pool of putative neurogenic precursors should reveal how environmental cues affect their specification, maintenance, and fate.

The Earliest A2B5⁺ Cells May Be Sensory Neurons

A comparison of the present results with previous in situ and in vitro experiments suggests to us that the initial postmitotic A2B5⁺ population, in vitro, represents primary sensory neurons, whereas the A2B5⁺ population that appears later in crest cell cultures includes other neuronal classes, some of which express autonomic traits. Thus, when migrating crest cells are transplanted heterochronically, the grafted cells give rise to all crest derivatives characteristic of the particular host environment, including sensory neurons (Weston and Butler, 1966; Erickson et al., 1980; Ayer-Le Lievre and Le Douarin, 1982). In contrast, heterochronic grafts of crest-derived tissues, placed in the crest migratory pathway of host embryos, give rise to most crest derivatives, but *not* sensory neurons. These crest-derived grafts included cultured neural crest cell clusters (Erickson et al., 1980), as well as sympathetic or parasympathetic ganglia of any age (Le Lievre et al., 1980; Ayer-Le Lievre and Le Douarin, 1982) and dorsal root ganglia from older (e.g., seven-day) embryos (Schweitzer et al., 1983; but see Rohrer et al., 1986). Since Ayer-Le Lievre and Le Douarin (1982) have demonstrated that differentiated neurons in grafts of neural crest–derived ganglia do not survive transplantation (see also Dupin, 1984), it seems likely that either the precursors of sensory neurons had been depleted or the surviving crest-derived cells in the grafted tissue had lost the capacity to generate this class of neurons.

Further, several lines of evidence suggest that the second A2B5[+] population to appear in neural crest cell cultures contains neurons with autonomic traits. Thus, the second wave of A2B5[+] cells can be detected shortly before the time when cells with adrenergic properties have been reported to appear in culture. Such properties include catecholamine storage (Sieber-Blum and Cohen, 1978) and release (Maxwell and Sietz, 1983) and the binding of soybean agglutinin (Sieber-Blum and Cohen, 1978). The latter property probably reflects the presence of a glycolipid found at higher levels on adrenergic neurons than on cholinergic neurons (Zurn, 1982). Additionally, Kahn and Sieber-Blum (1983) have reported that the appearance of adrenergic cells, but not melanocytes, was inhibited by blocking cell division during the first three days of neural crest culture. This correlates well with the present observations that the second wave of A2B5[+] cells arises from cells that are mitotically active only during this period in culture. Recent preliminary results indicate further that the proportion of A2B5[+] cells that express tyrosine hydroxylase (TH) immunoreactivity increases progressively in secondary cultures of 24-hour neural crest cell clusters (Figure 5) (Vogel and Weston, 1985) and that, consistent with this catecholaminergic trait, a few of the A2B5[+] cells have been observed to coexpress glyoxylic acid–induced catecholamine fluorescence. Finally, reports of discrete populations of substance P-, somatostatin-, serotonin-, and

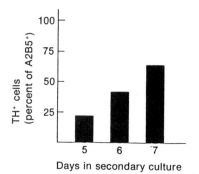

FIGURE 5. A subpopulation of A2B5[+] cells also exhibits tyrosine hydroxylase (TH) immunoreactivity. Subcultures of 24-hour quail neural crest cell clusters were double-labeled with monoclonal antibody A2B5 and a monospecific antiserum against TH. For each bar, 500 to 1000 cells were scored. The proportion of A2B5[+] cells that are also TH[+] increases between days 5 and 7 of secondary culture.

methionine enkephalin–immunoreactive cells in crest cell cultures (Maxwell et al., 1983; Sieber-Blum, 1984; Sieber-Blum and Patel, 1986) suggest that if all neurons in crest cell cultures are in fact A2B5[+], this immunoreactive neural crest population will contain a number of neuron subclasses.

Alternative Crest-Derived Phenotypes In Vitro

Since environmental cues affect the expression of various cellular phenotypes by the neural crest, it is clearly important to establish exactly when this control is exerted, so that the precise role of the environment can be assessed. The appearance of the second wave of A2B5-immunoreactive cells in crest cell cultures is likely to provide an appropriate paradigm for such an analysis because of its marked dependence on early culture conditions. Thus, large numbers of A2B5[+] cells appear when 24-hour crest cell clusters are removed from explanted neural tubes and subcultured on tissue culture substrata, whereas no *new* A2B5[+] cells arise in secondary cultures of clusters that remain associated with the neural tube explants for 32 or 48 hours (Girdlestone and Weston, 1985; and K. Vogel and J.A. Weston, unpublished data). In secondary cultures of older (32- to 48-hr) crest cell clusters, only the initial population of A2B5$^+$ cells, distinguished by their unipolar or bipolar morphology, can be detected, and the proportion of these A2B5-immunoreactive cells declines during the culture period, probably because of proliferation of A2B5$^-$ cells (see discussion earlier in chapter; Girdlestone and Weston, 1985). New A2B5$^+$ cells also fail to appear in cultures of older cluster-derived cells from which A2B5$^+$ cells have been ablated (Vogel and Weston, 1986; K. Vogel and J.A. Weston, in preparation).

Moreover, when 24-hour crest cell clusters are removed from explanted neural tubes, but subcultured on agar substrata to prevent cell dispersion for an additional 8 or 24 hours before then being allowed to disperse on plastic substrata, the pattern of cell differentiation resembles that of clusters maintained in association with the neural tube explant for 32 and 48 hours, respectively (Glimelius and Weston, 1981; K. Vogel and J.A. Weston, unpublished data). We conclude that the 32- and 48-hour clusters lack the precursors of A2B5-immunoreactive cells, and that this loss of developmental potential occurs within an 8-hour period in response to environmental conditions established within the cell clusters. Since this restriction occurs in the absence as well as in the presence of the original neural tube explant, it seems most likely that the effective environmental cues are established by the crest cells themselves.

Melanogenesis in these cultures exhibits a reciprocal pattern. Thus, the proportion of melanocytes that differentiate in secondary culture of 24-hour clusters (10 to 20%) is much lower than that in subcultures of 32- or 48-hour clusters (usually >90%) (Glimelius and Weston, 1981; Vogel and Weston, 1986). This reciprocal expression of neuronal and pigment cells might be reconciled by what has been observed in cultured sensory ganglia (see below; Nichols and Weston, 1977; Nichols et al., 1977).

Glial Cell Precursors Can Undergo Melanogenesis

Nascent sensory ganglia contain populations of non-neuronal cells, some of which may themselves be precursors of additional neuronal classes (Newgreen and Jones, 1975; Xue et al., 1985). Other non-neuronal cells of nascent avian sensory ganglia, which probably normally give rise to glial cells, are transiently able to undergo melanogenesis when interactions with neurons are disrupted or prevented (Nichols and Weston, 1977; Nichols et al., 1977). Thus, cultured ganglia from four-day (Hamburger-Hamilton stages 21 to 23) but not seven-day (e.g., stage 30) or older avian embryos produce populations of melanocytes. These melanocytes may be identified not only by cell morphology and the presence of melanosomes, but also by their ability to invade feather germs when cultured ganglia are grafted into the wing buds of unpigmented host embryos (Ciment et al., 1986a). Conversely, the differentiation of glial cell traits by the non-neuronal cells of the ganglion is promoted when interactions with neurons are established and/or maintained (Holton and Weston, 1982). Based on these observations, we have suggested that glia and melanocytes share a common progenitor cell type (Weston, 1981, 1982).

Recent immunocytochemical results support this suggestion. For example, the monoclonal antibody R24 (Dippold et al., 1980), originally raised against human melanoma cells, binds to cultured crest-derived avian melanocytes (Figure 6A,B). It is noteworthy, however, that this antibody also binds to non-neuronal cells from avian embryo spinal ganglia (Figure 6 C,D) (Girdlestone and Weston, 1985). This pattern of immunoreactivity is consistent with, but obviously does not prove, the notion that supportive and pigment cells share a specific lineage antecedent, as indicated in the hypothetical scheme presented previously (see Figure 1).

The reciprocal pattern of neurogenesis and melanogenesis in secondary cultures of neural crest clusters also may be reconciled by these observations. Thus, the choice between melanogenesis and gliagenesis in nascent ganglia appears to depend on the interaction between neurons and glial cell precursors. In some crest cell cultures, cells with neuronal phenotype can arise precociously, as suggested in the scheme of sequential segregation of crest phenotypes (see Figure 1), and few pigment cells appear. However, melanogenesis is enhanced in cultures of 48-hour clusters, where neurogenesis is minimal, as it is in cultures of dissociated nascent ganglia, where the interaction between neurons and the intermediate glial/melanocyte progenitors is disrupted.

Environmental Cues May Specify Neuron Type

Various neuronal subclasses may also be established by sequential restrictions within a

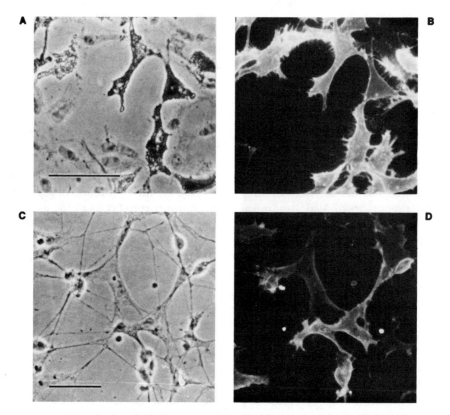

FIGURE 6. Monoclonal antibody R24 binds to neural crest–derived melanocytes and non-neuronal cells in vitro. (A, B) R24 immunolabeling of neural crest cells from clusters isolated from 48-hour explants of neural tubes and maintained in secondary culture for an additional 5 days. Both melanocytes and unpigmented cells exhibit varying levels of immunoreactivity. **(C, D)** R24 immunofluorescence of non-neuronal cells in 24-hr cultures of trypsin-dissociated, 6-day quail embryo dorsal root ganglia (DRG). Phase-contrast (A, C); epifluorescence (B, D). Bars = 50 μm. (From Girdlestone and Weston, 1985.)

neuronal lineage. Such sequential restrictions may have some interesting implications. For example, initial developmental restrictions may limit the kinds of environmental signals to which a given neuronal population can respond. If such differentially responsive subpopulations of neurons can be identified, then the nature of the environmental cues, the receptor(s) for these specific cues, and the nature of the cells' responses might all be analyzed in some detail (Davies, 1986).

To examine these issues, we have studied a subpopulation of sensory neurons that responds differentially to factors in the embryonic environment (Marusich et al., 1986a,b). A monoclonal antibody, SN1, which was raised against the membrane fraction of avian dorsal root ganglion neurons, binds to a subpopulation of sensory neurons, but not to other neurons of the peripheral or central nervous systems (Figure 7). SN1-immunoreactive fibers project to laminae I and II of

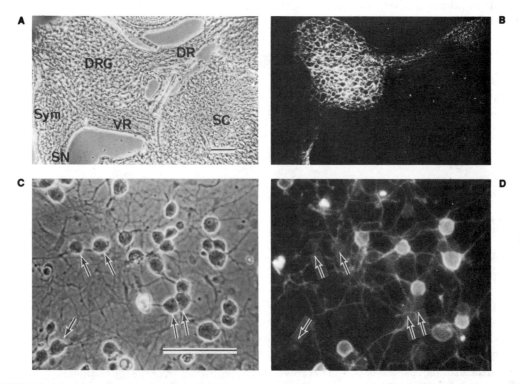

FIGURE 7. Monoclonal antibody SN1 binds to a subpopulation of sensory neurons. (A, B) Frozen transverse section at the lower thoracic level of an E15 quail embryo, showing SN1 immunoreactivity in dorsal root ganglion nerve cell bodies and fibers. In contrast, neurons of the sympathetic ganglion lack SN1 immunoreactivity, as does the spinal cord, except where afferent sensory fibers enter the dorsal horn. DRG, dorsal root ganglion; DR, dorsal root; VR, ventral root; SN, spinal nerve; Sym, sympathetic ganglion; SC, spinal cord. (C, D) Live culture of dissociated E12 quail DRG processed for SN1 immunoreactivity, showing that only a subpopulation of neurons are SN1[+], whereas others (some of which are indicated by arrows) are SN1[−]. Both the cell bodies and fibers are stained, and the pattern of fluorescence on individual living cells (rings and diffuse distribution) is characteristic of cell surface antigens. Phase-contrast (A, C); epifluorescence (B, D). Bars = 100 μm. (From Marusich et al., 1986a.)

the spinal cord dorsal horn and are found peripherally in the skin but not in deeper embryonic tissues. On the basis of their central and peripheral projections, we have suggested that SN1-immunoreactive neurons are cutaneous afferents with predominantly free and diffuse nerve endings.

The proportion of $SN1^+$ neurons in a spinal ganglion depends on axial level, so that expression of the SN1 immunoreactivity could be the consequence of axial level–dependent environmental cues. Consistent with this notion, SN1 immunoreactivity normally appears after neuronal processes have reached their peripheral targets (Marusich et al., 1986a). Moreover, experiments in which contact with

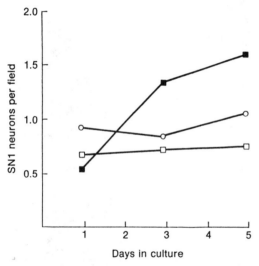

FIGURE 8. Trophic effects of embryonic skin-conditioned medium on the appearance of $SN1^+$ neurons. Purified neurons from E10 DRG were cultured in the various media for the times indicated and then scored for SN1 immunoreactivity. More neurons express SN1 immunoreactivity when they are cultured in medium conditioned by exposure to cultured embryonic skin (filled square) than when they are cultured in either fresh medium (open circle) or medium conditioned by exposure to cultured embryonic cartilage (open square). For each point, three cultures (20 fields per culture) were counted.

the periphery was prevented, either by explanation of ganglia in vitro or by ablation of peripheral targets, support the inference that contact with appropriate targets by fibers extending into the skin promotes the appearance of SN1 immunoreactivity (Marusich et al., 1986b). This inference has recently been further tested in experiments in which embryonic $SN1^-$ sensory neurons were cultured in medium previously conditioned by primary cultures of avian embryonic skin or other tissues. Results confirm that there are tissue-specific trophic effects on the appearance of this neuronal subpopulation (Figure 8). It is not yet clear, however, whether such tissue-specific cues can modulate expression in all sensory neurons, or if there are nonimmunoreactive precursors of $SN1^+$ cells that respond selectively to these environmental cues. Ablation experiments, such as those performed with other cell type–specific antibodies, may potentially elucidate these issues.

Summary

The general processes of cell differentiation and tissue interaction during nervous system development obviously involve the coordinated expression of phenotypic traits that are manifest at molecular, cellular, and tissue levels of organization. Some of the mechanisms that regulate these events appear to be autonomous, involving intrinsic processes that lead cells to make novel gene regulatory decisions and to relinquish the ability to regulate other genetic determinants.

Such autonomous regulatory mechanisms, coupled with rapid cell division, known to occur in neural crest populations during early development (Maxwell, 1976; J.A. Weston, unpublished data), would effectively generate significant cellular diversity within this embryonic tissue. The pattern of differentiation that is usually observed at the tissue level of organization seems likely to result, in part, from environmental factors acting on distinct developmentally restricted subpopulations to change their survival or proliferation rate (Brockes, 1984). In early embryos, where starting populations are relatively small and overall proliferation is rapid, even subtle changes in cell cycle parameters could radically alter the complexion of cell types in heterogeneous embryonic populations.

The recent studies of neural crest cell developmental behavior discussed above lead to the inference that some of the earliest developmentally restricted subpopulations arise autonomously. The demonstration that such distinct subpopulations exist provides a heuristic alternative mechanism by which environmental cues can effect diversity in the developing neural crest.

CHAPTER SIXTEEN

Genetic Mosaics as a Means to Explore Mammalian CNS Development

KARL HERRUP AND MICHAEL W. VOGEL

Introduction

A detailed understanding of central nervous system (CNS) development from the instructions encoded in the genome to the subtleties of function demonstrated by "the mind" is not likely to be achieved in our lifetime. In the quest for this understanding, one is forced to focus on small pieces of the process and hope that the knowledge gained in these restricted areas of research will offer new insights and perspectives on the larger problem that is developmental neurobiology. As a context for our own studies, we attempt to look at development from the standpoint of the DNA. What is the genetic logic of the nervous system? We view it as unlikely that the mental processes that lead to, say, the decision of what to have for breakfast this morning (i.e., function) can be read off the double helix directly. While this may seem self-evident, it is also our view that the picture of the brain described for the past century by anatomists and embryologists (i.e., structure) is not likely to be encoded directly in the genome either. What *is* present in the genome is DNA that makes gene products and DNA that regulates whether or not these gene products are made. What then are the links "from message to mind"?

One way of learning to understand the language of the genome is to explore the consequences of changing a single word in the lexicon. In other words, a powerful approach to the study of development in any system is the study of gene mutations that lead to arrests or abnormalities in CNS structure or function. In the mouse, several dozen mutations are known that fit this description (Green, 1981). Together, these mutants have provided a global picture of the genomic perspective on the CNS. The resolution of this approach is often limited because, in an organ in which cell-cell interaction and interdependence is the rule, the loss of a single gene function usually has a cascade effect throughout the brain. Once the initial pathological description of a given mutant has been achieved, therefore, it becomes essential to determine what Mullen (1977) has called "the primary site of gene action." There are many ways of approaching this goal. The one that has proved most useful to us is the analysis of aggregation chimeras made by joining one mutant and one wild-type preimplantation embryo. Studies of the composite nervous systems of the resulting mice frequently speed the unraveling of cause from effect by allowing the assessment of whether phenotype and genotype are separable at a cellular level. An added dividend of this understanding is that once it is achieved, the mutants and the chimeras become valuable experimental systems in which one can analyze the consequences of developmental manipulations that are difficult or impossible to achieve by standard experimental methods. Thus, instead of using the nervous system to understand genetics, we apply the genetics, in a defined manner, to help understand the epigenetics of the nervous system.

In this chapter we will explore two such applications of neurological mutant chimeras. The examples provide information about two very different times during development: (1) the earliest stages in the formation of the CNS and (2) the processes that do not unfold until the second and third week of postnatal life in the mouse. The two examples also differ in the developmental principles that they highlight: one leads to new insights into the role of cell lineage relationships in the formation of the mammalian CNS, while the second offers new information on the subject of target-related cell death in the brain. There are common themes that run through these examples, however. Both studies have added materially to our understanding of the importance of neuronal cell number regulation during mammalian CNS development, and each illustrates a different way in which the developing CNS achieves this regulation. An additional

common theme that the reader might consider is that both studies represent insights that could not easily have been achieved without the use of the experimental genetic approach.

Site of Gene Action of Two Neurological Mutants

A variety of neurological mutants of the mouse that affect cerebellar organization have been described. Two of these mutations, *staggerer* and *lurcher,* have proved to be valuable tools in the study of developmental interactions. Both mutants cause a pronounced ataxia and severe cerebellar atrophy, but they have different underlying defects. *Staggerer* (gene symbol: *sg*) is an autosomal recessive mutation that maps to chromosome 9 (Sidman et al., 1965). Homozygous animals (*sg/sg*) show uncoordinated behavior early in postnatal development. Their cerebella are reduced in size and in the extent of cortical folding. There is a reduced number of Purkinje cells, and those that remain are positioned ectopically, reduced in size, and have sparse, disoriented dendritic arbors (Sidman, 1968; Bradley and Berry, 1978; Herrup and Mullen, 1979a). Functional synapses are thought to be present between Purkinje cells and climbing fibers, and there appear to be immature Purkinje cell–parallel fiber contacts (Crepel and Mariani, 1975; Mariani and Changeux, 1980; Crepel et al., 1980; Sotelo, 1975; Landis and Sidman, 1978). However, the sites of Purkinje cell–granule cell synapses, the tertiary spines of Purkinje cell dendrites, are rarely, if ever, seen (Landis and Sidman, 1978; Sotelo and Changeux, 1974; Landis, 1971; Sidman, 1972). Virtually 100% of the granule cells die after migration (Sidman, 1968). There is some reduction in the rate of granule cell generation, but their subsequent degeneration is secondary to Purkinje cell defects (Yoon, 1972; Sotelo and Changeux, 1974; Herrup, 1983).

Lurcher (gene symbol: *Lc*) is an autosomal dominant mutation that maps to chromosome 6 (Phillips, 1960). Homozygous animals (*Lc/Lc*) do not survive past parturition, but heterozygotes are recognizable by the second or third postnatal week, when they begin to develop ataxia. Their loss of coordination is correlated with the degeneration of a large portion of the cerebellum; virtually 100% of the Purkinje cells, 90% of the granule cells, and 75% of the inferior olive cells degenerate with increasing age (Caddy and Biscoe, 1979). Development of the cerebellum appears morphologically normal until the beginning of the second postnatal week, when Purkinje cells begin to degenerate. Granule cell and inferior olive neuron loss begins at about the same time, and continues for a short while after Purkinje cell degeneration is complete.

The utility of these two mutations for developmental studies lies in knowing the primary site(s) of gene action in each mutant. Although many cell types are affected in *staggerer* and *lurcher*, the mutant gene is not necessarily directly responsible for the degeneration of each cell type. Cells that are intrinsically deficient as a result of the defective gene are considered a primary site of gene activity (Mullen and Herrup, 1979; Mullen, 1978). Other cell types may be affected by the mutation, but only indirectly if, for example, their existence is dependent on the normal functioning of the directly affected cell type. The site of gene action is studied with mutant ↔ wild-type chimeras (Mullen and Herrup, 1979; Herrup and Mullen, 1979b). Chimeras are constructed by aggregating, in vitro, two 8 cell stage embryos from different sets of parents. The embryos are cultured overnight while their cells aggregate to form an enlarged morula. The chimeric embryo is subsequently transplanted into a pseudopregnant host mother to complete gestation. (Mintz, 1962, 1965; Tarkowski, 1961; Mullen and Whitten, 1971). The rationale of these experiments is that the mix of wild type and mutant cells in the chimera will provide some elements of a normal environment. Those neurons of mutant genotype that are not the primary site of gene action may thus be restored to wild-type phenotype by interaction with normal (wild-type) environmental influences. Cells that are a primary site of gene action will retain their mutant phenotype despite normal environmental interactions.

Analysis of mutant ↔ wild-type chimeras has shown that in both lurcher (+/*Lc*) and staggerer (*sg/sg*) mutants, the Purkinje cell is a primary site of gene activity (Herrup and Mullen, 1979; Wetts and Herrup, 1982a). A cell-autonomous cytoplasmic enzyme, β-glucuronidase, was used in these studies as an independent cell marker to histochemically identify the genotype of Purkinje cells. β-Glucuronidase stained of the Purkinje cells in a C57BL/6J ↔ C3H/HeJ chimera is shown in Figure 1. C57BL/6J mice are homozygous for the normal activity (high-staining) allele for glucuronidase (*Gus^b/Gus^b*), while C3H/HeJ mice are homozygous for the reduced activity (low-staining) glucuronidase allele (*Gus^h/Gus^h*). Both the lurcher and staggerer mutant strains are homozygous for the high-staining glucuronidase allele. Therefore, strains that are homozygous for the low-staining but neurologically wild-type allele are aggregated with +/*Lc* or *sg/sg* embryos to make mutant ↔ wild-type chimeras. In lurcher chimeras, only wild-type Purkinje cells survive. In *sg/sg* ↔ wild-type chimeras, some abnormal, ectopic medium-to-large neurons of mutant genotype are found, but their phenotype, even down to the fine details, is not altered in the presence of a relatively normal cytoarchitecture. These results indicate that the defect in both mutations is intrinsic to the Purkinje cell. If it were not, it would be expected that some mutant Purkinje cells would be restored to a wild-type phenotype by normal environmental interactions within the chimeric animal.

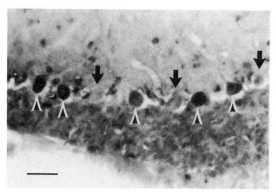

FIGURE 1. β-Glucuronidase histochemistry in the Purkinje cells of a C57BL/6J ↔ C3H/HeJ chimera. C57BL/6J mice are homozygous for the high-staining β-glucuronidase allele (*Gus^b^/Gus^b^*), while C3H/HeJ mice are homozygous for the low-staining allele (*Gus^h^/Gus^h^*). The perikarya of *Gus^b^/Gus^b^* Purkinje cells (arrowheads) become filled with a red reaction product following the β-glucuronidase histochemical reaction and appear dark in this black and white photomicrograph. The *Gus^h^/Gus^h^* Purkinje cells (indicated by arrows) are pale in comparison to the *Gus^b^/Gus^b^* Purkinje cells. Bar = 25 μm.

A similar type of analysis demonstrates that olivary neurons and granule cells are not directly affected by the staggerer or lurcher mutations. In the staggerer (Shojaeian et al., 1985; Blatt and Eisenman, 1985), and lurcher (Caddy and Biscoe, 1979) mutants, approximately three-quarters of the olivary neurons degenerate. Quantitative analysis of surviving olivary neurons in +/*Lc* ↔ wild-type chimeras demonstrates that more +/*Lc* olivary neurons survive than would be expected if degeneration of 75% of the olivary neuron population were an intrinsic characteristic of the lurcher mutation (Wetts and Herrup, 1982a). Some +/*Lc* olivary neurons are, therefore, rescued by the chimeric environment, and their degeneration in *lurcher* must be secondary to some other defect, probably the degeneration of their postsynaptic target, the Purkinje cells. A similar picture emerges for the staggerer muta-

tion (J. Mariani and K. Herrup, unpublished data).

Quantitative analysis of the granule cell population in *sg/sg* ↔ wild-type and +/*Lc* ↔ wild-type chimeras has shown that granule cells are also not intrinsically affected in these mutants (Herrup, 1983; Wetts and Herrup, 1982b). The β-glucuronidase marker does not work in granule cells because of their sparse cytoplasm. Instead, a nuclear marker based on the clumping of heterochromatin in ichthyosis mutants was used in these experiments (Goldowitz and Mullen, 1982). As in the case of olivary neurons, both wild-type (i.e., ichthyosis) and mutant (*sg/sg* or +/*Lc*) granule cells survived, and there were many more mutant granule cells than would be expected in a *sg/sg* or +/*Lc* animal. The mutant granule cells were apparently rescued by interactions with normal components of the chimeric cerebellum, indicating that the granule cells are not directly affected by the genetic defect. The most likely reason for granule cell death in *sg/sg* and +/*Lc* mutants is the loss of their postsynaptic target, the Purkinje cell.

Having established that the Purkinje cells, but not other cerebellar neurons, are a primary site of gene action in both the staggerer and the lurcher mutants, it becomes possible to use these mutants in chimeric studies of cerebellar development. In the next section we describe the use of lurcher ↔ wild-type chimeras to analyze cell lineage relationships among cerebellar Purkinje cells.

Cell Lineage Relationships

Genetic mosaics in any organism are a powerful approach to the analysis of cell lineage relationships. One of the problems that the developing CNS faces at many junctures during ontogeny is how to take an apparently homogenous group of cells and instruct a specific subset of them (1) to all behave alike and (2) to be different from all of the other cells around them. The two basic solutions to this problem are either to take an appropriate number of cells late in development and somehow instruct them and only them in the desired adult behavior pattern or to take one or a small number of cells early in development, imbue them with similar instructions, and then arrange for them to pass these instructions on faithfully to all of their cellular progeny. In this second solution, cell lineages become not just a brute force way of producing many cells out of few; instead, the dynamics of each family tree itself shape and mold the properties of the CNS in which the cells will ultimately reside.[1] The molecular problems of information transfer are probably similar in both cases, but the consequences for development and the implications for where researchers should look for controlling elements are very different.

For some time, attention in vertebrate development has focused on the first solution. Researchers concentrated their efforts at relatively late developmental times in search of cell-cell interactions that could transform postmitotic neuroblasts into the appropriate menagerie of cell types. Many of these searches were successful, and we now know of a number of physical and chemical interactions that materially change the behavior of developing nerve cells. Exploration of the role that cell lineage plays in guiding brain development was largely left to students of invertebrate neurobiology.

Experimental aggregation mouse chimeras are useful in the study of lineage relationships during all stages of mouse development (McLaren, 1976; LeDouarin and McLaren, 1984; Rossant, 1984). By independently marking cells according to their embryo of origin, one knows with certainty whether or not any one group of cells belongs to the same lineage.

Depending on their distribution, these family ties can be informative as to earlier developmental events. A recent series of papers provides an elegant and conceptually simple example of this. The cells of the intestinal epithelium were analyzed in chimeras using a specialized cell-marking system (Ponder et al., 1985; Schmidt et al., 1985). While there are many different cell types that make up each of the villi of the small intestine, these workers found that each of the intestinal crypts in a chimera are made up of cells of only one genotype. Thus each crypt is a single clone; one progenitor was specified to give rise to a family of cells adequate to form a crypt. This is an example of the second solution described above. Rather than organizing a large, indifferent sheet of cells into individual functional units based solely on microgradients or positional information, the mouse intestinal epithelium uses cell lineage information to augment the process.

In the CNS, this simple situation is not observed. Analysis of the brains of chimeric mice has shown that no one nucleus or neuronal cell type is a single clone. This is true for the cerebellum (Mullen, 1977, 1978; Oster-Granite and Gearhart, 1981; Wetts and Herrup, 1983b; Herrup and Sunter, 1986; Herrup, 1986), facial nucleus (Herrup et al., 1984a), retina (Sanyal and Zeilmaker, 1977; Mintz and Sanyal, 1970; West, 1978; Mullen and LaVail, 1976), and peripheral ganglia (Dewey et al., 1976), as well as other structures (Mullen, 1978; Herrup, in press). Further, the cells of the various lineages appear to intermingle extensively in the three spatial dimensions (Herrup, 1986; see especially discussion for Purkinje cells in Mullen, 1977, 1978; Oster-Granite and Gearhart, 1981; Wetts and Herrup, 1983b). This intermingling demonstrates that clones do not tend to remain as large patches (spatially contiguous cells of like genotype). A series of experiments from our laboratory has shown, however, that cells of the mammalian CNS do

---

[1]For an example of a system in which switches in adult fate go hand in hand with switches in lineage-specific cell division patterns, see Zackson (1984).

exist in large, well-defined "numerical patches." These patches, or quanta, as they are more properly called, serve as evidence that certain cell types within the brain can be correctly described as a small number of clones of cells with several hundred to several thousand cells per clone.

Wetts and Herrup (1983b) counted the number of Purkinje cells in a series of $+/Lc \leftrightarrow$ wild-type chimeras following the degeneration of the $+/Lc$ Purkinje cells. The inbred strain used as the source of the wild-type embryos was C3H/HeJ, and, since the lurcher gene acts to remove all of the $+/Lc$ Purkinje cells, only C3H/HeJ Purkinje cells survive in the chimeric cerebella. The resulting Purkinje cell counts were not distributed randomly from 0 to 100% of wild-type values, but rather, the numbers of Purkinje cells in all of the $+/Lc \leftrightarrow$ C3H/HeJ chimeras were integral multiples of a single value, 10,200 Purkinje cells. The interpretation given to the integral value, or numerical quantum, is that it represents a single developmental clone of cells arising from a single progenitor cell. Since the C3H/HeJ strain has 81,600 Purkinje cells per half cerebellum, this interpretation suggests that a wild-type C3H/HeJ mouse has eight clones of 10,200 Purkinje cells per half brain. The assumptions underlying this interpretation are diagramatically represented in Figure 2. In this hypothetical example four progenitor cells (circles) in chimeric embryos are selected to generate 64 Purkinje cells in the adult cerebellum. The model assumes that, once selected, each progenitor cell produces the same number of descendants, and cells may not enter or randomly leave the pool of progenitor cells once it is established. To produce 64 cells each progenitor cell must divide four times, producing a total of 16 descendants. Analysis of the adult chimeric cerebella in Figure 2 shows that cerebellum A contains 16 labeled Purkinje cells, while the cerebella in B and C contain 32 and 48 labeled Purkinje cells, re-

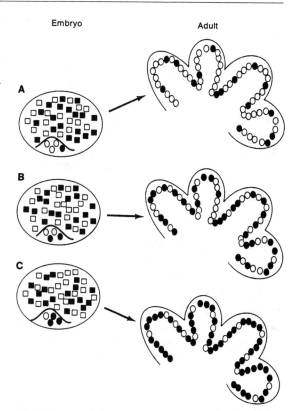

FIGURE 2. Clonal development of cerebellar Purkinje cells. In this hypothetical example, three chimeric embryos are shown on the left, each containing equal numbers of light and dark cells (represented as circles and squares). The distribution of 64 Purkinje cells is shown in the representation of adult cerebella on the right. In each embryo, four cells (circles) have been selected to become Purkinje cell progenitors, but the ratio of dark to light cells differs in each case (A = 1:3, B = 2:2, C = 3:1). If each progenitor cell divides four times to produce 16 progeny cells, a total of 64 descendants will be produced by the four progenitor cells. However, the number of progenitor cells in the embryo can be determined by analyzing the number of filled Purkinje cells in the adult cerebella.

spectively. The highest integral multiple of these values is 16, which equals the number of cells generated by one progenitor cell. Since the total number of Purkinje cells is known (64), it can be calculated that four progenitor cells were needed to produce all of the Purkinje cells.

A clone implies the existence of a founder cell or progenitor, and some of the properties of the Purkinje cell progenitors can be deduced. First, if symmetrical (geometric) cell divisions are assumed, for one cell to generate 10,200 requires between 12 and 13 cell generations. Previous work has defined the day in gestation when the Purkinje cell mitotic precursors cease division (approximately embryonic day 12.5; Miale and Sidman, 1961; Inouye and Murakami, 1980) and average cell cycle times are known for the mouse neural tube (Kauffman, 1968; Atlast and Bond, 1965; Wilson, 1981). Putting these estimates together, it is possible to calculate when in gestation the first cell division of the progenitor took place—roughly embryonic day 7.5 to 8.5. Since neural tube closure in the mouse does not begin until embryonic day 8.5, this suggests that the selection of progenitors takes place during the earliest events in the founding of the nervous system. It has been known for many years that during these early times in development the neural plate becomes a functional mosaic of cells committed to many different morphological fates and that these fates are, in large part, unresponsive to surgical changes in their developmental environment. Our quantitative analysis of adult chimeras leads to the same picture of the embryo.

Analysis of additional chimeras has refined our view of this process even further. First, a similar clonal organization has been shown to describe the motor neurons of the facial nerve nucleus (Herrup et al., 1984a). This conclusion was reached using a slightly different form of data analysis, but the results point to similar early times in development as moments when the precursor cells for this motor neuron population are identified. Second, it is clear that the selection of both facial nucleus and Purkinje cell progenitors occurs independently on each side of the developing CNS (Herrup et al., 1984b), in agreement with experimental embryological studies (Jacobson, 1959).

Finally, recent studies have implicated these cell lineage relationships as part of the mechanism used by the mammal to regulate neuronal cell numbers during development. Comparison of the data from lurcher chimeras made with the C3H/HeJ strain with data from chimeras made using two other inbred strains as the source of wild-type embryos (for C57BL/6, see Herrup, 1986; for AKR/J, see Herrup and Sunter, 1986) has shown that the size of a clone is an autonomous property of the lineage. That is to say, the number of descendents that a single Purkinje cell progenitor will produce is intrinsic to that cell. By contrast, the number of progenitors that will be selected (i.e., the number of clones) is not prespecified in any strain-specific way and most likely results from cell-cell interactions in the early CNS. This result is also predicted by the view derived from experimental embryology.

If the size of a clone of nerve cells is intrinsic to each progenitor cell, the concept of the functional mosaic of the neural plate/early neural tube takes on deeper meaning. The implication is that both the morphological and the numerical fates of the neuroblasts are laid down early in development. We feel that this is a true "rough draft" of the CNS, with size, balance among the component parts, and some of the morphological features of the adult neuronal cell types sketched in. Each of these properties is refined during subsequent development through cell-cell interactions. We believe this draft is closer to the code that is actually inscribed in the genome, since it consists not of anatomy and physiology, but of potentials and simple rules for interaction and cellular behavior.

This concept may provide a glimpse of one of the intermediates in the transition from gene to brain, but we know that this is very far from the whole story. The "potentials and simple rules" need to be translated into real cells and synapses. This translation—which is what the embryologist actually sees as CNS development—is essential to building the brain, but is probably not guided directly by gene expression. For example, one could speculate that the pattern of cerebellar cortex folding arises from interaction of previously specified components (much as T4 bacteriophage heads and tails come together to form virus particles spontaneously, without any requirement for additional gene expression). Thus, although cerebellar folial pattern is genetically determined (indeed, it is constant within any one inbred strain and unique to it; Inouye and Oda, 1980), its construction may well be directed primarily through intercellular communications rather than directly by gene expression. Naturally-occurring cell death during CNS development is an example of an important developmental phenomenon that is regulated primarily by this intercellular communication. The chimeras allow the study of this epigenetic phenomenon through genetic means, as described in the next section.

Naturally Occurring Cell Death in Cerebellar Development

A period of naturally occurring cell death is a common feature in a variety of neural circuits, including spinal motoneurons in the mouse and chick (Lance-Jones, 1982; Oppenheim, 1981) and the isthmo-optic nucleus and ciliary ganglion of the chick (Clarke and Cowan, 1976; Landmesser and Pilar, 1976). The period of cell death is usually coincident with early synaptogenesis. In the chick neuromuscular system, the first neuromuscular contacts are made on embryonic day 5 (E5) and E5.5, and the peak of cell death occurs between E6.5 and E9, during which 40 to 50% of the lumbosacral motoneurons die (Landmesser and Morris, 1975; Hamburger,

1975; Oppenheim et al., 1978). Target extirpation and transplantation experiments have shown in a variety of neuronal systems that neuron survival is dependent on target contact. Target removal increases the extent of cell death (Hamburger, 1958; Landmesser and Pilar, 1974; Lamb, 1981), while expansion of the target or reduction in the presynaptic neuronal population rescues a portion of the cells that would normally die (Hollyday and Hamburger, 1976; Narayanan and Narayanan, 1978; Boydston and Sohal, 1979; Lance-Jones and Landmesser, 1980; Pilar et al., 1980).

The function of target-related cell death has been hypothesized both as a mechanism for correction of inappropriate developmental projection errors (Lamb, 1980, 1981) or as a mechanism for matching the numerical size of the pre- and postsynaptic populations (Hamburger, 1975; Hamburger and Oppenheim, 1982). The latter hypothesis is supported by the target removal and supernumerary target experiments described above. However, these experiments do not directly address the precision of neuron-target size matching; target removal causes the death of almost all presynaptic neurons (Hamburger, 1958; Prestige, 1967; Oppenheim et al., 1978; Lamb, 1981; Lanser and Fallon, 1984), while supernumerary targets do not rescue as many neurons as might be expected on the basis of amount of target added (Hollyday and Hamburger, 1976). The first direct evidence that size matching can be very precise has come from quantitative analysis of neuron-target populations. In the chick neuromuscular system, McLennan (1982) has described a correlation between motoneuron number and myotube number during the cell death period. In an analysis of chick-quail chimeras, Tanaka and Landmesser (1986) found a linear relationship between the number of motoneurons surviving the period of cell death and the number of myotube clusters present during the period of cell death. The precise matching of neuron-target numbers has also been found in the mouse cerebellum. Wetts and Herrup (1983a) analyzed Purkinje and granule cell numbers in four different wild-type strains of mice and found a linear relationship between neuron and target.

The availability of genetic cerebellar mutations such as *staggerer* and *lurcher* makes the granule cell–Purkinje cell circuit an excellent model for the analysis of target-related cell death. In the wild-type cerebellum, pyknotic cells are observed in the external granule cell layer from days 3 to 10 in the mouse and in the internal granule cell layer until day 21 in the rat (Landis and Sidman, 1978; Lewis et al., 1976). Immature synapses on Purkinje dendritic spines and mature parallel fibers are visible by three days after birth, suggesting that granule cell death does not begin until the first synaptic contacts are made (Landis and Sidman, 1978; Larramendi, 1969). In the staggerer mutant, virtually 100% of the granule cells die during the period of Purkinje cell–granule cell synaptogenesis (Sidman, 1968). The granule cells die presumably because the synaptic spines that are the sites of parallel fiber–Purkinje cell synapses are missing, making the staggerer cerebellum equivalent to a cerebellum without a granule cell target (Sotelo and Changeux, 1974). Similarly, 90% of the granule cells in the lurcher mutant die subsequent to the degeneration of 100% of the Purkinje cell population (Caddy and Biscoe, 1979). As described above, the degeneration of granule cells in both mutants is secondary to the intrinsic defect in the Purkinje cells.

The intrinsic Purkinje cell degeneration characteristic of both mutants confers unique qualities on chimeras made between mutant and wild-type embryos and makes them powerful models for analysis of neuron-target interactions. The percentage of the contributions of the two embryos to a chimera varies almost uniformly from 0 to 100%. Furthermore, there is extensive mixing of cells of both genotypes to produce a fine-grained mosaic (Oster-Granite and Gearhart, 1981; Mullen, 1978). In wild-type ↔ staggerer or wild-type ↔ lurcher chimeras, the mutant Purkinje cells either do not develop normally or degenerate early in postnatal development, leaving behind a mosaic cerebellum with a variable number of wild-type Purkinje cells scattered throughout the cerebellar cortex (Wetts and Herrup, 1983b; Herrup and Sunter, 1987). Therefore, if enough chimeric embryos are generated, a cell dose experiment may be performed to examine the role of target size in regulating cell death.

The correlation of Purkinje and granule cell numbers found by Wetts and Herrup (1983a) in nonchimeric inbred mouse strains was based on relatively few data; only four data points were used in the correlation. The numerical relationship between granule and Purkinje cell number has been further studied in staggerer ↔ wild-type chimeras (Herrup and Sunter, 1987). Because of the intrinsic defect in *sg/sg* Purkinje cells, the granule cell target size in staggerer chimeras ranges from 0 to 100% of wild-type values. Fourteen hemicerebellums in nine chimeric animals were analyzed, and a linear relationship was found between Purkinje and granule cell numbers with a correlation coefficient (r^2) of 0.969. These results indicate that there is little variability in the interactions between Purkinje cells and granule cells during the establishment of their neural circuit. The ratio of granule cells to Purkinje cells appears to be fixed, so that a Purkinje cell population of a given size can support only a defined number of granule cells, and a given granule cell population needs a certain number of Purkinje cells to persist into adulthood.

We have attempted to repeat this experiment by using the lurcher mutation, instead of the staggerer, to remove Purkinje cells. In the lurcher mutation 100% of the Purkinje cells die by the end of the second postnatal month. The mutation has been shown to act in a cell in an autonomous fashion (Wetts and Herrup, 1982a). When lurcher chimeras were made and analyzed, Purkinje cell numbers were found to range from zero to wild-type values and the qualitative appearance of the cerebellar cortex indicated a reduced granule cell population as well.

In lurcher ↔ wild-type chimeras, however, increased numbers of granule cells are found to be supported by proportionately fewer adult Purkinje cells than would be predicted by the staggerer chimera data discussed above (Wetts and Herrup, 1983a; Vogel and Herrup, 1986). The explanation offered by Wetts and Herrup (1983a) for the nonlinear granule cell–Purkinje cell relationship in lurcher ↔ wild-type chimeras is that the extra granule cells are rescued by the transient presence of lurcher Purkinje cells. In the lurcher mutant, Purkinje cells do not begin to degenerate until one week postpartum. By contrast, in the staggerer mutant, most Purkinje cells fail to develop and the remaining Purkinje cells never make productive contacts with granule cells. Prior to their degeneration, lurcher Purkinje cells are competent to form mature granule cell synapses (Caddy and Biscoe, 1979). We assume that this timing applies to the $+/Lc$ ↔ wild-type chimeric brain as well. Therefore, some granule cells may be stabilized by their initial interactions with $+/Lc$ in addition to $+/+$ Purkinje cells. These granule cells may then survive the subsequent degeneration of the lurcher Purkinje cells, thereby increasing the ratio of granule cells to Purkinje cells in adult chimeric animals. This interpretation supports the suggestion that there is a "critical period" for granule cell development during which the granule cells must be supported by Purkinje cell interactions (Changeux and Danchin, 1976).

The hypothesis that transient $+/Lc$ Purkinje cells stabilize supernumerary cells would seem to predict that early-generated granule cells will have a temporal advantage over late-generated granule cells in the competition for target. To test this hypothesis, we have analyzed the frequency of early-generated (postnatal day 4; P4) and late-generated (postnatal day 13; P13) granule cells in various experimental circumstances. Wild-type and $+/Lc$ control animals and $+/Lc$ ↔ wild-type chimeras were injected with tritiated thymidine ([^{3}H] thymidine) at either P4 or P13. Dividing granule cell precursors incorporate the radiolabeled thymidine, and those that are in the S-phase of their final cell division at the time of [^{3}H]thymidine injection will retain the ^{3}H label into adult life. Precursor cells that continue to divide after [^{3}H]thymidine injection will dilute the amount of incorporated radiolabel and will appear unlabeled in autoradiographic tissue sections from adult animals. The total number of granule cells, the number of [^{3}H]thymidine–labeled granule cells, and the number of Purkinje cells were then counted in sagittal sections of the adult animals. We have analyzed 18 hemicerebella, 11 of them chimeric, with Purkinje cell numbers ranging from 0 to 108% of wild-type values. In wild-type animals, approximately 3% of the total granule cell population is labeled by thymidine injection at either P4 or P13 (Figure 3). In $+/Lc$ animals, however, over 35% of the granule cells surviving to adulthood are generated at P4, while less than 1% of the surviving granule cells are generated at P13. In chimeric animals with intermediate numbers of wild-type Purkinje cells, the percentage of granule cells labeled by thymidine at P4 is dramatically reduced, compared to $+/Lc$ values, while the percentage of thymidine-labeled granule cells at P13 is only modestly increased compared to the mutant.

Examination of the total number of surviving early- and late-generated granule cells (as opposed to the ratios of these neurons) reveals additional complexities in the developmental interactions (Figure 4). In chimeras with fewer Purkinje cells, P4 granule cells appear to have

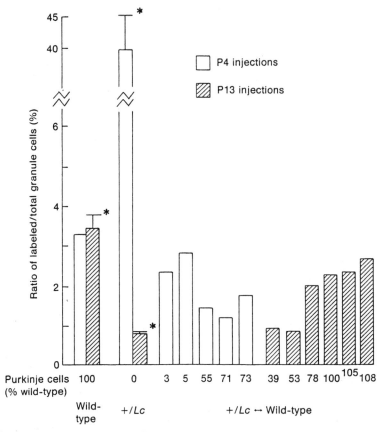

FIGURE 3. Ratio of granule cells labeled by [³H]thymidine at P4 or P13 to the total number of granule cells in the right and left brain halves of +/*Lc* ↔ wild-type chimeras and control animals. The percentage of wild-type Purkinje cells present in each half brain is shown underneath each histogram bar. Lighter bars represent results from animals injected at P4, while darker bars show results from animals injected at P13. *, *n* = 2; all others, *n* = 1.

a slight competitive advantage over P13 granule cells. For example, our counts of cells in two hemicerebella, each with approximately 50% of the wild-type number of Purkinje cells, demonstrated that the brain injected with [³H] thymidine at P4 had twice as many labeled granule cells as did the brain injected at P13. However two additional cerebella, with 73% and 78% of the wild-type number of Purkinje cells, respectively, had equal numbers of P4- and P13-generated granule

cells. As the number of Purkinje cells increases in chimeric animals, it appears that early-generated granule cells lose their competitive advantage. As expected, the number of surviving granule cells is sensitive to Purkinje cell number. The number of surviving P13 granule cells increases monotonically as a function of Purkinje cell number. The total number of surviving P4 granule cells, however, is sensitive to the number of wild-type Purkinje cells in a complex fashion. Over

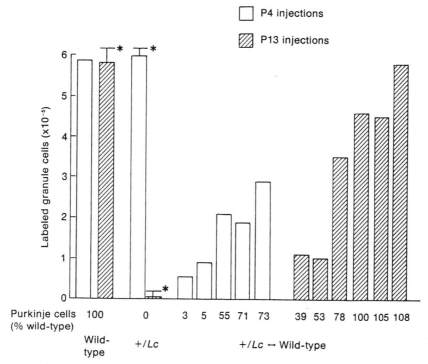

FIGURE 4. The total number of surviving labeled granule cells in the right and left brain halves of $+/Lc \leftrightarrow$ wild-type chimeras and control animals injected with [^{3}H]thymidine at either P4 or P13. The percentage of wild-type Purkinje cells remaining in each half brain is shown underneath each histogram bar. Lighter bars represent results from animals injected at P4, while darker bars represent results from animals injected at P13. *, $n = 2$; all others, $n = 1$.

500,000 granule cells labeled at P4 survive in both adult wild-type animals, with a full complement of Purkinje cells, and $+/Lc$ mutants, with no adult Purkinje cells. Yet in chimeras, the presence of very few wild-type Purkinje cells (5 to 8% of wild-type values in one case) dramatically reduces the total number of surviving granule cells. While the number of surviving P4 granule cells does increase with further increases in the numbers of wild-type Purkinje cells, this rise only serves to return the number of labeled granule cells to non-chimeric $+/Lc$ values.

We propose that our data are consistent with a model of granule cell competition that involves two phases of competitive interactions (Figure 5). The first phase is an acute event in which recently born granule cells must be stabilized by their initial Purkinje cell contacts in order to survive. The second phase is a chronic process wherein granule cells compete with other granule cells, born at different times, for stabilization by Purkinje cells. This hypothesis suggests that in $+/Lc$ animals, the early-generated granule cells are stabilized by their initial contacts with $+/Lc$ Purkinje cells during the first, acute phase. The subsequent degeneration of these Purkinje cells deprives later-generated granule cells of this stabilization and they cannot survive. The

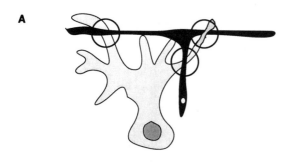

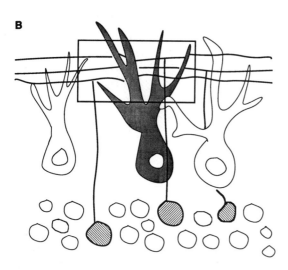

FIGURE 5. Schematic diagram of the proposed acute and competitive phases of granule cell competition. (A) Newly-generated granule cells must be stabilized by their initial contacts with Purkinje cells, possibly when the granule cells descend through the Purkinje cell layer to the internal granule cell layer. Potential sites for the early, stabilizing interactions are enclosed in circles. (B) At later stages, granule cells of different ages compete among themselves for Purkinje cell target space. In this hypothetical example, the three shaded granule cells were generated at different times and they are competing for target space on the same Purkinje cell (shaded dark). The competition presumably occurs at the level of parallel fiber–Purkinje cell dendritic spine synapses (shown within the rectangle), but the full parameters mediating the competition are unknown. Our results indicate that factors other than their date of generation will determine which granule cells are successful in the competition for target space.

early-generated granule cells therefore face no additional, chronic competition from later-generated neurons, and they survive by benefit of the early stabilizing influences.

The situation becomes more complicated in chimeric animals. The addition of wild-type Purkinje cells to the cerebellum means that some later-generated granule cells may be stabilized by first-phase interactions. The number of stabilized granule cells will increase as the number of wild-type Purkinje cells increases. These later-born granule cells would then enter into competition for target support with early generated granule cells for second-phase stabilization. This would account for the drop in the number of P4 labeled granule cells from $+/Lc$ to the chimera values and for the rise in the number of labeled granule cells, with increasing numbers of Purkinje cells, in both the P4 and the P13 granule cell populations. As more target becomes available, more early- and late-generated granule cells will survive. At low Purkinje cell numbers, proportionally more early-generated granule cells may survive because the early presence of $+/Lc$ Purkinje cells will have allowed more early-generated granule cells to survive the acute phase of competition. As Purkinje cell numbers approach wild-type values, P4 and P13 granule cells may be placed on equal competitive footing. We must emphasize that the hypothesis states that the acute phase is both a necessary and sufficient condition for neuronal survival, while the second phase is necessary only in situations where competition for target is a limiting factor on presynaptic neuron number.

We propose this model for two phases for target-related interactions on the basis of an analysis of our data, yet this hypothesis may have relevance to other model systems for naturally occurring cell death. Two types of degenerative changes have been observed in a variety of presynaptic neuronal populations during natural cell death and target removal–

induced cell death (for review, see Oppenheim, 1981). Type I cell death is characterized by regressive nuclear changes, while type II cell death is characterized by the dilation of the rough endoplasmic reticulum and the disruption of the cytoplasm, followed by nuclear degeneration (Pilar and Landmesser, 1976; Chu-Wang and Oppenheim, 1978). In the chick ciliary ganglia, neurons undergo type II degeneration during natural cell death, but when neurons are deprived of their target, they undergo type I cell death (Pilar and Landmesser, 1976). According to Pilar and Landmesser, neurons deprived of their target do not receive a critical inductive signal, and therefore they do not differentiate normally. The neurons undergo a series of morphologically distinct degenerative changes, and eventually die. The authors also suggest that natural cell death is a consequence of forming either too few synapses or not enough stable synapses. Although in other systems, both types of cell death may be present during natural as well as induced cell death (Oppenheim, 1981), type I and type II cell death may represent different underlying mechanisms. In reference to the model we have presented here, neurons that are not stabilized by early target interactions may undergo type I cell death during the first, acute phase of competition. Neurons that are stabilized initially but subsequently fail to compete successfully may then die, undergoing type II degeneration.

Summary

Viewing central nervous system development through the lens of mouse genetics has led to many new insights, including the elucidation of the developmental principles illustrated here. Cell number in the CNS appears to be regulated during development by the harmonic interplay of cell lineage and cell death (as well as by other factors yet to be analyzed). It is worth emphasizing that neither the control of the cell lineage relationships nor the specification of the rules for the neuron–target interactions are likely to be the result of the expression of a single gene. However, both are examples of seemingly complicated developmental issues being separable into component parts that have simple rules for cellular behavior and cell-cell interaction. We believe that the discovery of the workings of these components allows a more focused approach to the underlying molecular bases of brain development. As only one example, the demonstration that the number of Purkinje cells produced by each progenitor is a cell-autonomous property suggests that it is necessary to begin thinking, at a mechanistic level, about how a neuroblast counts cell divisions.

Finally, this chapter has offered examples of the tremendous power that genetic mutations and mutant ↔ wild-type chimeras can bring to bear on the study of developmental neurobiology. The tetraparental animals help to unravel cause from effect in the etiology of a given mutant phenotype and allow the researcher to perform manipulations during development that are not possible through other means. The analysis of this material encourages adoption of an approach that asks, "How does the genome see the nervous system?" Both this perspective and these data help to fill in links in our search to complete the chain that runs from message to mind.

VI

MECHANISMS OF BRAIN PLASTICITY

MICHAEL V. JOHNSTON

Genetic programs direct the assembly and maintenance of neuronal circuitry in the developing brain but cellular interactions, hormones, neurotransmitters, neuronal activity, and many other influences determine its final form. The interactions between genetic programs within immature neurons and their "environments" can be studied most easily in simpler nervous systems or in vitro. However, despite the complexities involved, it is tempting to search for neurobiological models of genetic–environmental interactions in intact higher brains. The chapters in this section represent novel examples of what will become an expanding list of models available for study and manipulation.

One example of the force of powerful external influences on brain development is sexually dimorphic behaviors. The developing brain is both a regulator of the reproductive system, controlling neuroendocrine signals and gonadal function, and an important endocrine target, responding to hormones. Pervasive behavioral changes are accompanied by striking neurochemical and morphological sex-related differences in the mature brain. Even before puberty, early exposure of the brain to gonadal hormones has profound lasting effects, which are reviewed in the chapters by Gorski and McEwen et al. Gorski's Chapter 17 describing the influence of gonadal hormones on the volume of the sexually dimorphic nucleus of the preoptic area (SDN-POA) in mammals complements and extends observations on sexually dimorphic nuclei in avian brains (Nottebohm, 1976). Gonadal hormones appear to serve a major organizational function, and play a role in regulating neurogenesis, programmed cell death, neurite outgrowth, and neurite sprouting in response to injury in certain sensitive neurons. McEwen's experiments with receptor-mediated hormonal effects on brain neurotransmitter systems suggest neurochemical correlates and, possibly, mechanisms of hormone action. Despite the limited understanding of the relationships between sexually dimorphic behaviors and the morphologic and neurochemical changes, these approaches should yield information about basic control mechanisms.

Many other hormones, such as thyroid hormone, have similarly powerful influences on gene expression in brain development. At present, probably only a small fraction of influential hormone-like actions on brain development are recognized. For example, it was only recently discovered that nerve growth factor (NGF), an active molecule in the developing peripheral nervous system, stimulates and regulates the development of specific populations of cholinergic neurons in the brain (Mobley et al., 1986). It is noteworthy that NGF acts as a "survival" factor for neurons in dorsal root ganglia and for developing cholinergic neurons in the brain. Gonadal hormones exert similar effects on neurons in the developing SDN–POA, suggesting that sex hormones may direct organizational changes in the brain by mechanisms which resemble actions of growth factors (Thoenen and Edgar, 1985).

Sensory experiences mold the developing mind but neurobiological studies have as yet provided little information about the morphologic, neurochemical and electrophysiological mechanisms. Greenough's work on morphologic synaptic alterations in response to an enriched environment (Chapter 19) suggests that there are structural correlates of memory formation. Synaptic reorganization may continue in the mature brain and common mechanisms may control synaptic plasticity in the immature as well as in the aging brain. These studies point to metabolic synaptic responses to neuronal activity which have been suggested also by experiments from much simpler systems. These results also indicate that synaptic changes are coordinated with metabolic and structural alterations in glial support structures and blood vessels, probably mediated by as yet unknown trophic factors. Although there is no information yet, it is reasonable to speculate that similar structural synaptic changes will be found to correlate with the electrophysiological plasticity demonstrated in barn owls in Knudsen's experiments with sound localization (Chapter 20).

These chapters describe provocative, exciting examples of nervous system plasticity in mammalian brain. Application of new molecular techniques (for example, in situ hybridization of brain messenger mRNA) and refined hypotheses to these brain preparations will lead to important advances. A general theme suggested by the work presented in this section is that the diverse signals (neurotransmitters, growth factors, hormones, sensory inputs) modify gene expression and act through a discrete number of final common molecular pathways. Common mechanisms may mediate the effects of both an enriched environment and gonadal hormones on neuronal growth, survival and connectivity. Evidence already exists that several "conventional" neurotransmitters may play a role to regulate neuronal plasticity during development. Such dual purpose (short-term information transfer and long-term trophic influence) signals may continue to operate in a modified mode in the adult brain. In a sense, the word "message" in the title *From Message to Mind* refers to the interaction of two types of messages: the genetic programs delivered into the cell cytoplasm by messenger RNA, and the multiple hormonal, experiential, and chemical messages that influence them. Understanding of these processes should greatly accelerate in the next few years, and help to decode the signaling mechanisms which influence the assembly of neuronal circuitry.

CHAPTER SEVENTEEN

Sexual Differentiation of the Brain: Mechanisms and Implications for Neuroscience

ROGER A. GORSKI

Introduction

Sexual differentiation of the brain is a fundamental concept of neurobiology in general and of developmental neurobiology in particular. In this chapter, I present evidence to suggest that the mammalian brain is inherently female, or at a minimum bipotential, and that aspects of brain function and structure characteristic of the male are imposed on the developing brain by the hormonal environment during the perinatal period by testicular activity. After describing the evidence for this statement in terms of brain function, the evidence for the structural sexual differentiation of the brain will be reviewed and with one marked structural sex dimorphism—the sexually dimorphic nucleus of the preoptic area (SDN-POA)—used as the model system, the possible mechanisms of steroid action will be considered.

To introduce the topic further, two fundamental concepts must be presented. In the process of the sexual differentiation of the reproductive system, as is currently understood, fertilization of the ovum with an X or Y sperm determines whether the individual will be female or male and, thus, have ovaries or testes. Subsequent development of the reproductive system in terms of both the extragonadal internal reproductive organs and the external genitalia depends exclusively on the production of hormones by the testes of the male and on the response of the tissues of the individual to these hormones (Wilson et al., 1981). As will be reviewed below, the brain plays an integral role in the process of reproduction and can conceptually be considered an important component of the reproductive system.

The second concept relates to the nature of steroid action on the central nervous system. Gonadal steroids can exert either *activational* or *organizational* effects on the central nervous system (Phoenix et al., 1959). By activational is meant the transient stimulation (or inhibition) of neuronal function induced by the steroid modification of genomic activity. Examples of this type of action of gonadal steroids include the facilitation of reproductive behavior and the feedback alteration of pituitary hormone secretion, including the surge of luteinizing hormone (LH) responsible for ovulation in the female. Important for the present consideration, however, is the fact that during certain periods of development, steroids also exert permanent or organizational effects on the brain. Although it is not known whether the precise biochemical mechanisms of the activational and organizational actions of gonadal steroids are different or perhaps equivalent, this concept is very useful in describing the process of the sexual differentiation of the brain.

Sexual Differentiation of the Brain

The concept of the sexual differentiation of the brain states that the numerous functional and more recently discovered structural sex differences in the mammalian brain are not due to autonomous sex differences in neuronal gene expression, but rather are imposed on what appears to be an inherently female brain by the gonadal hormonal environment during a critical phase of development (Goy and McEwen, 1980; DeVries et al., 1984a; Gorski, 1985a). Two functional sex differences in the rat brain that have been the most thoroughly studied and serve to illustrate both the concept and its experimental validation are the neural control of ovulation and that of reproductive behavior.

Although the precise neuroanatomical components and neurochemical interrelationships responsible for ovulation in the female rat are still obscure, it is generally accepted that rising titers of plasma estradiol interact with a cyclic neural system presumably resident within the preoptic area of the hypothalamus, resulting in the activation of this system. This event in turn leads to a dis-

charge of luteinizing hormone–releasing hormone (LHRH), which produces the surge of LH responsible for ovulation (Kalra and Kalra, 1983; Gorski, 1985b). This "facilitatory feedback" effect of estradiol is one example of the activational effects of gonadal steroids. In the laboratory situation, it is possible to gonadectomize adult male rats and attempt to induce facilitatory feedback by the administration of exogenous ovarian hormones or even to use ovarian grafts to measure the ability of the brain of the genetic male rat to support the cyclic release of LHRH ultimately responsible for ovulation. However, when the male rat gonadectomized as an adult is so treated, no evidence for the existence of a functional cyclic neural system is obtained.

Reproductive behavior is another example of a system dependent in adulthood upon the activational effects of steroids (Gorski, 1974, 1985a). Feminine sexual behavior is the result of the sequential activational effects of estradiol and progesterone, whereas masculine reproductive behavior is testosterone dependent. Upon exposure to estradiol and progesterone, the gonadectomized adult female rat becomes sexually receptive and her behavior can be readily quantified by indices such as the lordosis reflex: when the receptive female is mounted by a male she arches her back and her head is elevated, as is the perineal region, which permits intromission by the male. Although the male rat gonadectomized as an adult is capable of showing this lordosis reflex in response to the activational effects of estradiol and perhaps progesterone, the frequency of such a response by the genetic male is markedly and significantly less than that of the female. These two functional sex differences in the rat brain are shown in Table 1.

The basic experimental approach that has been used to document the concept of the sexual differentiation of the brain has been to modify the hormonal environment perinatally, either by exposing females to exogenous steroids or by preventing exposure of the perinatal male to endogenous testicular hormones (e.g., gonadectomy [Feder and Whalen, 1964] or drug treatment [McEwen et al., 1977a]), and then to measure the effects of such treatment on subsequent functional activity of the neuroendocrine brain in adulthood. As shown in Table 1, the injection of exogenous gonadal steroids (within the physiological dose range) does not alter the neuroendocrine characteristics of the adult male. However, similar treatment of the female produces an irreversible and permanent masculinization and defeminization of the brain. Thus, a single injection of testosterone propionate (TP), given within the first 10 days of postnatal life, permanently blocks the ability of the female to exhibit facilitatory feedback, as indexed either by LH measurement in plasma or ovulation. Moreover, this androgen-exposed female exhibits the lordosis reflex much less frequently than does the normal female rat.

The result of gonadectomy neonatally has equally dramatic and permanent effects, but this time basically only in the genetic male. Although the precise temporal window may be dependent on strain, if the testes are removed within about the first three days of postnatal life, when the genetic male reaches adulthood, he will exhibit facilitatory feedback, again as indexed either by the direct measurement of plasma titers of LH in response to steroid injection in adulthood or by the occurrence of ovulation and corpora lutea formation in ovarian grafts. In addition, following the exposure to ovarian hormones in the activational mode, the neonatally gonadectomized male also exhibits the lordosis reflex at levels comparable to those displayed by the normal genetic female. Thus, the ability of the adult rat to exhibit the cyclic release of LH or lordosis behavior depends not on genetic sex, but on the presence or absence of gonadal steroids during the early postnatal period of life (Goy and McEwen, 1980; Gorski, 1985a,b).

Although female rats may normally demonstrate some of the elements of masculine copulatory behavior such as mounting (Whalen, 1968), exposure perinatally to exogenous gonadal steroids clearly increases this and other components of masculine copulatory behavior exhibited by the adult female (Christensen and Gorski, 1978). It

TABLE 1. Influence of Gonadectomy or Exposure to Exogenous Gonadal Hormones Perinatally on Two Exemplary Functional Sex Differences in the Adult Rat Brain[a]

Genetic Sex	The Normal Adult	Gonadectomy Perinatally	Exposure to Gonadal Steroids Perinatally
Female	Cyclic release of luteinizing hormone	Cyclic release of luteinizing hormone	*Tonic release of luteinizing hormone*
	High levels of lordosis responding	High levels of lordosis responding	*Lordosis rarely displayed*
Male	Tonic release of luteinizing hormone	*Cyclic release of luteinizing hormone*	Tonic release of luteinizing hormone
	Lordosis rarely displayed	*High levels of lordosis responding*	Lordosis rarely displayed

[a]Italics indicate sex-reversed phenomena.

should also be emphasized that the functional systems that appear to undergo sexual differentiation in one species or another are not restricted to overt sexual functions (Goy and McEwen, 1980; Gorski, 1985a). Thus, aggressive behavior, the regulation of food intake and body weight, social and play behaviors, territorial marking, urination posturing, learning, the lateralization of cognitive function, and perhaps even human psychosexual identity are examples of brain functions that also appear to be sensitive to gonadal steroids during development and that undergo, at least to a degree, the process of sexual differentiation. It should also be emphasized that sexual differentiation occurs prenatally in those animals that are born relatively mature; even in the rat, in which the sexual differentiation of the brain can be modified rather dramatically in the postnatal period by the hormonal environment, this process probably begins prenatally (Gorski, 1985b). The general hypothesis, therefore, is that for many, perhaps all, brain functions that are sexually dimorphic in the adult, the sex differences are created by the hormonal environment during critical periods of development.

Possible Mechanisms of Steroid Action

The question of the possible mechanism by which gonadal steroids exert their organizational effects is intriguing but yet unexplained. Alternative and not mutually exclusive potential mechanisms include an alteration in fundamental neuronal properties such as membrane characteristics, resting potential, or general reactivity. Since many of the reported functional sex differences in the brain depend upon the activational effects of gonadal steroids in adulthood, it is possible that early exposure to gonadal hormones alters the manner by which neurons in the adult ultimately process or respond to the steroid hormonal signal (Olsen and Whalen,

1980; Rainbow et al., 1982). In addition, there are a number of reported sex differences in neurotransmitter activity (DeVries et al., 1984b) and a vast body of literature that relates certain neural transmitters to specific functions such as ovulation or reproductive behavior. Therefore, an alteration in neurotransmitter production or response mechanisms induced by the perinatal exposure to steroids might well underlie functional sex differences in the adult brain. Finally, as will be documented below, there are structural sex differences in the brain that again could underlie sex differences in brain function (Arnold and Gorski, 1984).

Structural Sex Differences in the Brain

Although it is not possible to present a thorough review of structural sex differences in the mammalian brain here, it may be helpful to place this concept in some historical perspective. Although several structural sex differences in the brain had been reported previously, certain experiments had a particularly important impact on this field. Raisman and Field (1973) demonstrated at the ultrastructural level that a sex difference in a specific region of the medial preoptic area (MPOA) of the rat brain was determined by the hormonal environment during the temporal period of the sexual differentiation of brain function. Just three years later, it was reported that in two species of songbirds, there are a series of marked hormone-dependent sex differences at the level of nuclear organization in the brain (Nottebohm and Arnold, 1976). This observation suggested the possibilities that similarly marked sex differences might exist in the mammalian brain and that other investigators studying the rat had focused on too fine a level of resolution (i.e., individual neuronal nuclear or nucleolar size; Pfaff, 1966; Dörner and Staudt, 1968), which indeed turned out to be true.

Previous data had suggested that the MPOA might be one site of the organizational action of hormones in the rat, and when sections through the male and female MPOA were compared side by side, an obvious sex difference was identified (Figure 1) (Gorski et al., 1978). The intensely staining cell group in the MPOA, from the volume of which we could determine sex, was found to be an area of increased neuronal density in relation to the immediate surroundings (Gorski et al., 1980).[1] On this basis, we labeled this cell group the sexually dimorphic nucleus of the preoptic area (SDN-POA). It is important to emphasize that although there is a marked difference in volume of the SDN-POA in males compared to that of females, there is no sex difference in neuronal density within the SDN-POA (Jacobson and Gorski, 1981); therefore, this nucleus is composed of more neurons in the male than it is in the female. If it could be shown that the development of the SDN-POA is sensitive to gonadal steroid hormones, then one potential mechanism of steroid-induced sexual differentiation of the brain could be put forward and challenged experimentally, namely, that gonadal hormones control the number of neurons that form a given nucleus.

Before this question of the hormonal sensitivity of the developing SDN-POA is considered further, it should be emphasized that numerous structural sex differences are now known to exist. In addition to the SDN-POA, sex differences in regional nuclear volume have been reported for the ventromedial hypothalamic nucleus (Matsumoto and Arai, 1983) and the bed nucleus of the stria terminalis (Hines et al., 1985). Structural sex differences are clearly not restricted to the rat or songbird, but have been identified in the guinea pig (Hines et al., 1985), gerbil (Commins and Yahr, 1984), and ferret (Tobet et al., 1983). An

SDN-POA has been reported to exist in the human brain (Swaab and Fliers, 1985), although in our studies of the human hypothalamus we have failed to confirm a sex difference in the volume of what we believe is the same nucleus (Allen et al., 1986). However, there are sex differences in volume in two nuclei that we have labeled interstitial nuclei of the anterior hypothalamus.

Sex differences in nuclear volume are not limited to the brain, since Breedlove and Arnold (1981) have reported that the spinal nucleus of the bulbocavernosus (SNB) in the lumbar region of the spinal cord of male rats is absent in females. Sex differences in synaptic organization have been reported in the MPOA (Raisman and Field, 1973; Greenough et al., 1977), the arcuate (Matsumoto and Arai, 1980), ventromedial (Matsumoto and Arai, 1986), amygdaloid (Nishizuka and Arai, 1981) and suprachiasmatic (Güldner, 1982) nuclei and the lateral septum (DeVries et al., 1984b). It is likely that the list of structural sex differences in the central nervous system and the species in which such sex differences exist will grow as more studies are conducted. Moreover, it may be predicted that as the analyses of the brain include the neurochemical specificity of individual neurons or neuronal groups, this list of sex differences will continue to grow markedly. The results of studies of Loy and Milner suggest the existence of sex differences at a more dynamic level. Although it is known that the medial septum projects a cholinergic pathway to the hippocampus and that following lesions of this cholinergic pathway, sympathetic noradrenergic fibers from superficial blood vessels innervate the denervated hippocampal cells, these investigators have demonstrated that this adult plastic response is sexually dimorphic (Loy and Milner, 1980). In the female, the quantitative and qualitative degree of reinnervation is dramatically greater than that which occurs in the male. Moreover, this sex difference is sensitive to the perinatal

[1]This observation does not rule out a possible sex difference in glia as well.

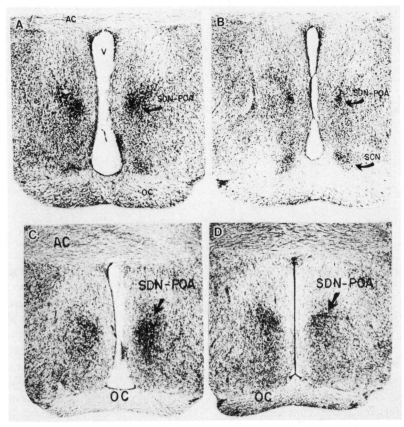

FIGURE 1. Representative coronal sections through the center of the SDN–POA of young adult male (A) and female (B) rats. The lower figures illustrate the sex reversal of the SDN–POA in genetic females that were exposed daily from embryonic day 16 through postnatal day 10 to testosterone propionate (C), or diethylstilbestrol (D). All at the same magnification. AC, anterior commissure; OC, optic chiasm; V, third ventricle. (From Döhler et al., 1984.)

hormonal environment (Milner and Loy, 1982).

The Sexually Dimorphic Nucleus of the Preoptic Area

The mere existence of the SDN-POA does not demonstrate that this neuronal structure undergoes sexual differentiation. To address this question we first studied the development of the SDN-POA. Does this structural sex difference become established only after puberty, or is it present throughout the life of the individual rat? As shown in Figure 2, the sex difference in the volume of the SDN-POA is first detectable on day one of postnatal life (Jacobson et al., 1980). Thereafter, the SDN-POA volume continues to increase in the male, while there is no statistically significant change in SDN-POA volume in the female. Moreover, SDN-POA volume essentially reaches its adult level by about postnatal day 10. Thus, the development of the SDN-POA within the first week or so of postnatal life

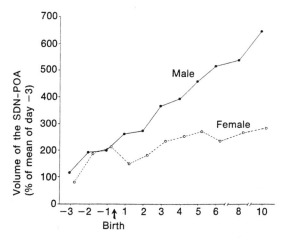

FIGURE 2. The ontogeny of the sex difference in SDN–POA volume related to the day of birth. The data are expressed as the mean volume observed in both male and female rats sacrificed at embryonic day 20 [3 days (−3) before expected birth]. (Data from Jacobson et al., 1980; reprinted from Gorski, 1984.)

coincides closely with the period of the functional sexual differentiation of the brain.

The fact that the SDN–POA develops during the postnatal period of sexual differentiation of brain function still does not establish that its development is hormone sensitive. Initially we applied the now classic approach: the administration of a single injection of testosterone propionate (TP) to female rats or gonadectomy of the male on postnatal day one (Gorski et al., 1978). As shown in Figure 3, both treatments affected the volume of the SDN–POA as measured in the 30-day-old animal. In the male, castration caused a 50% reduction in volume attained by the nucleus (Figure 3C). Interestingly, this reduction could be prevented completely by the injection of 100 µg TP one day after castration, i.e., on postnatal day two (Figure 3D; Jacobson et al., 1981). Thus, 50% of SDN–POA volume in the male is dependent on the postnatal hormonal environment. In females, the postnatal administration of exogenous TP significantly in-

creases SDN–POA volume (Figure 3E) (Gorski et al., 1978).

Two important points can be made from these observations. First, a significant component of the volume of the SDN–POA is sensitive to gonadal hormones postnatally. Second, although the male castrated on postnatal day one exhibits neuroendocrine functions characteristic of the female, his SDN–POA volume is significantly greater than that of the genetic female. Similarly, although the androgen-treated female exhibits functional characteristics typical of the male sex, her SDN–POA volume is much smaller than that of the genetic male. Thus, nuclear volume per se does not correlate with the ability of the animal to exhibit lordosis behavior or to release the cyclic surge of LH required for ovulation. In addition, the failure to "sex-reverse" SDN–POA volume suggests that neuronal genomic activity might contribute to the sex difference in the SDN–POA or that these hormonal manipulations might have been inadequate. For example, in the case of the male, the process of the sexual differentiation of the brain may actually begin before birth, and in the female, a single injection of TP on postnatal day four may just be too little hormone exposure or perhaps too little too late.

To address this issue, hormonal exposure was maximized: injections of 2 mg TP per day were given to pregnant rats beginning on day 16 postfertilization. After birth the pups were injected subcutaneously with 100 µg TP daily through postnatal day 10. These animals were sacrificed at 30 days of age and SDN–POA volume determined (Döhler et al., 1982a, 1984). As shown in Figure 3G, this prolonged exposure to exogenous androgen did not affect SDN–POA volume in males (Döhler, 1984). However, this treatment did completely sex reverse SDN–POA volume in females (Figure 3F; see also Figure 1C) (Döhler et al., 1982a). This observation clearly demonstrates that hormones alone can completely determine

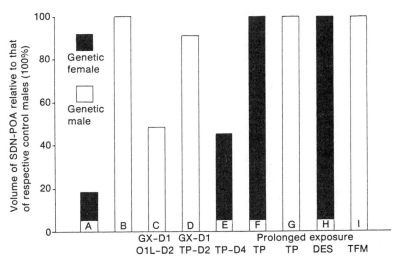

FIGURE 3. Highly schematic summary of the sex difference in SDN–POA volume (compare A and B) and the influence of the perinatal hormonal environment. SDN–POA volume is expressed as a percentage of the volume of this nucleus in control males from several independent studies. In C, males were gonadectomized on postnatal day 1 (GX-D1), and in D, similar males were then injected with 100 µg TP on day 2 (TP-D2) (Jacobson et al., 1981). In E, females were injected with 1 mg TP on day 4 (TP-D4) (Gorski et al., 1978). Although prolonged exposure to TP (from embryonic day 18 through postnatal day 10) has no effect on SDN–POA volume in males (G) (Döhler et al., 1984), this treatment sex-reverses SDN–POA volume in females (F) (Döhler et al., 1982a). Prolonged exposure to DES also sex reverses SDN–POA volume in females (H) (Döhler et al., 1982b). Finally, male rats with the testicular feminizing mutation (TFM) have a normal SDN–POA in terms of volume (I) (Gorski et al., 1981). See individual studies for statistical analyses.

SDN–POA volume and the final number of neurons in this nucleus. This does not rule out possible neuronal gene involvement under more physiological circumstances, but clearly indicates the importance of the hormonal environment for the development of this nucleus.

Before the possible mechanisms by which gonadal hormones influence the development of the SDN–POA are considered, it is necessary to digress slightly and to consider the precise molecular species of gonadal hormones that act to masculinize the brain. As shown in Figure 3H (see also Figure 1D), a similarly prolonged exposure perinatally to diethylstilbestrol (DES), a synthetic estrogen, is equally effective in sex reversing SDN–POA volume

in the female (Döhler et al., 1982b). Since testosterone can be metabolized, it is possible that testicular testosterone may act in the brain as testosterone itself, as dihydrotestosterone (DHT) following reduction by 5-alphareductase, or as an estrogen following aromatization by aromatase.

At the present time, it can be stated that the balance of the evidence suggests that in the rat hypothalamus, estrogen masculinizes the brain. Estradiol benzoate is more potent than TP when administered systematically (Gorski, 1971), DHT—which cannot be aromatized—is much less effective if at all (Whalen and Rezek, 1974; Korenbrot et al., 1975), the enzyme aromatase is present in the hypothalamus (Naftolin et al., 1975; Selmanoff et al.,

1977), inhibitors of this enzyme block normal masculine differentiation in male rats (McEwen et al., 1977a,b), and antiestrogens also inhibit endogenous sexual differentiation (Booth, 1977; Doughty and McDonald, 1974).

In one sense, therefore, neurons appear "to produce" the hormone to which they or others respond. Further confirmation of the role of estrogen in the development of the SDN–POA comes from the study of rats with the testicular feminization mutation (TFM) (Bardin et al., 1970). In this animal, androgen receptors are markedly deficient (Naess et al., 1976), although estrogen receptors appear to be normal, as judged from the TFM mouse (Attardi et al., 1976). The volume of the SDN–POA in adult TFM male rats is equivalent to that of the normal male (Figure 3I), an observation that is consistent with the view that estrogen determines the masculine differentiation of the SDN–POA (Gorski et al., 1981). Interestingly the SNB has been evaluated in the same animals. The development of this spinal nucleus seems to be androgen dependent, since in the TFM male rat the SNB does not exist (Breedlove and Arnold, 1981). This observation is mentioned to emphasize the fact that although this discussion focuses on the SDN–POA, which appears to be influenced by estrogen, this does not preclude a role for testosterone and/or DHT in sexual differentiation of other anatomical or functional components of the central nervous system.

Possible Mechanisms of Steroid Action

Although exposure to gonadal steroids of whatever molecular species during development might influence the way the adult neuron responds to hormonal signals and/or modify neural transmitter regulation and/or production, the demonstration of marked structural differences such as those in the SDN–POA clearly indicate that gonadal steroids can regulate the number of neurons that form a given neural structure. There are several potential mechanisms for this action of gonadal steroids: These hormones may regulate neurogenesis itself or, when neurons become postmitotic, may modify the process of neuronal migration, may interact with or induce a cell surface recognition factor that promotes aggregation of neurons into the nucleus, may specify the neurochemical or morphological characteristics of the neurons, and/or may influence the process of developmental cell death either during migration or after the neurons have reached the region of the SDN–POA.[2]

The results of studies of the neurogenesis of the MPOA suggest that it is unlikely that the sole mechanism of action of gonadal hormones is to modify the formation of the neurons of the SDN–POA, since most of the neurons of the MPOA become postmitotic by around day 16 of gestation (Ifft, 1972; Altman and Bayer, 1978; Anderson, 1978; Jacobson and Gorski, 1981) yet the volume of the SDN–POA can be altered substantially more than a week after the neurons have become postmitotic. However, as illustrated in Figure 4, the genesis of those neurons that specifically form the SDN–POA is surprisingly prolonged (Jacobson and Gorski, 1981). Approximately 25% of the neurons in the SDN–POA of the 30-day-old rat are labeled by exposure to tritiated thymidine on embryonic day 18, and the genesis of SDN–POA neurons continues for several additional days (C. Jacobson, personal communication). In addition, the analyses of these data revealed two statistically significant sex differences. The percentage of cells labeled on embryonic day 14, and present in the SDN–POA on postnatal day 30, is significantly greater in females than in males. The reverse is true for label given on day 17. The physiological significance of these statistical sex differences remains unknown. It is not

_____

[2]In these processes, any possible role of glia remains unknown.

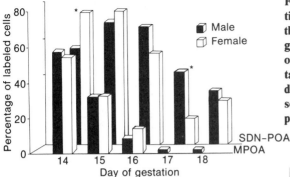

FIGURE 4. The influence of embryonic age, at the time of injection of [³H]thymidine to pregnant rats, on the percentage of labeled cells determined autoradiographically in the SDN-POA and in a control region of the MPOA of their offspring sacrificed on postnatal day 30. Asterisk indicates statistically significant differences within the SDN-POA. (Data from Jacobson and Gorski, 1981; reprinted from Gorski, in press.)

even possible to conclude that these represent sex differences in neurogenesis per se. Since these animals were not sacrificed until they were 30 days of age, potential sex differences in other developmental processes (e.g., cell death) might have contributed to the observed sex difference. Nevertheless, these data suggest that gonadal steroids could be mitogenic to neuroblasts that form SDN-POA neurons. This question, which would be of considerable biological significance, can best be addressed in vitro.

Because the genesis of neurons forming the SDN-POA is significantly prolonged with respect to that of the surrounding region, it is possible to label specifically and permanently this late-arising group of cells (Figure 5) and to take advantage of that fact in order to identify both the origin of these neurons and the path of migration they follow to reach the SDN-POA. To accomplish this, rats were exposed to tritiated thymidine on embryonic day 18 and sacrificed after different intervals. As summarized in Figure 6, these late-arising neurons of the SDN-POA are formed in the subependymal lining of the third ventricle, migrate to the base of the ventricle and the surrounding brain tissue, and then migrate upwards and laterally to reach the SDN-POA (Jacobson et al., 1985). Labeled neurons between the base of the brain and the SDN-POA clearly disap-

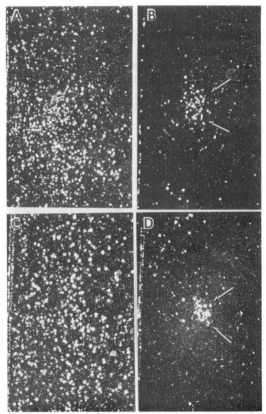

FIGURE 5. Dark-field autoradiograms of the region of the MPOA including the SDN-POA in 30-day-old male (A, B) and female (C, D) rats exposed to [³H]thymidine on embryonic day 15 (A, C) or 18 (B, D). Arrows indicate the SDN-POA. Note the relatively specific labeling of SDN-POA neurons when exposed to [³H]thymidine on embryonic day 18. (After Jacobson and Gorski, 1981; reprinted from Gorski, 1984.)

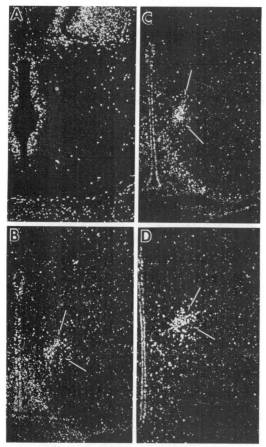

FIGURE 6. Illustration of the migration of the cells of the SDN–POA during the perinatal period. Animals were all exposed to [³H]thymidine on embryonic day 18 via an injection to their mothers. The rats were then sacrificed at 18 days plus 2 hours (A), 22 days (B), 26 days (C), or 32 days (D) postfertilization. (Data from Jacobson et al., 1985; modified from Gorski, 1985c.)

pear. Whether they continue to migrate beyond the SDN–POA or die is currently under investigation in our laboratory.

In this study three groups of animals were compared; males, females, and males that were castrated on day one of postnatal life in an attempt to detect a possible sex or hormonal influence on this process of migration.

Analysis of the results of this study, however, did not reveal any statistically significant differences. There are three possible explanations for this negative finding. First, it may well be that the process of migration is not modified by the hormonal environment. Second, it is possible that the analysis performed was inadequate to detect what may be subtle differences in the timing of migration or the precise location of labeled cells. Finally, it is also possible that these late-arising neurons of the SDN–POA, which can be permanently and specifically labeled by exposure to tritiated thymidine, are not representative of the SDN–POA as a whole and are not in fact sensitive to the hormonal environment!

Since SDN–POA volume in females can be sex reversed by the prolonged perinatal exposure to TP, this provides a means to determine whether or not the late-arising SDN–POA neurons are sensitive to gonadal steroids. The results of this preliminary study clearly indicate that these late-arising neurons of the SDN–POA are indeed sensitive to the hormonal environment (R. Dodson and R.A. Gorski, unpublished data). As shown in Figure 7, four groups of animals exposed to tritiated thymidine on embryonic day 18 were evaluated: males, females, and females who were exposed to TP for a prolonged period either beginning on embryonic day 16, as previously reported, or on embryonic day 20, two days after exposure to tritiated thymidine. Both hormonal treatments were continued through postnatal day 10 and the animals were sacrificed at postnatal day 30. It is important to note that the second injection scheme, beginning on embryonic day 20, was also effective in sex reversing SDN–POA volume in the female. Regardless of whether the sex-reversing hormonal regime was initiated two days before or two days after pulse labeling with tritiated thymidine, the number of labeled cells within the SDN–POA of these females was increased and comparable to that in the males.

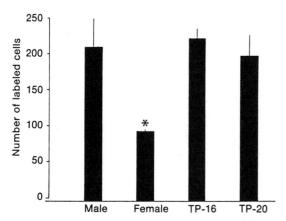

FIGURE 7. The influence of the perinatal hormonal environment on the number of labeled cells in the SDN-POA. All animals were exposed to [³H]thymidine on embryonic day 18 and sacrificed on postnatal day 30. Some females were exposed to TP for a prolonged period beginning on either embryonic day 16 (TP-16) or embryonic day 20 (TP-20), and in both cases continuing through postnatal day 10. Asterisk indicates significantly less than all other groups.

The influence of gonadal hormones on mouse hypothalamic neurons in explant culture has been studied in a series of elegant studies (Toran-Allerand, 1984). The results of these studies demonstrate that estradiol can stimulate the growth of neurite processes from estrogen-responsive neurons in vitro (Toran-Allerand et al., 1980) and, in fact, that estradiol seems to be required for such outgrowth (Toran-Allerand, 1980). Thus, at least in vitro, estradiol is a neurite outgrowth–promoting factor. Another phenomenon of potential significance for an understanding of SDN-POA formation is that of developmental cell death (Purves, 1977; Cowan, 1978; Silver, 1978; Varon and Adler, 1980; Hamburger and Oppenheim, 1982). It is well documented that in many regions of the central nervous system there is during development a frank overproduction of neurons followed by the death of many. Although the mechanisms of neuronal survival during this period of develop-

mental cell death are unknown, it has been suggested that migrating neurons may initially depend on some neuronotrophic substance, but that when these neurons reach their target site and begin to extend processes, they become sensitive to another neuronotrophic substance that is produced by their target neurons apparently in limited quantity.

Putting these observations together, one can suggest the following potential mechanism for the development of sex differences in the SDN-POA: in the male or in females exposed to exogenous steroids, the neurite outgrowth–promoting effect of these hormones causes more neurons to extend processes to their target sites and thus to obtain the neuronotrophic substance necessary for survival. In females or in the neonatally gonadectomized male, without this hormonal stimulus fewer neurons survive either because they fail to extend processes to the appropriate target site or because the temporal pattern of such growth is altered. However, another explanation is also possible. From the study of the migration of the late-arising SDN-POA neurons, it is clear that the cells between the base of the brain and the SDN-POA disappear (see Figure 6). If these neurons are actually dying, and if there is a sex difference in this process, gonadal steroids may be directly neuronotrophic.

Among the possible mechanisms of steroid action illustrated in Figure 8, at the present time it appears most likely that gonadal steroids promote neuronal survival either by virtue of a neurite outgrowth-promoting effect or perhaps by a direct neuronotrophic effect. At the same time, it may be premature to preclude a possible regulatory role of gonadal steroids on neurogenesis per se, or on migration, aggregation, or cell specification. Nor is it possible to exclude an additional and indirect effect of estrogen: hormones may stimulate the development of afferent connections that promote the survival of SDN-POA neurons. Model systems, such as the SDN-POA, offer

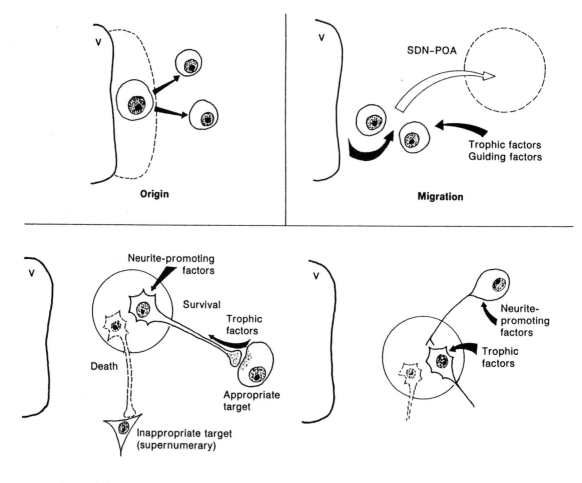

FIGURE 8. Highly schematic illustration of the formation of the SDN–POA and some of the points at which gonadal steroids (or other chemical substances) might act to produce the observed sex difference in the SDN–POA. These include the formation of neurons and their migration and/or their survival due to the development of critical efferent or afferent connections. V, third ventricle.

the opportunity to elucidate more clearly the fundamental process of the sexual differentiation of the brain.

Another model, another mechanism? As valuable a model system as the SDN–POA may be, one must question whether a neurite outgrowth-promoting or neuronotrophic effect is the *only* mechanism of gonadal hormone action on the developing brain. Bleier and colleagues (1982) have reported a sexual dimorphism in another nucleus of the preoptic region of the rat brain. In this case, however, the nucleus appears to be greater in volume in the female and, according to the results of lesion studies, may be involved in the neural regulation of ovulation (Wiegand and Tera-sawa, 1982). Although these authors have labeled this nucleus the medial preoptic nucleus (MPN), we have adopted another terminology and call this nucleus the anteroventral periventricular nucleus (AVPV) (Simerly et al., 1985a). [In fact, we have used the term MPN to describe a nucleus in the MPOA, of which the SDN–POA is a subcomponent (Simerly et al., 1984).] We have studied the AVPV immunohistochemically using antibodies directed against tyrosine hydroxylase and dopamine betahydroxylase. In terms of the latter, there is little staining and there is no obvious sexual dimorphism. However, in the case of antibodies to tyrosine hydroxylase, there is a marked sex dimorphism that we attribute to dopaminergic elements. As illustrated in Figure 9, there are significantly more "dopaminergic" positive fibers and somata in the AVPV of the female than in the male. Once again, perinatal exposure to TP reverses this sex difference, and such females have fewer dopaminergic positive fibers and cell bodies than do control females (Simerly et al., 1985b). Since the AVPV of the male is smaller than that of the female, it may be that in this nucleus steroid action actually results in cell death. However, at the present time it cannot be excluded that in the development of this nucleus steroids act to alter the neurochemical specification of its neurons and/or afferent input. Given the complexity of the mammalian brain, it may be simplistic to assume that there is only one mechanism of hormone action during development.

Are Morphological Effects of Steroid Hormone Action Restricted to Development?

The results of the studies of the SDN–POA clearly indicate that gonadal steroids can exert morphological effects during development. An important question in neurobiology is whether this ability is restricted to develop-mental critical periods. Data are accumulating that suggest that the answer to this question is no. The administration of exogenous estrogen at the time of puberty increases synaptogenesis in rats (Clough and Rodriguez-Sierra, 1982). After deafferentation of the hypothalamus of the adult rat, exogenous estrogen again promotes synaptogenesis (Arai et al., 1978). Gonadal steroid administration facilitates regeneration of the hypoglossal nerve in adult rats (Yu and Yu, 1983). Following lesions of the lateral septum in adult male rats, exposure to exogenous steroids leads to apparently permanent changes in behavioral responsiveness to steroids (Nance et al., 1975) that appear to have a neurochemical (Gordon et al., 1979), and perhaps even a neuroanatomical, basis. In the gerbil, the volume of the apparent equivalent of the SDN–POA, the sexually dimorphic area, changes in volume with alterations in the steroid environment in the adult (Commins and Yahr, 1984). Finally, and most convincingly, in the songbird there is a hormone-dependent seasonal variation in nuclear volume (Nottebohm, 1981) as well as in the growth of its dendrites (DeVoogd and Nottebohm, 1981) and the formation of synapses (DeVoogd et al., 1982).

It is tempting to speculate that gonadal steroids always have the capacity to induce neuronal growth but that in the adult this action cannot be expressed unless lesions are made that presumably make synapses available for new terminals. However, the results of the studies in the gerbil and the songbird suggest a far more dynamic interrelationship between the hormone environment and the fine structure of the central nervous system. It is possible that the synaptology of the adult brain is modulated by gonadal hormones.

Summary

It is now clear that one mechanism by which gonadal hormonal action leads to the

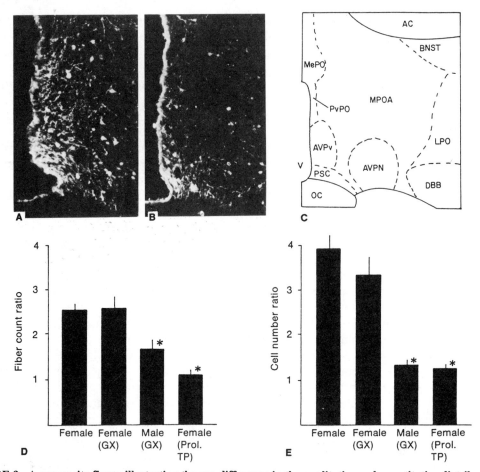

FIGURE 9. A composite figure illustrating the sex difference in the qualitative and quantitative distribution of putative dopaminergic cells and fibers in the AVPV and the influence of prolonged exposure perinatally to TP on these distributions. (A, B) Fluorescence photomicrographs of the AVPV of colchicine-treated female (A) and male (B) rats. (C) The general neuroanatomical relationships of the AVPV. (D, E) The lack of an effect of gonadectomy (GX) of young adult animals, and the marked influence of the prolonged exposure of the genetic female to TP (prol. TP), is shown on an index of fiber density (D) and on the number of immunoreactive cells (E). In both cases, data are presented as a ratio between each treatment group and an intact reference male. Asterisks indicate mean ratios that were significantly lower than those of intact males. AC, anterior commissure; AVPN, anteroventral preoptic nucleus; BNST, bed nucleus of the stria terminalis; DBB, nucleus of the diagonal band; LPO, lateral preoptic area; MePO, median preoptic nucleus; OC, optic chiasm; PSC, suprachiasmatic preoptic nucleus; PvPO, periventricular preoptic nucleus; V, third ventricle. (Nomenclature from Simerly et al., 1984; A, B, from Simerly et al., 1985a; C, after Simerly et al., 1984; D, E, after Simerly et al., 1985b.)

sexual differentiation of the central nervous system involves some fundamental processes governing neuronal production, migration, and/or survival. Moreover, the ability of gonadal hormones to modify the subtle neuroanatomy of the central nervous system in the adult appears to be a distinct possibility. It is interesting to point out that in this book, there has actually been minimal reference to sex or to hormones. Sex differences in brain structure and function do exist and, as represented by the SDN–POA, can be relatively marked. Although clearly an overstatement, this point needs to be made: until proven otherwise, it may be safest to assume that there may be a sex difference in any parameter under investigation. In addition, the existence of structural sex differences in the brain such as those in the SDN–POA can serve as valuable model systems for the study of developmental neurobiology. As exemplified by the SDN–POA, sexual differentiation involves a very pervasive modulation of development by a physiological environmental factor external to the nervous system, a factor that can be controlled reasonably well experimentally depending on the animal species. Further study of this model can be expected to yield new insights into the environmental modulation of normal neuronal growth and differentiation.

With respect to the field of sexual differentiation of the brain itself, we can surely expect more phenomenology, that is, descriptions of additional sex differences in brain structure and function at the neuroanatomical, neurochemical, and species levels. However, again using the SDN–POA as an example, the application of in vitro and molecular techniques can be expected to elucidate the basic mechanisms by which gonadal hormones modify neuronal genomic expression and lead to functional and structural sex differences in the brain.

CHAPTER EIGHTEEN

Developmental Actions of Hormones: From Receptors to Function

BRUCE S. McEWEN, VICTORIA N. LUINE, AND CHRISTINE T. FISCHETTE

Introduction

In birds and mammals, there is a clearly defined period during development when hormonal signals trigger cellular events that lead to the sexual differentiation of the reproductive tract, brain, and other secondary sex characteristics. Such a sequence of events requires incredible coordination, entailing hormonal secretion at a time when the target cells of the developing brain, reproductive tract, and other tissues have expressed receptors enabling them to respond to the hormone.

It would have been much simpler, perhaps, had these differences been programmed into the sex chromosomes so that they could have been established as invariant traits. What the hormone mediation sequence achieves, however, is a remarkable degree of plasticity for realization of environmentally controlled individual differences that would not be possible via genetic control. Such individuality is possible because of variations in the magnitude and timing of secretion of the hormonal signals. In addition, the secretion of other hormonal factors, such as thyroid and stress hormones, can further help to increase distribution of phenotypic traits in the fully mature individual. This is illustrated by the range of masculine characteristics found among female mice because of variations in proximity in utero to testosterone-producing male fetuses (vom Saal and Bronson, 1978) and by the ability of prenatal stress to suppress testosterone production and thus reduce the magnitude of masculinization among male rat offspring (Ward, 1984).

In mammals, the actions of testosterone on the developing brain lead to sex differences in brain structure and function that alter behavior and neuroendocrine function as well as the way the brain responds to hormones and drugs related to neurotransmitters. Such sex differences have implications in the realm of disease and abnormal behavior, in that such a disorder as dyslexia is more prevalent among males (Hier, 1979), while endogenous depressive illness is more common among females (Robins et al., 1984). Yet while morphological sex differences provide a tangible aspect of brain sex differences, these structural differences do not tell us about function. On the other hand, biochemical sex differences may or may not have a demonstrable morphological basis. In this chapter we begin by reviewing the overall plan of sexual differentiation in mammals and then address the relationship between morphological and functional sex differences in the brain.

Organization of Brain Sexual Differentiation

To provide perspective for the discussion that follows, it is useful to look at the overall plan by which the reproductive tract and brain of mammals become sexually differentiated. Most importantly, the sex chromosomes only determine the sex of the gonad (Jost, 1970). Once this is established, mammals follow a plan whereby testosterone (T) is secreted by the testes during a limited period of early development, and T interacts with receptors that have been elaborated as part of developmental programs within reproductive tract anlage and specific parts of the brain. Both male and female embryos have such receptors; if T is administered to the female embryo, sexual differentiation will ensue. Sexual differentiation of the type that we shall discuss involves two separate processes: masculinization and defeminization. Masculinization is the enhancement of male-typical traits such as androgen-dependent aggressive and mating behavior. Defeminization is the suppression of female-typical traits such as the estrogen-dependent surge of luteinizing hormone (LH) and mating behavior. Masculinization begins somewhat earlier in the rat than defeminization and involves different receptors and T-metabolizing enzymes. Let us briefly consider the receptors and associated T-metabolizing enzymes and their role in this process.

Enzymes exist regionally localized in brain that convert T to estradiol (aromatizing

enzymes) and to 5α-dihydrotestosterone (5α-reductase). In the rat brain, aromatizing enzymes, estrogen, receptors, and androgen receptors are all initially expressed prenatally (McEwen, 1983; George and Ojeda, 1982; Vito and Fox, 1982). After the initial appearance of receptors for estrogens, we know that subsequent expression and increase of receptor number can occur in the absence of circulating gonadal and adrenal hormones, as judged from experiments on fetal hypothalamic transplants into adult hosts (Paden et al., 1985). Perinatally, males and females appear to have the same levels of androgen and estrogen receptors (Figures 1 and 2), but males show greater levels of occupancy of these receptors as a result of substantially higher levels of circulating T. Figure 3 shows that in males, estrogen receptors are occupied by estradiol derived from aromatization of T. Similarly, androgen receptors are occupied by androgens derived by 5α-reductase from T (Meaney et al., 1985).

Androgen and estrogen receptors are both important for brain sexual differentiation (Baum, 1979; McEwen, 1983). Whereas androgen receptors are linked to masculinization of rough and tumble play behavior (Meaney et al., 1983), spinal motor control of penile muscle (Hart, 1967; Breedlove and Arnold, 1983), and certain other aspects of masculine sexual behavior (Baum, 1979), estrogen receptors that receive estradiol from aromatization of T are associated with defeminization of both the lordosis reflex and the ability to display cyclic ovulation (McEwen et al., 1982; McEwen, 1983). In the rat and mouse, masculinization of sexual and aggressive behavior begins before birth and therefore can be influenced in females by the proximity of male fetuses in utero through the passage of T from males to females (vom Saal and Bronson, 1978). Defeminization, however, begins only postnatally; otherwise, females would be exposed to estradiol derived through aroma-

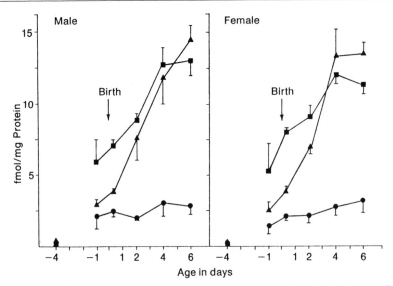

FIGURE 1. Perinatal ontogeny of soluble cytoplasmic estrogen receptor sites in the brains of male and female rats. Cytosols were prepared from either midbrain and brain stem (●), cerebral cortex (▲), or pooled hypothalamus, preoptic area, and amygdala (■) and were labeled in vitro with 2nM [³H]moxestrol, a synthetic estrogen that binds to estrogen receptors and not to the alpha fetoprotein that abounds in the neonatal rat blood. Receptor-bound radioactivity was measured by Sephadex LH-20 gel filtration and is expressed as femtomoles (fmol) of [³H]moxestrol bound per milligram of cytosol protein. Each data point represents the mean (+/− S.E.M.) of four determinations at each age. (From MacLusky et al., 1979.)

tization from T and some of them might be born incapable of ovulating and mating and thus incapable of propagating the species.

The question of defeminization raises the important issue of protection of the developing fetus from gonadal hormones arising from the mother, from other fetuses in the womb, or from the environment. Because nature established androgen and estrogen receptors as mediators of sexual differentiation and placed them more or less equally in male and female embryos, it was also necessary to establish protection mechanisms for the developing organism during the critical period. In the rat, there are two types of protection: the first operates in utero at the placental barrier, and the second operates in the vulnerable early postnatal period when defeminization may occur. In utero, the placental barrier protects the fetus from circulating hormones. The barrier appears to be particularly effective toward natural estrogens, although synthetic estrogens such as diethylstilbestrol (DES) bypass this mechanism (McEwen, 1983) and produce unfortunate teratological and behavioral effects in humans (Meyer-Bahlburg et al., 1984; Ehrhardt et al., 1985). In the rat, the proximity of male fetuses in utero (see above) would defeminize females. Therefore, defeminization is a postnatal event, but the early postnatal period must also be protected. A special estrogen-binding version of the ubiquitous alpha fetoprotein is thus present in high levels in the rat and mouse during late fetal and early postnatal ages (Nunez et al., 1974; Ruoslahti and Engvall, 1978). This protein has enormous capacity to bind estradiol but not synthetic estrogens like DES and mox-

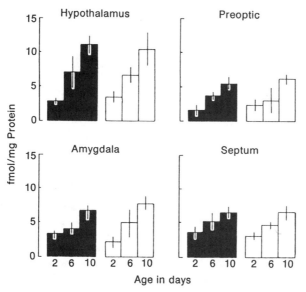

FIGURE 2. Cytosolic androgen receptor concentrations as a function of developmental age in days, measured by the binding of a synthetic androgen, 3H R1881, and expressed as bound radioactivity in fmol/mg protein for specific brain regions in gonadectomized, neonatal male (open bars) and female (filled bars) rats. Mean (+/− S.E.M.) based on four to six replications. (From Meaney et al., 1985.)

estrol (RU 2858); therefore, these substances cause defeminization even at extremely low doses (Plapinger and McEwen, 1978). Figure 4 depicts the postulated role of alpha fetoprotein (fEBP) in relation to T and estradiol (E_2) action. T does not bind to fEBP and enters target cells in brain, where it is converted to estradiol; the conversion can be blocked by the competitive antagonist 1,4,6-androstatriene-3,17-dione (ATD), and the binding of E_2 can be blocked by the nonsteroidal antiestrogen CI 628. Both ATD and CI 628 interfere with the defeminization caused by T (McEwen, 1983).

FIGURE 3. Testosterone levels in male and female rats (picograms/ml serum; bottom panel) are compared with occupation of cell nuclear estrogen receptor sites in brain by estradiol derived from the aromatization of testosterone (fmol/mg DNA; top panel) at indicated postnatal ages. (From McEwen, 1983.)

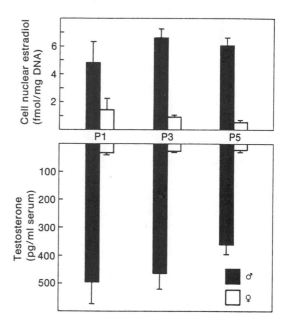

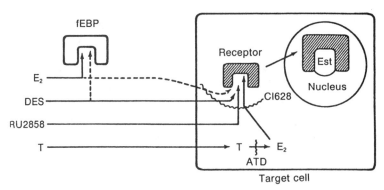

FIGURE 4. Schematic summary of events involved in the interaction of testosterone (T) and estrogens with the rat brain. fEBP, alpha-fetoprotein; RU 2858, moxestrol; DES, diethystilbestrol; E_2, estradiol; ATD, aromatase inhibitor; CI 628, antiestrogen; Est, receptor-bound estrogens in nucleus. The fEBP protects target tissues from E_2, but not from DES or RU 2858, which therefore defeminize at very low doses compared to E_2. T is not bound by fEBP and gains access to brain, where it is converted by target cells of hypothalamus and limbic brain to E_2; the E_2 binds to estrogen receptors in the nucleus. Conversion of T to E_2 can be blocked competitively by ATD, and CI 628 blocks estrogen receptors from binding E_2. Both ATD and CI 628 prevent the defeminizing actions of T to suppress the development of feminine sexual behavior in male rats. (Reprinted from McEwen et al., 1975.)

What makes the critical period unique? In other words, what do the gonadal hormones initiate via their receptors during the critical period that is different from what goes on in the same neurons at maturity? The answer appears to be that hormones initiate growth of neuronal processes and differentiation of neuronal characteristics that would otherwise not occur at that time. The question of differentiation will be discussed later in the chapter. As for growth, we know from the work of Toran-Allerand (1976, 1978, 1980) that testosterone or estrogens stimulate neurite outgrowth in developing mouse preoptic area and hypothalamus. Such outgrowth constitutes a means by which neurons can establish a wider range of synaptic contacts, leading to increased chances of survival and larger cell body size, which are two of the features associated with morphological sex differences in the adult brain.

What determines the end of the critical period? It would appear that the establishment of stable synapses and myelin deposition are associated with the end of T effects on brain sexual differentiation. Administering thyroid hormone, which speeds up synapse formation and myelination, tends to shorten the critical period (Kikuyama, 1966; Phelps and Sawyer, 1976), whereas a hypothyroid condition prolongs it (Kikuyama, 1969). When in adulthood, lesions reduce synapse number in at least one hypothalamic area (the arcuate nucleus), subsequent estrogen treatment enhances collateral sprouting within this nucleus, which results in the formation of larger numbers of new synaptic connections on vacant synaptic sites (Matsumoto and Arai, 1979). Thus, developmental programs, which may well underlie the developmental effects of estradiol derived from T through aromatization, can apparently be reactivated in adulthood by lesions. It remains to be established whether estrogens are also able to enhance synaptic turnover in the absence of lesions, for example, in the adult hypothalamus during the course of the estrous cycle.

Sex Differences in Neurotransmitter and Hormone Receptor Systems

Thus far, we have discussed only the developmental actions of hormones related to brain sexual differentiation. These actions are sometimes referred to as "organizational;" they occur during a sensitive or critical period of development and are permanent for the life of the organism. Another type of hormone action is called "activational;" these effects are reversible and occur in differentiated cells after the organizational phase is completed. Sometimes, the capability to show activational effects is subject to influences of hormones during the organizational period, e.g., the ability of estradiol to induce an enzyme in a male brain may be increased or suppressed by the developmental effects produced by T during sexual differentiation.

The existence of morphological sex differences in the brain (Gorski, Chapter 17) implies that there are also related functional sex differences that may be detected by biochemical means. These may be of two types: (1) those based on differences in quantity of a neurochemical irrespective of the presence or absence of hormone and therefore not involving any activational effects, but rather reflecting the number of neurons or terminals of a particular type; and (2) those based on differences in quantities of neurochemicals that depend on hormone priming, implying sex differences in the activational effects of the hormone and not necessarily in the number of neurons or terminals of one type or another. In this section we shall summarize some of the information pertaining to both types of biochemical sex differences.

Sex Differences in Gonadal Hormone Receptors

Contrary to initial expectations, brains of males contain receptors for female as well as male hormones, and brains of females contain receptors for male as well as female hormones (McEwen, 1983). Figures 1 and 2 illustrate the developmental pattern of estrogen and androgen receptors, respectively, in male and female brains. Brains of both sexes are similar not only during postnatal development (Lieberburg et al., 1979; MacLusky et al., 1979; Meaney et al., 1985), but also in adult life (Maurer, 1974; Lieberburg and McEwen, 1977). In fact, as noted above, because aromatization of T is important for defeminization of the developing male brain, estrogen receptors are occupied by estradiol (E) derived from T during early development in the male brain (Figure 3) (Lieberburg et al., 1979). Likewise, androgen receptors, which are involved in masculinization, are occupied in the male by androgens derived from T during the early postnatal period (Meaney et al., 1985).

Aromatization also plays a role in governing sexual behavior and other brain functions in the adult male brain (Davis and Barfield, 1979; Krey et al., 1982), and therefore estrogen receptors function in the male as well as female brains in adulthood. The same may be said for androgen receptors. Androgen receptors in the female rat brain and pituitary show occupancy by endogenous androgens, and cyclic changes in androgen receptor occupancy are noted in female pituitary throughout the estrous cycle (Handa et al., 1986). Moreover, progestin receptors are present in male and female brains and are inducible by estrogens in the pituitary, hypothalamus, and preoptic area of males and females (Moguilewsky and Raynaud, 1979; Rainbow et al., 1982). In the intact male brain, the level of progestin receptors is comparable to that of the proestrus female brain (Krey et al., 1982), but their exact function is unknown.

Given the overall similarity of gonadal hormone receptor levels and distribution in male and female brains, what is the evidence that there may be some sex differences in receptor level, however subtle or regionally localized? Existence of such sex differences might begin to explain sex differences in response to hormone, such as the reduced sensitivity of males to E or progestin, and the reduced sensitivity of females to androgens. However, the differences that have been reported are small and are not always easy to replicate from laboratory to laboratory.

Estrogen receptors have been reported in some laboratories to be reduced in adult male versus female hypothalamus and preoptic area, whereas others have not found consistent differences (Whalen and Massici, 1975; Lieberburg et al., 1980; Nordeen and Yahr, 1983). One study, using Palkovitz punch dissection of rat brain to achieve higher resolution, reported a quantitative reduction of E receptors in males in the medial preoptic area but not in other preoptic and hypothalamic nuclei (see Figure 5) (Rainbow et al., 1982). Thus it is conceivable that sex differences in E receptors may be highly localized and therefore not easily detected using larger pieces of brain tissue. The high degree of localization may also create problems of replicability of the anatomical dissection. For example, recent research failed to find a statistically significant sex difference in medial preoptic area E receptor content (Kranzler et al., 1984).

Varying results have also been reported for sex differences in progestin (P) receptor (Etgen, 1985; Kato, 1985). Using gross dissection of hypothalamus and preoptic area (HPOA), Moguilewsky and Raynaud (1979) found greater E induction of P receptors in HPOA of adult female rats than in male rats without any sex difference in noninducible receptors. In the adult guinea pig HPOA, Blaustein and colleagues (1980) found a similar sex difference in inducible P receptors that tended to be reduced by higher doses of E priming, as if male HPOA might be generally less sensitive to E than females. In the adult hamster, I. Fraille and colleagues (1987) have

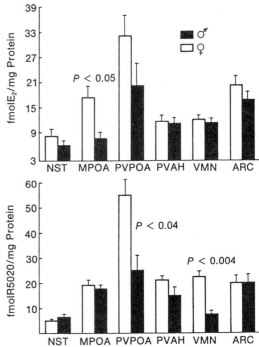

FIGURE 5. Estrogen and progestin receptor levels in hypothalamic nuclei of male and female rats. Rats were gonadectomized for at least 1 week and not more than 3 weeks before assays were performed. Males and females were assayed together and were matched for age and length of time after gonadectomy. Steroid receptor levels were measured in hypothalamic nuclei by the specific binding to cytosol fractions of [^{3}H]estradiol (E$_2$) (top panel) and [^{3}H]R5020, a synthetic progestin (bottom panel). Results are expressed as femtomoles (fmol) of radioactive steroid bound per milligram of cytosol protein. NST, nucleus of stria terminalis; MPOA, medial preoptic area; PVPOA, periventricular preoptic areas; PVAH, periventricular anterior hypothalamus; VMN, ventromedial nucleus; ARC, arcuate nucleus. (From Rainbow et al., 1982.)

found a sex difference in E induction of P receptors that reaches a peak as a function of E priming dose and then is reduced with higher E priming. Etgen (1985) reports that, for rats, low doses of E priming do not result in a sex difference in hypothalamic P receptors, but that repeated injections of E over several days result in higher levels of P receptors in female hypothalamus. Using a similar repeated E-priming regimen, Rainbow and coworkers (1982) reported higher P receptor levels in female ventromedial nucleus (VMN) and periventricular preoptic area (PVPOA) (Figure 5). However, the sex difference was not widespread and did not exist in arcuate nucleus or in other preoptic area nuclei besides PVPOA. It is thus unclear to what extent sex differences in E induction of P receptors in the brain are highly localized within discrete E-sensitive cell groupings or are the result, at least in part, of more widespread sex differences in E sensitivity.

Given the existence of a provisional sex difference in E induction of P receptors in the ventromedial hypothalamic area, what is the evidence that such a difference can be reversed by neonatal hormonal manipulations? Administration of the aromatase inhibitor ATD to newborn male pups elevated P receptor induction in VMN in adulthood to that of control females; neonatal E treatment of females produced the opposite effect, namely, a reduction of P receptor induction in VMN in adulthood to that of control males (Parsons et al., 1984). Figure 6 summarizes these results. In parallel with these results, ATD blocked defeminization of lordosis behavior, while E defeminized lordosis (Parsons et al., 1984). Figure 7 presents these results in terms of three measures of lordosis behavior. Because males are insensitive to P as far as activation of lordosis reflexes in adulthood, even after extensive E priming, these results support the no-

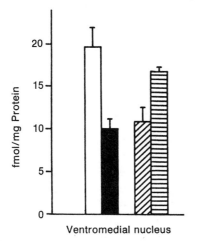
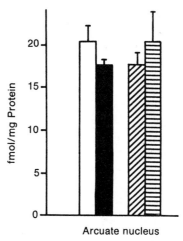

FIGURE 6. The effects of hormonal manipulation during perinatal development on the induction of progestin receptors by estradiol benzoate (EB). Progestin receptor levels expressed as fmol of [^{3}H]R5020 bound per milligram protein, +/− S.E.M. Open bars, control females; filled bars, control males; diagonal hatch bars, neonatally estrogenized females; horizontal hatch bars, neonatally ATD-treated males. Analysis of variance tests revealed a significant treatment effect of perinatal EB and ATD administration on P receptor levels in the ventromedial nucleus (VMN) ($p < 0.001$), but not in arcuate nucleus. Comparisons of individual means in the VMN using the Newman-Keuls test revealed that P receptor induction in normal females and ATD-treated males differed significantly from those in normal males and EB-treated females ($p < 0.05$). Each treatment group consisted of 24 animals; results were pooled to generate 8 independent determinations. (Data from Parsons et al. 1984).

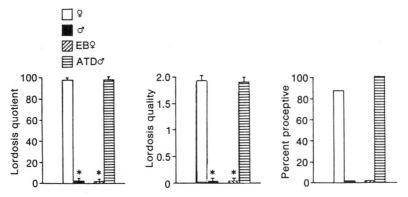

FIGURE 7. The effects of hormonal manipulation during perinatal development on activation of feminine reproductive behavior, as measured by the lordosis quotient, lordosis quality score, and percentage of animals showing proceptivity, a measure of behavior which attracts the male. Analysis of variance revealed highly significant treatment effects of perinatal estradiol benzoate (EB) and ATD administration. Comparisons of individual means using the Newman-Keuls test revealed that all three measures for normal females and ATD-treated males differed significantly from those of normal males and EB-treated females. Each treatment group consisted of 12 animals. All results are presented as the mean ±S.E.M.; * = results significantly different from those of controls. (From Parsons et al., 1984.)

tion that the lesser VMN P receptor induction in males limits the sensitivity of the male to P. However, it is not the only aspect of hypothalamic response to E that is less, or at least different, in males compared to females; thus the entire explanation for the sex difference in behavioral and neuroendocrine response to P does not rest on E induction of P receptors. In the next section, we shall consider some other aspects of sex differences that occur in response to gonadal hormones.

In summary, estrogen, androgen, and progestin receptors are present in brains of both sexes. Sex differences in E receptors have been difficult to find consistently, and if they exist, they are not large. Sex differences in E-inducible P receptors have been more consistently reported, and they appear to reflect a reduced sensitivity to E in the male brain. One of the areas of the rat brain where this difference in sensitivity is most evident is the VMN. In one case studied thus far, such a sex difference, found in the VMN, is reversed by

neonatal manipulations that reverse the defeminization of lordosis behavior.

Sex Differences in Preoptic–Hypothalamic Neurochemicals

In the search for neurochemical changes associated with gonadal hormone action, monoaminergic and cholinergic neural systems have been investigated because they are implicated in sexual behavior and generation of the LH surge (McEwen and Parsons, 1982). Because sexual behavior and gonadotropin regulation are sexually dimorphic, it comes as no surprise that neurochemical sex differences have been found in cholinergic and monoaminergic systems in hypothalamus and preoptic area. In addition, sex differences are found in the aromatization of testosterone and in the action of the neuropeptide angiotensin II on thirst. Table 1 summarizes these sex differences.

Of the eight neurochemical endpoints

TABLE 1. Summary of Sex Differences in Neurochemicals of Hypothalamus and Preoptic Area of the Rat Brain

Neurochemical	Sex Difference	Neonatal T[a]	Reference
Tyrosine hydroxylase (TH)	After puberty, TH levels in females are higher and show cyclicity in basal hypothalamus (HYP); E and P regulate TH in female HYP.	?	Porter, 1986 Beattie et al., 1972
Monoamine oxidase (MAO)	Females generally show higher activity than males; E and P regulate differently in males and females (see Table 2).	?	Skillen et al., 1961 Kamberi and Kobayashi, 1970 Luine and Rhodes, 1983
Monoamine turnover	In female rat preoptic area (POA), norepinephrine (NE) turnover cycles, but the male's does not; E and P regulate NE turnover in females differently than in males.	Yes	Kalra and Kalra, 1983 Rance et al., 1981 Hiemke et al., 1985 Lookingland et al., 1982 Wise et al., 1981
Serotonin-1 receptors (5HT1)	5HT1 receptors higher in male POA; E treatment regulates 5HT1 receptors in different male and female brain regions (see Figure 8).	?	Fischette et al., 1983
Cholinergic enzymes	Choline acetyltransferase and acetylcholinesterase are induced by E in female basal forebrain and not in male basal forebrain.	Yes	Luine et al., 1975 Luine and McEwen, 1983 Luine, 1985 Luine et al., 1986 Libertun et al., 1973
Muscarinic receptors	E treatment induces muscarinic receptors in female HYP and POA but not in male HYP and POA.	?	Rainbow et al., 1980 Dohanich et al., 1982 Rainbow et al., 1984
Angiotensin II (AII)	AII infusion into POA increases drinking; E suppresses this effect in females but not in males.	Yes	Jonklaas and Buggy, 1983
Aromatizing enzymes (ARE)	In POA of male quail brain, T elevates ARE much more than in female; however, T induces 5 α-reductase equally in male and female POA.	?	Schumacher and Balthazart, 1986

[a]Indicates whether there is any evidence regarding the effects of neonatal manipulations on the development of sex differences.

summarized in Table 1, three have been shown thus far to be reversed by neonatal hormonal manipulations, that is, neonatal castration of males has promoted development in the direction of the feminine phenotype, whereas T administration to newborn females has promoted development in the direction of the masculine phenotype. The other five endpoints shown in Table 1 have not as yet been subjected to such manipulations.

Another important feature of the neurochemical sex differences listed in Table 1 is that all of them involve differences in the response to gonadal hormones, i.e., in the activational effects of these hormones. Some of the sex differences involve qualitative differences in hormone effect: the opposite and regionally localized effects of E on serotonin-1 receptors (Figure 8); the opposite responses of male and female hypothalamic MAO activity

to E (Table 2); and the response of basal forebrain cholinergic nuclei to E treatment (Luine and McEwen, 1983). The other sex differences shown in Table 1 consist of a response in one sex that is diminished or absent in the other, very much like the sex difference seen in response of P receptor induction to E treatment (see Figures 5 and 6).

The existence of sex differences in the ability of gonadal hormones to induce neurochemical changes in the brain suggests that male and female brains differ in their ability to respond to psychoactive drugs. Moreover, such sex differences in drug response are likely to be accentuated by the effects of circulating gonadal hormones through effects like those summarized in Table 1. Indeed, there are sex differences in drug response. Table 3 lists some of these sex differences that appear to involve CNS mechanisms. Like the

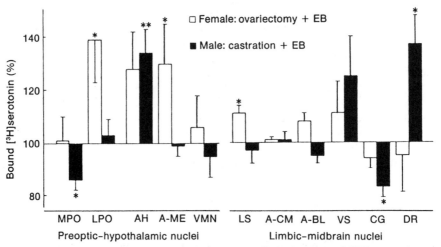

FIGURE 8. Effect of estrogen (EB) priming on specific high-affinity binding of [³H]serotonin to serotonin-1 receptors obtained from gonadectomized male and female rats in microdissected preoptic, hypothalamic, and limbic-midbrain nuclei. Results are expressed as percent of controls that did not receive estrogen. MPO, medial preoptic area; LPO, lateral preoptic area; AH, anterior hypothalamic nucleus; A-ME, arcuate-median eminence; VMN, ventromedial nucleus of the hypothalamus; LS, lateral septum; A-CM, corticomedial amygdala; A-BL, basolateral amygdala; VS, ventral subiculum of the hippocampal formation; CG, central gray; DR, dorsal raphe nucleus. In comparison with control rats subjected to gonadectomy, * = $p < 0.05$, ** = $p < 0.01$ (two-tailed paired t-test). (From Fischette et al., 1983.)

TABLE 2. Sex Differences in Estrogen and Progesterone Control of Ventromedial Hypothalamic and MAO Activity

Dose of Estradiol	Female Response		Male Response	
	E	E + P	E	E + P
Low	Up	Down	Down	None
High	Down	Up	Down	None

Data from Luine and Rhodes (1983) and V.N. Luine (unpublished data).

differences outlined in Table 1, they appear to reflect quantitative differences in drug response as well as at least one instance of a qualitative difference in hormone effect (myoclonus in guinea pig). Other sex differences have been noted in phencyclidine-induced stereotyped behavior (Nabeshima et al., 1984), in barbiturate protection from cerebral ischemia (Nakatomi et al., 1983); in behavioral sensitization to cocaine (Glick and Hinds, 1984); and in response to anxiolytic drugs (Rodriguez-Siera et al., 1986) and antidepressants (Wilson and Roy, 1986). In the case of these effects, differences in peripheral metabolism of drugs appear to make some contribution, but sex differences in central nervous system mechanisms may also be involved.

Relationship of Neurochemistry to Morphology

Quantitative Predictions

The relationship of neurochemical sex differences to reported morphological sex differences is intriguing and complex. The simplest prediction is that when there are more terminals and/or cell bodies of a given neurochemical system in a particular brain area, that area should contain more of that par-

TABLE 3. Summary of Sex Differences in Response to Drugs

Sex Difference	Neonatal T[a]	Reference
"Serotonin behavioral syndrome" shows greater sensitivity in female rats than in male rats; sex difference can be reduced or eliminated by castrating of males as adults. Mediated by 5HT1 receptors.	There may be a developmental influence on androgen sensitivity.	Fischette et al., 1984 Lucki et al., 1984
Myoclonus in guinea pigs is enhanced by E treatment, which is much more effective in females than in males. P causes response to increase in E-treated males and to decrease in E-treated females.	?	O'Connor and Feder, 1984, 1985 O'Connor and Fischette, 1986
Rotational response to amphetamine is greater in female rats than in male rats. Response varies with estrous cycle in females and is enhanced by E treatment in females. Parallels in human dopamine system, i.e., women are more susceptible to Parkinsonism and to tardive dyskinesia caused by neuroleptic drugs.	?	Robinson et al., 1982 Becker et al., 1982 Beer and Becker, 1984 Donlon and Stenson, 1979

[a]Indicates whether there is any evidence about influence of testosterone in producing the sex difference.

ticular neurochemical. Thus, the higher density of vasopressin fiber staining in lateral septum of male rats (DeVries et al., 1984) should be reflected in higher vasopressin levels in males. Larger numbers of dopamine neurons in female periventricular preoptic nucleus (Simmerly et al., 1985) should be reflected in higher tyrosine hydroxylase levels in females. The problem is that no one has yet biochemically measured vasopressin and tyrosine hydroxylase in these areas. One instance where biochemistry and morphology coincide is for α-bungarotoxin binding in mouse amygdala; in this case, a larger area of α-bungarotoxin binding, determined autoradiographically, corresponds to a larger binding capacity for bungarotoxin, determined biochemically (Arimatsu et al., 1981).

In the case of demonstrated biochemical sex differences, one would predict that morphology would follow suit. Thus, sex differences in γ-aminobutyric acid and glutamate (Frankfurt et al., 1984), substance P (Yoshikawa and Hong, 1983), and met-enkephalin (Hong et al., 1982) levels should be reflected in larger number of terminals and/or cell bodies. Again, such observations have not yet been made.

Where there has been some attempt to relate morphology and biochemistry, there are some surprises and disappointments. Simmerly and coworkers (1984) showed that serotonin innervation of the medial preoptic nucleus (MPN) is more extensive in the male because the MPN is larger in the male. However, Renner and colleagues (1985) found serotonin levels to be the reverse, that is, higher in the female than in the male, although the area sampled was not identical in these two studies. Consistent with the Simmerly et al. observations, however, is the fact that serotonin-1 receptor levels are higher in medial preoptic area of males (Fischette et al., 1983). With respect to cholinergic enzymes in the horizontal limb of the diagonal band of Broca, biochemi-

cal sex differences in cholinergic enzyme activity in the rat are not readily reflected in the number of AChE-stained neurons (G. Dohanich, unpublished data).

Another factor entering into comparisons between morphology and biochemistry is the fact that tissue stains may reveal an aspect of morphology that is not reflected in particular biochemistry. For example, within the sexual dimorphic nucleus (SDN) of the preoptic area (Gorski et al., 1978), the female rat actually has a more extensive area of opiate receptor staining than the male, even though the SDN of the male is larger than that of the female (Hammer, 1984).

Sex Differences in Regulation

Biochemical sex differences also involve regulation of neurochemistry in what we have referred to as the activational effects of hormones. Except for those regulatory processes that can be studied using morphological techniques such as quantitative autoradiography, the strictly biochemical approach contributes something that cannot be predicted or studied by morphological techniques, namely, an insight into function. At the same time, the underlying basis for biochemical sex differences may depend heavily on morphology and the pattern of connections to a particular brain region of males and females. Because a phenotype is the visible characteristic of an organism, tissue, or cell, we call the characteristic that concerns regulation of cellular events a "regulatory phenotype" (McEwen et al., 1984; McEwen, 1984). In this section we shall examine possible underlying mechanisms for sex differences in the regulatory phenotype.

At first glance, sex differences in response to a hormone, such as those enumerated above, may be due to sex differences in hormone receptor levels or distribution or sex differences in the program of cellular response to the hormone. Presumably, both types of sex differences would be subject to determination by the perinatal presence or absence of testosterone, as discussed previously. With respect to hormone receptors, it was seen that sex differences do exist but that they are either not large or discretely localized within the brain. A discretely localized reduction in androgen receptors in female Japanese quail brain might, for example, account for the less sensitivity of the female to elevating aromatase activity (Schumacher and Balthazart, 1986). However, in order for this to be a viable explanation, the equal induction of 5α-reductase by T in males and females would have to occur in cells in which there was no sex difference in hormone receptors. Likewise, the failure of E to inhibit angiotensin II–induced drinking in male rats (Jonklaas and Buggy, 1984) might reflect a local deficit in males in E receptors such as was reported for PVPOA by Rainbow and coworkers (1982). Finally, the reduced induction of P receptors in VMN of both males and neonatally E-treated female rats is not accompanied by a reduction in VMN estrogen receptors (Rainbow et al., 1982). What then might be the additional explanatory factors?

First, we have seen that actions of E in males and females may be intricately related to dose and duration of E exposure. For example, studies on progestin receptor induction by E reveal that sex differences in inducibility may only occur at certain doses or durations of E exposure (Blaustein et al., 1980; Etgen, 1985; I. Fraille et al., unpublished data). The effects of E on MAO activity in VMN and arcuate nucleus are also dose dependent, in that low doses of E decrease MAO in males but increase or have no effect on MAO activity in females, whereas higher doses of E decrease MAO in both sexes (Luine and Rhodes, 1983; V.N. Luine, unpublished data). Similarly, P effects on MAO are dependent on prior E-priming dose: with low-dose E priming, P increases MAO activity in females and again has no effect in males (V.N. Luine, unpublished

data). At this stage, we have no explanation for these complex actions of E. However, we would note that variations in time course and dose of E action are not unknown, in that within the female brain differences in time course and degree of genomically mediated E effects on cell nuclear morphology and ribosomal RNA levels are noted among E-sensitive cell groupings (Meisel and Pfaff, 1985; Jones et al., 1985; Jones et al., 1986).

Second, certain hormone effects noted above show qualitative differences in males and females that are simply not explainable on the basis of differences in hormone receptor levels or distribution. For example, sex differences in the direction of E effects on serotonin-1 receptors (Fischette et al., 1983), MAO activity (V.N. Luine, unpublished data), and choline acetyl transferase (ChAT) activity (Luine and McEwen, 1983) require a radically different explanation. We present two alternative, but not mutually exclusive, models of what may be occurring (Figure 9). In model A, it is postulated that the innervation of hormone-sensitive neurons plays a role in determining the regulatory phenotype and that sex differences in innervation result in different patterns of response. For example, recent observations of sexually dimorphic patterns of shaft and spine synapses in the hormone-sensitive ventrolateral VMN (Matsumoto and Arai, 1986) may be related through model A to the sex differences in response to E described above. We know from a number of studies with blocking drugs that certain neurotransmitter inputs to E- and P-sensitive neurons alter their hormone-binding capacity (Nock et al., 1981; Nock and Feder, 1984; Clark et al., 1985; Gietzen et al., 1983; Thompson et al., 1983). It is conceivable that such effects may be involved in sex differences in response to hormone. Recent observations from our laboratory indicate that removal of serotonin innervation of mediobasal hypothalamus by 5,7-dihydroxytryptamine (DHT) lesions el-

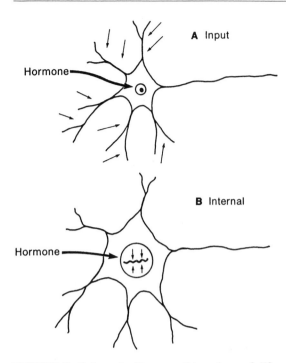

FIGURE 9. Schematic diagram of two views of differentiation of response to hormonal stimulation. (A) Response of neuron to hormone is controlled by effects of neural input via secondary effects of neurotransmitters. (B) Response of neuron to hormone is controlled by developmentally established differentiation of genome.

evates female sexual receptivity in both male and female rats (Luine et al., 1984; J. Moreines et al., unpublished data). The effect in males is particularly noteworthy, in that the lesions cause them to respond to P after E priming, which they do not normally do. In the future, we will attempt to find out whether biochemical responses in the basal hypothalamus that are sexually dimorphic, such as E induction of P receptors, are also reversed in 5,7-DHT–lesioned males.

Alternatively, model B in Figure 9 postulates that sex differences in the regulatory phenotype are largely the result of intracellular and, perhaps, intragenomic factors

that differentially influence how the hormone receptor exerts its effects. An example would be sex-specific chromosomal proteins that mask or modify the acceptor and effector sites through which the steroid hormone–receptor complex acts to regulate transcription (Buller and O'Malley, 1976; Spelsberg et al., 1983; Yamamoto, 1985). There is very little direct evidence for model B thus far in the nervous system. We know that progestin receptors can be induced in some, but not all, E-sensitive neurons; for example, they are induced in hypothalamus and preoptic area, but not in amygdala (MacLusky and McEwen, 1980). What accounts for this difference? It is possible that additional regulatory factors, such as those postulated in model B, determine that P receptors are produced in amygdala neurons on a constitutive basis, but these regulatory factors also determine whether E-receptor complexes can or cannot induce them. Only the study of specific gene products at the mRNA level will allow us to test the model directly.

Finally, it is important to note that aspects of both models might be correct. Neural inputs (model A) might help to determine the responsiveness of the genome to hormones (model B).

Summary

Sexual differentiation of the brain is a remarkable phenomenon and one that is useful for the developmental neurobiologist, because it involves a circulating hormone that interacts with specific receptors produced at the correct time of development in specific brain regions. Consequences of sexual differentiation are far-reaching and involve some of the most challenging problems in the study of the neural basis of behavior, as well as questions of clinical and sociological significance.

The underlying mechanisms involved in the development of brain sex differences clearly involve differential regulation of gene expression. We have discussed the degrees to which this differentiation involves both morphological and biochemical alterations which can be studied in parallel. Our conclusion is that although morphological sex differences must have a biochemical counterpart, and vice versa, the relationship between the two is far from obvious. Some of the problems encountered to date are due to the fact that coordinated morphological and biochemical studies have not frequently been carried out. Where they have, methodological problems arise involving the units of measurement and interpretation of results. At the level of the regulatory phenotype, the question of biochemical sex differences transcends morphology and enters the realm of function. In considering sex differences in hormone action, it is possible to see fruitful connections between how neurons are "wired" and how these connections may influence the neurons' response to hormonal signals. In future studies of these types of interactions, the precision of the steroid hormone as a tool to find specific brain regions where sexual differentiation is taking place and to study genomic events in those brain areas offers great advantages to the experimenter.

The Turned-on Brain: Developmental and Adult Responses to the Demands of Information Storage

WILLIAM T. GREENOUGH

Introduction

As other chapters of this book have suggested, the functional organization of the mammalian brain can be dramatically affected by experience during its development. One mechanism by which this may occur is the overproduction of synaptic connections, followed by the selective loss (or selective retention) of many of them, such that a functional pattern of brain organization remains. In parallel to this, there are often changes in the morphological characteristics of synapses that may reflect either the status of a synapse in this process (i.e., stabilized or not) or specific changes in its character (i.e., altered strength) that would affect its role in the overall organizational pattern. Thus, evidence strongly suggests two general types of developmental synaptic plasticity—the plasticity of pattern suggested by Ramon y Cajal (1893) and the plasticity of strength suggested by Tanzi (1893)—which together determine the functional organization of the developing brain and which reflect the preserved influence of past experience. This chapter reviews evidence for specific changes of both types and argues for their continuing role in adulthood and aging. In addition, the following questions about synaptic plasticity are considered.

1. Does synaptic plasticity merely reflect experiential driving of synaptic activity, or is it a selective response to experience?

2. Do proposed neuromodulatory substances regulate plastic synaptic phenomena?

3. What is the metabolic cost of synaptic plasticity? Does the brain adjust its metabolic capacity, on a long-term basis, to the demands of the environment?

4. How does synaptic plasticity differ in development and adulthood?

5. How well do various facets of brain plasticity persist in the absence of the eliciting phenomenon? Persistent effects seem more likely to underlie long-term changes in functional organization.

Why Use Complex (or "Enriched") Environments?

We have for some time been involved in examining effects of experience differences on indices of synaptic connectivity. We have used three conditions, modeled on the original home and laboratory conditions arranged for rats by Hebb (1949) and his students (Hymovitch, 1952; Forgays and Forgays, 1952) (Figure 1). Environmental complexity (EC) involves housing rats in groups of about 10 to 14 in a large cage with subsets of a large set of play objects that are changed daily. These animals are additionally treated to a 30- to 60-minute daily period in a 4-foot square box also equipped with a new set of toys daily. Individual cage (IC) animals are housed singly in standard laboratory cages; social cage (SC) animals are housed in pairs in standard laboratory cages. Male littermate sets of Long Evans hooded rats, matched for body weight, coat color, and coat pattern (hooded vs. fully pigmented), are randomly assigned, one per set, to the environments.

Using similar rearing conditions, Rosenzweig and colleagues (1972; Rosenzweig and Bennett, 1978) reported that regions of dorsal neocortex were heavier and thicker in EC rats than in IC rats, particularly in occipital cortex. Values in SC rats were intermediate, but often closer to those of the IC rats. In addition, they reported that nerve cell bodies were larger and glial cells more numerous in EC than in IC occipital cortex.

A question that is often asked is why the complex (or "enriched") environment is a key element of this research. The goal of our research has been to detect possible substrates of experience-induced plasticity. The two most common manipulations used in quantitative neuroanatomical studies of this sort have been visual deprivation and the use of complex environments. In both cases, the goal has been to use a heavy experiential "sledgehammer" to maximize the detection of plastic change. The

FIGURE 1. A representative photo of part of the complex environment (EC) home cage (top), and of individual (IC) cages (below). SC rats are housed in pairs in cages identical to those in the lower photo.

effects of visual deprivation are limited to restricted sensitive periods in development, and the associated plastic phenomena are potentially unique to this developmental process, as we shall see later in this chapter. In contrast, the complex environment approach works, at least to some extent, throughout the life of the animal. This is one reason for the suggestion below that the brain's response to differential environmental complexity (and to formalized behavioral training) may be different from what occurs after manipulation of the postnatal visual environment.

It is also important to address what I believe to be a widespread misperception of the relative merits of various approaches to differential experience: that visual deprivation is somehow "clean" because only one modality is affected (by extension, stripe rearing is cleaner than dark rearing, since it is restricted to channels within modalities), whereas differential environmental complexity is "dirty" because the elements of experience responsible for differences in the brain cannot be isolated. In fact, neither manipulation is free of extraneous factors; every change in the environment, no matter how seemingly specific, has nonspecific consequences. For example, dark rearing may involve maternal and infant endocrine imbalance due to the acyclic lighting (Mos, 1976) and may reduce body weight (Eayrs and Ireland, 1950). In addition, dark rearing damages the retina noticeably in some species (Rasch et al., 1961). Pattern deprivation by eyelid closure alters optical axis length, producing myopia (Wiesel and Raviola, 1977). At least some components of reported dark-rearing effects on behavior and on the brain may result from these non-CNS and nonvisual causes. The differential environmental complexity procedure involves similar confounds, of course. For example, body weight differences are often reported between individually housed rats and those housed in complex environments, although there is no systematic relationship between body weight differences and the morphological differences described below.

Thus, ignoring the developmental limitations, there is no reason to assume that effects of various forms of visual deprivation will be routinely more interpretable than those of differential environmental complexity. In addition, the complex environment condition more closely approaches the stimulation of the natural environment, and, in rats, differences after postweaning environmental complexity versus isolation (Greenough and Volkmar, 1973; Juraska, 1984) are larger than differences between rats reared in the light (in cages) and those reared in darkness (Borges

and Berry, 1978; Fifkova, 1968; Rothblat and Schwartz, 1979). In either the dark-rearing or the differential environmental complexity paradigm, one ultimate goal is to determine the functions of the differences in brain structure that are detected. This can proceed in two directions: examining the effects of seemingly more specific experience differences (e.g., stripe rearing or learning tasks) and ruling out possible causes of the brain differences, such as stress-induced hormone effects. Examples of both approaches are presented later in this chapter.

A final note of caution: one would probably not see this sort of very large effect if one merely placed a few toys in a cage with some rats and left them alone. We expend significant effort arranging elaborate arrays of toys, observing the animals to be certain that they actively explore their surroundings, and even shopping (mostly at garage sales) for the toys themselves. Several studies support our informal observations of the importance of active participation by the animals in generating differences in brain and behavioral measures between animals reared in a complex environment and animals reared in simple cages (Forgays and Forgays, 1952; Ferchmin et al., 1975).

Interpreting Morphological Frequency Data

Theories of the organization and modification of the nervous system by experience have focused upon the synapse, proposing that either changes in the strength of existing synaptic connections or the formation and/or loss of connections could alter the functional circuitry of the brain in development and adult memory. Thus, morphological studies of experientially induced plasticity have focused upon measures indicative of synapse formation or loss and measures indicative of synaptic efficacy change. However, these measures are not independent. Qualitative changes in existing synapses are inferred either directly from relative frequencies, such as the percentage of synapses with a particular morphological characteristic (Greenough et al., 1978; Vrensen and Cardozo, 1981), or from mean values, such as size, that will change if new synapses differing in mean values are added. Thus, if synapses with certain characteristics are formed at higher rates, qualitative changes would appear to have occurred in the population of synapses. Thus both the number and the characteristics of synapses must be measured in order to detect possible changes, and if numbers change, conclusions about changes in pre-existing synapses must be tentative.

A second, very critical point has to do with the way in which the number changes are assessed. It is now widely understood that if synapses differ in size, this will affect the probability of their appearing in tissue sections, and thus group differences in synapse size can affect synapse density estimates (Weibel, 1979). There is less agreement on the best method to correct for such differences. What is not so widely understood is that synapse density may have very little meaning with regard to the parameter that one wants to evaluate— the number of synapses in a given functional system—if one wants to draw conclusions about synapse formation or loss. The reason for this is that, as described below, new synapses may require varying amounts of axon, dendrite, glia, and so on, all of which take up some space in the tissue. Thus the *volume* added to the tissue with each new synapse may be substantially greater than the volume of the synapse itself. In the extreme, each new synapse could theoretically require the same volume of tissue as any other, in which case the density of synapses would not change at all with addition of new synapses. While this extreme may be unlikely, it illustrates why synapse density estimates by themselves are unlikely to reflect changes in synapse numbers accurately. More

accurate indicators are (1) the number of synapses per neuron, used in the more recent studies described below, or (2) judicious use of quantitative dendritic field analyses, as with the Golgi-impregnation technique.

Effects of a Complex Rearing Environment on the Brain

Given the theoretical focus upon the synapse, our morphological studies of the effects of rearing environments on brain structure have concentrated on two aspects of cerebral synaptic connectivity: (1) group differences in the frequency of morphological characteristics that might indicate synaptic efficacy differences, and (2) group differences in the number of synapses per neuron. More recently, we have begun to examine nonsynaptic tissue elements as well.

Changes in Synapse Morphology

The core of the set of findings that I will address clearly indicates both changes in morphology that might reflect alteration of pre-existing synapses and changes in the number of synapses. These studies compared occipital, or visual, cortex—in Rosenzweig and colleagues' work (1972), the area showing the greatest differences in weight and thickness— in two or all of the groups described earlier: EC rats, reared communally in cages with a changing set of toys; SC rats, reared in pairs in standard cages; and IC rats, reared individually in identical or similar cages. With regard to changes in average morphological measurements, which are depicted in Figure 2, the following have been reported:

1. *Size.* While no CNS data are available, data from peripheral synapses suggest that larger synapses are stronger (Grinnell et al., 1983). If some brain synapses in members of an experimental group had

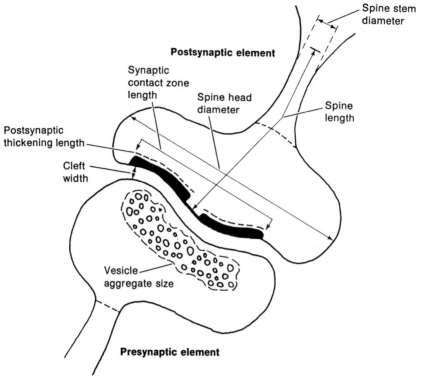

FIGURE 2. Schematic depiction of measurements of synapses. (From Sirevaag and Greenough, 1985.)

been altered in this way to increase efficacy, then the average size of synapses would be greater. In layer IV of occipital cortex, synapses (postsynaptic thickenings or synaptic contacts) are larger in EC than in IC rats, and intermediate in size in SC rats (West and Greenough, 1972; Diamond et al., 1975; Sirevaag and Greenough, 1985). A report of very large synapse size differences in layer III (Mollgaard et al., 1971) was later repudiated by the junior authors (Diamond et al., 1975). Measures of the various other presynaptic and postsynaptic parameters depicted in Figure 2 (excluding cleft width; see below, under "Miscellaneous") do not differ, except that EC rats have

larger maxima of presynaptic terminal size (Sirevaag and Greenough, 1985) and larger presynaptic vesicle aggregates (Sirevaag and Greenough, in press) in layer IV.

2. *Shape.* The curvature of synaptic contacts has been proposed as an index of synapse maturity (Dyson and Jones, 1980) or activation level (Cooke et al., 1974), although what role this plays in synaptic function remains uncertain. In visual cortex, EC synapses are more presynaptically concave than IC synapses (Wesa et al., 1982). Spine shape measures (Sirevaag and Greenough, 1985) reveal no differences of the sort proposed to affect synaptic resistance in long-term potentiation studies

(Fifkova and Van Harreveld, 1977; Chang and Greenough, 1984).

3. *Miscellaneous.* In the course of these studies, other aspects of synaptic structure have been found to differ. The relative frequency of perforations in the postsynaptic density is higher in upper visual cortex layers in EC than in SC or IC rats (Greenough et al., 1978). In aldehyde-osmium fixed preparations, the pre- and postsynaptic membranes are more closely apposed, that is, cleft width is smaller, at these perforations (Sirevaag and Greenough, 1985). Spine heads and stems of synapses in the upper visual cortex of EC rats are more likely to contain polyribosomal aggregates (PRA), presumed sites of protein synthesis, than in SC or IC rats (Greenough et al., 1985a).

A number of parallel reports suggesting synaptic efficacy alterations as a result of experience have arisen from other developmental work. Effects of visual deprivation on synapse curvature (Tieman, 1984) have been reported, while effects of visual deprivation on synapse size have been inconsistent (for review, see Greenough and Chang, 1985). The frequency of perforated synapses in adults has been reported to be increased by visual training (Vrensen and Cardozo, 1981) and motherhood (Hatton and Ellisman, 1982). In contrast to our findings with postweaning rats, the housing of adults in EC or IC environments did not affect synapse contact size in layer IV or the relative frequency of perforated synapses, although the absolute number of perforated synapses was greater due to the greater number of synapses overall in EC rats (Hwang and Greenough, in preparation). There was evidence for a parallel effect on synapse curvature in the adults.

The increase in PRA in spines also occurred in adult EC rats and was largely maintained throughout a subsequent period of IC rehousing (Table 1). Thus, to the extent that these effects represent qualitative changes in pre-existing synapses, our data suggest that only changes in synaptic contact curvature and PRA location occur after the postweaning period of development, while others' findings argue similarly for continued plasticity of synapse perforations under at least some conditions. The PRA finding is particularly interesting, since it suggests a lasting alteration in the metabolic state of synapses resulting from behavioral experience in adults. As a consequence of the animals' experience in the complex environment, the synapses appear to have been "turned on" or "turned up" on a relatively permanent basis.

Changes in Synapse Number

Changes in the number of synapses are caused by both developmental and adult experience manipulations:

1. In light microscopic studies of Golgi-impregnated occipital cortex, many neuronal populations have more extensive dendritic fields in EC rats than in IC rats, with SC rats intermediate (for review, see Greenough and Chang, 1985). One study, in which differential housing began at weaning, reported that in the upper layers (I to IV), differences in the amount of dentrites per neuron between EC and IC rats were as large as 20 to 25% (Greenough and Volkmar, 1973). Another report indicated that the density of spines along dendrites was also higher in EC rats than in IC rats (Globus et al., 1973).

2. Electron microscopic studies similarly indicate more synapses per neuron and more dendrites per neuron in the visual cortex of EC rats (Bhide and Bedi, 1984; Turner and Greenough, 1985; Sirevaag and Greenough, in press). Across layers I to IV, differences in the number of syn-

TABLE 1. Persistence of Experience-Induced Synapse Differences in Adult Rats

	Duration of Housing (days)							
	10		30		60			
	EC	IC	EC	IC	EC	EC/IC	IC/EC	IC
Synapse number per μm^3	0.570* ±0.02	0.497 ±0.02	0.596* ±0.02	0.488 ±0.03	0.596*** ±0.02	0.556^x ±0.01	0.539^x ±0.02	0.479^y ±0.02
Synapse number per neuron	10081** ±183	8738 ±357	11417*** ±152	9079 ±290	11474**** ±264	10951^x ±374	10291^x ±326	8742^y ±132
PRA in spine head, % of total spine heads	6.5 ±0.4	5.2 ±0.4	5.8** ±0.4	3.5 ±0.3	6.6*^{x} ±0.6	6.3^{xy} ±0.6	6.6*^{x} ±0.6	4.2^y ±0.3

Note: Rats were placed in EC or IC housing for durations of 10, 30, or 60 days or transferred from 30 days EC to 30 days IC housing (EC/IC) or from 30 days IC to 30 days EC (IC/EC). PRA, polyribosomal aggregate. For within-age comparisons: $*p < 0.05$, $**p < 0.01$, $***p < 0.001$; within the 60-day group, means identified by the same letter (x or y) do not differ statistically; means identified by different letters differ by Tukey's studentized range test, $p < 0.05$.

apses per neuron between EC and IC rats averaged 20 to 25%, with SC rats intermediate (Figure 3) (Turner and Greenough, 1985). Synaptic density estimates were much closer together for the two groups. The parallels between these light and electron microscopic findings strongly supports the use of quantitative measures of Golgi-impregnated dendrites to infer changes in the number of synapses per neuron.

3. Both the dendritic field differences and the synapse per neuron differences also occur in the visual cortex when *adult* animals are placed in EC and IC housing. Some neuronal populations may be less affected in adults than in postweaning animals, but in the upper visual cortex, dendritic length and synapse per neuron differences between EC and IC rats are not dramatically less in adult rats than in weanlings (Uylings et al., 1978; Juraska et al., 1980; Green et al., 1983; Hwang and Greenough, in preparation). There is reason to believe that, at least in the aggregate, this plasticity declines in some brain regions of aging rats (Connor et al., 1980; Greenough et al., 1986; Black et al., 1986), as has been suggested in humans (Flood et al., 1985). Nonetheless, the fact that synapses can be added to the brain throughout much of adulthood has made synapse addition a possible mechanism for brain memory storage.

4. The memory storage interpretation of the role of synapse addition is supported by the persistence of complex environment–produced changes well beyond the termination of the initiating events. In both an adult-onset study of synapse per neuron numbers in visual cortex layer IV and a postweaning study of dendritic field differences in pyramidal and non-pyramidal neurons in layers III and IV of visual cortex, differences between EC and IC animals persisted for at least 30 days

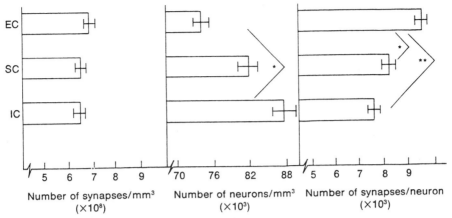

FIGURE 3. Synaptic and neuronal density and synapses per neuron, corrected for group differences in size, in upper visual cortex of rats reared for 30 days after weaning in environmental complexity (EC), social cages (SC), or individual cages (IC). The lower density of neurons in EC and SC rats reflects increases in neuropil and associated tissue elements that move neurons farther apart. The greater number of synapses per neuron in the EC group confirms predictions from Golgi-impregnation studies showing greater dendritic length per neuron. *, $p < 0.05$; **, $p < 0.002$. (From Greenough, 1984. Copyright 1984, Elsevier Science Publishers.)

following the removal of the animals from EC housing and IC cages (Tables 1 and 2) (Camel et al., 1986; Hwang and Greenough, in preparation). Thus if the added synapses are involved in altered functional organization of the visual cor-

tex, the functional alterations are relatively permanent.

Complex environment effects of virtually identical magnitude have been described in cats (Beaulieu and Colonnier, 1985), and par-

TABLE 2. Persistence of Experience-Induced Dendritic Field Differences in Young Rats[a]

	EC	IC	EC/IC	IC/IC
Pyramidal Cells (Layer III)				
n	6	4	5	6
Total Length (μm)[a]	1435.4	1242.5	1301.6	1193.8
Number of Branches[b]	39.0	37.6	38.7	34.6
Nonpyramidal Cells (Layer IV)				
n	6	5	8	8
Total Length (μm)[b]	953.8	905.3	931.0	771.9
Number of Branches[b,c]	24.2	23.3	22.4	20.6

Note: Rats were killed after 30 days postweaning housing in a complex environment (EC) or in individual cages (IC), or after 30 days postweaning housing in an EC environment, followed by 30 days housing in an IC environment (EC/IC), or after 60 days housing in an IC environment (IC/IC). In the analysis of variance, "environment" refers to the initial housing environment and "age" to when they were killed. An age x environment interaction would be the strongest evidence for nonpersistence of the effects of the first 30 days of EC housing, and there were none. From Camel et al., 1986.
[a]Environment, $p < .001$. [b]Environment, $p < .05$. [c]Age, $p < .001$.

allel results in cerebellum have been described in mice and monkeys (Pysh and Weiss, 1979; Floeter and Greenough, 1979). Studies of persistence and of adult change are presently limited to rodents. Extensive parallels have been reported in visual deprivation studies (for review, see Greenough and Chang, 1985), but these effects are restricted to relatively brief developmental periods.

Cellular Mechanisms of Information Storage

The tendency of individual researchers to focus on one cellular mechanism in their studies of memory should not be misconstrued as an indication that there is but one cellular mechanism by which nervous systems store information. Not only is it quite likely that vastly different mechanisms will have arisen among lines of adaptive radiation as diverse as the various invertebrates and vertebrates that dominate the discussions in this volume, but even within vertebrates it seems quite likely that the differences between mammals and most other vertebrates may be equally vast, given the importance of memory to the survival of mammals. Most of the information that we, and many other mammals, use in daily life was probably not present in the zygotes from which we developed. The central role of memory of one's individual environment in the adult life of mammals, as well as the central role of experience in early mammalian brain/behavior development, may render it quite likely that multiple cellular mechanisms for the storage of information have evolved in this class. Thus the evidence for changes in the structural or metabolic characteristics in pre-existing synapses need not be incompatible with evidence for changes in synapse numbers. While they may be different aspects of the same process, they may also be separate cellular processes that work independently or in tandem in brain information storage.

How Do Synapse Number Differences Arise?

Our primary hypothesis regarding the differences in the number of synapses per neuron is that increases in synapses underlie the functional reorganization resulting from experience, or, loosely speaking, memory. However, we must keep in mind that terms like "memory" have been invented to explain certain subjectively or objectively evident phenomena and they need not map isomorphically to brain processes. Thus, synapse formation may be a cellular mechanism by which the nervous system is organized or reorganized, one use of which is in some aspects of the memory formation process. The important aspect of synapse number changes would be the changes in the *pattern* of connections, but except when gains and losses are equal, these are grossly measureable as changes in their numbers.

Alternative interpretations for these data exist and can be grouped into concepts of the behavioral/neural *causes* of these synapse number differences and concepts of the behavioral/neural *consequences* of the differences.

The Role of Neural Activity

A simple hypothesis regarding causation is that neurons, like muscles, grow with use and that the number of synapses is a positive function of neural activity. That is, action potentials are the necessary and sufficient cause of synapse formation (or reduced loss). Few would argue that nerve cell activity is unnecessary for synapse formation, but the obverse, that synapse formation automatically follows nerve cell activity, seems unlikely. Moreover, there is at least some evidence against this possibility. Cerebellar Purkinje cell spiny branchlets, for example, are altered by rearing conditions that do not affect the dendrites of the granule cells afferent to them

(Floeter and Greenough, 1979). In addition, certain patterns of electrical stimulation are associated with synapse formation in hippocampal subfield CA1, while identical amounts of stimulation in different patterns are without effect (Lee et al., 1980, 1981; Chang and Greenough, 1984).

An interesting class of experiments suggests that the same external stimuli in a sensory channel may have different consequences for recipient neurons, depending upon the use that is made of the input. For example, auditory cortex neurons increase in both spine density and dendritic field size in visually deprived or blinded animals (Ryugo et al., 1975; Burnstine et al., in preparation). Auditory stimulation is equivalent in deprived and sighted animals, so the presumed synapse number increase must involve some aspect of the animal's increased reliance upon audition in the absence of vision. That is, the brain's differential use of the same auditory information determines the information's effect on brain structure. This result also indicates that the relationship between neural activity and synapse formation is not a simple, positive function, but appears to depend upon the purpose of the neural activity.

The Functional Consequences of Synapse Differences

The functional consequences of experientially altered patterns of synaptic connections have been very well worked out for the developing visual system (LeVay et al., 1981). With regard to the effects of a complex environment, this area has historically received very little research attention, possibly because it is not clear what would constitute a demonstration of a specific effect of changes in synapse number. It has often been noted that EC and IC animals differ behaviorally, with EC rats better at certain types of learning tasks (Greenough, 1976), but the relationship of this

to specific synaptic connections or even to numbers of connections in general would be difficult to show. Alterations in gross electrophysiological characteristics, rarely examined, suggest functional alterations (Edwards et al., 1969), but again the relationship to specific morphological changes is unclear. A description of parallel changes in connectivity, demonstrated anatomically and physiologically to occur as a consequence of an experience, would be a good first step in this direction.

Higher-Level Causation: Behavioral Events Eliciting Dendritic Change

The occurrence of synapse number changes in adult brain in response to behavioral experience led to a somewhat different approach to dissection of the experiential sledgehammer's effects. The question, quite simply, was whether more restricted experiences that resulted in demonstrable learning/memory formation, such as training on mazes or other learning tasks, also resulted in alterations of our light microscopic predictor of synapse number change, dendritic field dimensions. The following findings have been reported:

1. Adult rats trained on a series of changing maze patterns over 3.5 weeks had, in the visual cortex, more extensive branching from the upper region of the apical dendrite of layer IV and V pyramidal neurons (Greenough et al., 1979).

2. When visual input from maze training was directed largely to one hemisphere through use of unilateral eye occlusion and split-brain surgery (in rats, over 90% of optic nerve axons decussate), the same region of layer V pyramids was more highly branched in the hemisphere opposite the nonoccluded eye (Chang and Greenough, 1982).

3. In adult rats that learned to reach into a tube for bits of food, apical dendrites of

layer V pyramids in motor–sensory fore-limb cortex opposite trained forelimbs were more highly branched than those opposite nontrained forelimbs (Greenough et al., 1985b). This and the preceding result indicate that the effects of training are not likely to have resulted from non-specific metabolic consequences of the training procedure.

4. Physical activity not associated with a great deal of opportunity for learning/memory (running in a wheel) may have marginally measureable effects on these and other brain parameters, but the effects are far smaller than those of either the modest physical demands of the training procedures above or exposure to a complex environment (Krech et al., 1960; Zolman and Morimoto, 1965; Black et al., 1986).

5. Stress and accompanying endocrine responses do not seem likely to bring about training effects, since freely circulating hormones should not produce lateralized effects in the trained animals. Similarly, there is no consistent relationship between adrenal weight, an indicator of stress response history, and changes in the visual cortex in EC animals (Wallace et al., 1986).

These studies indicate that mental activity, of the sort occurring in learning tasks, seems far more important than physical activity in producing differences in synapse number. Moreover, the effects occur in brain regions demonstrated in brain lesion studies to be critical to the performance of the learned tasks (Peterson and Devine, 1963) and, in the case of the reach training, demonstrated to be metabolically active during performance of the task (Fuchs et al., 1983). Alternative sources of these effects such as general metabolic or hormonal effects have been largely ruled out by the lateralization of brain effects of lateralized training as in the reaching experiment. These observations are compatible with the interpretation that the additional synapses mediate memory.

Evidence for Neuromodulatory Neurotransmitter Effects on Development of Morphological Patterns

A question that may have answers at several levels is what determines whether neural activity has consequences for neural structure. Norepinephrine (NE) was proposed to regulate developmental plasticity by Kasamatsu and Pettigrew (1976, 1979), who reported that 6-hydroxydopamine (6-OHDA) destruction of noradrenergic innervation will prevent the shift in ocular dominance of visual cortical neurons that normally occurs in young kittens when one eye is deprived of vision. More recent studies suggest that the blockage may result from toxic effects of 6-OHDA administration to local visual cortex, since the ocular dominance shift does occur following systemic or more general treatments or with other drugs that deplete NE (Bear and Daniels, 1983; Daw et al., 1984, 1985). Bear and Singer (1986) argued that the effect may have been the result of 6-OHDA's damage to both NE and acetylcholine projections to the visual cortex. Using a different procedure, they found that combined damage to both systems was necessary to block the ocular dominance shift. However, manipulations thought to be selective to NE systems affect adult learning (Gold, 1984), late developmental plasticity (Mirmiran and Uylings, 1983; O'Shea et al., 1983), and the morphology of developing neurons (Felten et al., 1982), suggesting a role for NE independent of acetylcholine. Electrophysiological studies on adult brain suggest that NE may truly act as a "modulator" of neural responsiveness, suppressing "spontaneous" activity while enhancing responses to inhibitory or excitatory afferent activity (Waterhouse and

Woodward, 1980; Foote et al., 1983). This sort of role suggests that NE might potentiate effects of afferent activity in organizing developing systems by increasing the influence of active afferents while suppressing less systematic background activity.

To examine this, Loeb and coworkers (1987) studied the effects of neonatal 6-OHDA administration on the somatosensory cortical barrels that receive input from the large facial whiskers, or mystacial vibrissae, in mice (Woolsey and Van der Loos, 1970). Each barrel receives primary input from one vibrissa, and deletion of the vibrissa and its follicle during early development disrupts the organization of the barrel (Van der Loos and Woolsey, 1973; Harris and Woolsey, 1981). Barrels can be visualized because the nerve cells constituting them are more concentrated in the walls of a cylindrically shaped structure in layer IV of the cortex and extend much of their dendrite into the central hollow, where most thalamic afferent fibers terminate (Killackey and Leshin, 1975; White, 1978; Woolsey et al., 1975). We reasoned that if NE acted in development to sharpen the distinction between vibrissal afferent and background activity, then its absence would lead to a more loosely organized barrel, since the activity-based influence of these afferents would be reduced. To test this, we examined Golgi-impregnated neurons in barrels of adult mice treated neonatally with 6-OHDA (which reduced somatosensory cortical NE by 96 to 98%). There was no difference in dendritic field *size* in the treated and the control groups, suggesting that NE does not promote synaptogenesis, at least of a type that is permanent. In addition, the location of cell bodies in barrels in 6-OHDA–treated mice was basically normal, as Lidov and Molliver (1982) previously reported. However, barrels had a less rigidly organized dendritic pattern in the treated mice than in controls. For example, dendrites of the spiny nonpyramidal cells that receive most specific thalamic afferent (i.e., whisker) input in barrels simply showed less "respect" for the boundary between the wall and the hollow with 6-OHDA treatment (Figure 4). Thus the results suggest that one role of NE in development is to promote afferent-dependent stabilization of synapses. The result is a somewhat fuzzy contribution to a very fuzzy area, but it strengthens the case for the importance of neuromodulatory substances in brain development and points towards one possible answer to the problem of why only *some* neural activity affects neural structure. Norepinephrine may be involved at the cellular level, indicating the significance of an event at the behavioral level (Gold, 1984; Sparenborg et al., 1986).

The Tissue Cost of a Synapse

Our comparisons of various tissue parameters in rats reared from weaning in a complex environment have provided additional insight into the reason for the relatively small effects of complex environment experience on synaptic density compared to the effects of the number of synapses per neuron—the additional synapses are accompanied by a considerable amount of additional tissue. (Preliminary evidence suggests that the additional tissue per synapse may be substantially less in rats first exposed to a complex environment as adults, because synaptic density was significantly greater in adults placed in a complex environment than in adults placed in isolation; Hwang and Greenough, 1986.) The major difference is in the density of neurons, which is lower in the EC rats. The density is lower, of course, because there is more tissue between them. This tissue has been found to consist of the following:

1. *Vasculature.* In a study just being completed, postweaning EC rats had about 21% more peak vascular capacity (the

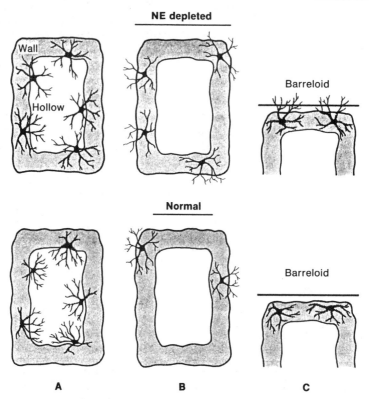

FIGURE 4. Schematic depiction of dendritic organization of spiny nonpyramidal neurons in mouse somatosensory cortex barrels of controls or after 6-hydroxydopamine (6-OHDA) treatment that destroys norepinephrine-containing nerve terminals. Normally, dendrites exhibit a strong tendency to grow into the hollows of the barrels whether the cell bodies from which they arise are in the hollow or the cell-rich walls. After 6-OHDA treatment, this tendency is reduced. The effect is overemphasized for clarity in this depiction. (From Loeb, Chang, and Greenough, in press. Copyright 1987, Elsevier Science Publishers.)

volume fraction of fully distended visual cortex blood vessels) per synapse, or about 42% more per neuron, than IC rats (Black et al., 1985; Sirevaag and Greenough, in press). This arose both from the addition of new capillaries and from the average enlargement of capillary diameter. The increase per synapse suggests greater metabolic demand from the EC rats' synapses, which is again an indication that their visual cortex is turned on or up relative to the cortex of the IC rats.

2. *Dendrites and axons.* The volume of den-

drite per neuron is about 12% greater in the upper visual cortex of EC rats than in that of IC rats (Sirevaag and Greenough, in press). This reflects the differences in the dendritic field measures in quantitative Golgi-impregnation studies. The volume of axonal material is also greater per neuron in EC rats (A.M. Sirevaag, unpublished data).

3. *Neuronal nuclei.* Neuronal nuclei are also larger in the visual cortex of EC rats (Diamond et al., 1966; Turner and Greenough, 1985). Presumably this increase is in-

volved in support of the additional synapses and associated neuronal tissue, as is evident in Betz and Purkinje cells.

4. *Glia.* As initially determined using thymidine incorporation by Altman and Das (1964) and described anatomically by Diamond et al. (1966), glial tissue is present in a higher proportion to neurons in EC rats. A recent study found, in upper visual cortex, about 34% more glial nuclear volume per neuron in ECs or about 14% more glial nuclear volume per synapse (Sirevaag and Greenough, in press). While significant details of the metabolic functions of glial tissue remain to be determined, these findings suggest some sort of general metabolic activation.

Two separate conclusions from these results should be emphasized: (1) because so many things can affect it, synaptic density alone may not provide an accurate assessment of changes or differences in synapse numbers associated with neuron populations or functional systems in the brain, and (2) the brain of a rat reared in a complex environment shows a pattern of metabolic activation that is evident not only in neuronal nuclear volume and the postsynaptic location of ribosomal aggregates, as discussed earlier, but also in the amount of vasculature and other supportive tissue per neuron. Since the EC situation probably falls far short of the natural environment in the amount of stimulation it provides, this might give some caution to those who use laboratory cage-reared animals in certain kinds of research, e.g., brain aging (Black et al., in press; for a discussion of the dependence of cerebral sex differences on postweaning housing conditions, see Juraska, 1986).

Are All Synapse Number Increases Equal?

Some authors have noted that a pattern of overproduction of both cells and synapses is commonly followed by selective retention of

some of them. The increase in the number of synapses per neuron in EC rats seems to be different, in that the synapses are not lost thereafter, at least over a time period equal to that in which they are gained. However, an overproduction process could be involved in the gain, and the overproduction could take either of the following forms:

1. The developmental overproduction process could simply continue, at reduced rates, in later life. Selective retention of a portion of these synapses that were appropriate in some way to the effects of experience would result in accumulation of synapses at higher rates in EC rats.

2. The formation of new synapses could occur specifically in response to experience, triggered in some way by its neurophysiological consequences.

Distinguishing between these alternatives would be greatly facilitated by a method of reliably detecting newly forming synapses, since only the second alternative predicts a difference *in the rate of synapse formation* between EC and IC rats. We and others had thought that the location of polyribosomal aggregates in postsynaptic elements might be an indicator of newly forming synapses (Steward, 1983; Hwang and Greenough, 1984; Greenough et al., 1985a; Steward and Falk, 1986), but the persistence of this change in rats removed from EC renders PRA location less likely as a marker (Hwang and Greenough, 1986). Thus, while there is evidence that regions of the brain can generate new synapses quite rapidly in response to neural activity (Lee et al., 1980, 1981; Chang and Greenough, 1984), the role of such processes in experience-related increases in synapse number remains uncertain.

Summary

The data discussed in this chapter make it clear that synaptogenesis is not merely a characteristic of early development but continues in the intact adult mammalian brain. Moreover, associated with the effects of experience on synaptogenesis is a set of effects upon the metabolic state of the brain, which are reflected in the persistent alteration of the positions of some polyribosomal aggregates, in increased vascularization, and in increased nuclear volume of glial cells and neurons. While the extent of some of these metabolically related changes may be less in adult than in developing brain, they do occur in adults. The cause of these changes at the behavioral level appears to be more the elicitation of mental than of physical activity, judging from studies in which these factors have been independently manipulated. The consequences of the changes, for brain activity or behavior, remain unconfirmed at this point, although it seems quite probable, given parallels between synaptic and functional organization in earlier brain development, that they mediate, at least in part, the incorporation of experientially based information into the functional organization of the brain.

While more specific behavioral and electrophysiological methods of eliciting plastic neural responses have been of major value, the sledgehammer approach of differential environmental complexity continues to serve an important role. The very massiveness of the effects of these environments upon selected brain regions such as the rodent or feline visual cortex speaks to the value of this manipulation in the detection of plastic neural phenomena at the cellular level. The reliability of these phenomena has aided us in sharpening our anatomical tools as well.

Important future tasks include more precise definition of the eliciting conditions for these phenomena at both the behavioral and the cellular levels. While we know quite a bit about the transformation from physical to cellular signals at the sensory surface, we know very little about either the translation of the resulting activity into structural change or the manner in which the nervous system determines which neural activity will result in structural change. Equally important is progress in determining specific functional roles of synapse addition and other morphologically detectable effects of behavioral experience. In addition to specifying the roles of new and modified connections, the meaning of the associated changes in metabolic indicators must be determined. The fact that early behavioral experience may irreversibly determine such factors as the quality of the brain's vascular supply, for example, has enormous potential importance if situations in which vascularization limits brain activity are likely to occur in adult life. The consequences for the aging brain, when vascular quality may be of critical importance (Smith, 1984), must also be carefully examined. While the complex environment paradigm will certainly not provide the answers to all of these questions, it has done a very good job of providing the questions.

CHAPTER TWENTY

Sensitive and Critical Periods in the Development of Sound Localization

E.I. KNUDSEN

Introduction

The nervous system has evolved to cope with an environment that is largely predictable. Therefore, much of the connectivity in the brain can be preprogrammed. However, not all aspects of the environment are certain: microhabitats and social conditions vary, as do details of the physical characteristics of individuals. To deal with such contingencies, the nervous system maintains a degree of plasticity, which allows experience to fine-tune otherwise predetermined pathways. Thus, the patterns of connectivity present in the adult nervous system reflect the interplay of innate and experiential influences.

A neural circuit that clearly exhibits the effects of this interplay is the auditory pathway that subserves sound localization (Knudsen, 1984a). Because the locations of acoustic stimuli do not map directly onto the sensory epithelium of the ear, the auditory system must localize sounds by associating sets of acoustic cues with appropriate locations in space. Localization cues, such as interaural

differences in timing and intensity, vary markedly with sound frequency and, more importantly, vary with the size and shape of the head and ears. As a consequence, to localize sounds accurately, the auditory system must form frequency-specific associations of cues with locations, and it must modify these associations according to the physical properties of the individual and adjust them as the animal grows.

This chapter discusses the results of behavioral and neurophysiological experiments on barn owls that reveal the effects of innate and experiential influences on the circuitry underlying sound localization. The results show that innate factors predispose the auditory system to form associations based on normal correspondences of cues with locations in space, while sensory experience exerts a powerful guiding influence that, within limits, directs the formation and maintenance of these associations. This interplay of genetic and experience-based forces leads to a neural circuit that is fine-tuned and customized for the individual.

The Experimental Model

The barn owl is a good experimental model for studying the development of sound localization for a number of reasons. First and foremost, the owl is an expert at sound localization, capable of localizing sounds to within 1 to 2° in both the horizontal (azimuthal) and the vertical (elevational) dimensions (Payne, 1971; Knudsen et al., 1979). Second, owls are altricial: the ear canals first open at about two weeks of age, the feathers that form the acoustically reflective surfaces of the external ears (called the facial ruff; Figure 1) grow in during the first seven weeks of life, and the skull doubles in size over this same period (Knudsen et al., 1984a). This long postnatal period of maturation facilitates observation and experimental manipulations during the development of sound localization. Finally, the barn owl's range of audible frequencies approximates that of humans, and the external ears

are stationary on the head, as in humans (Konishi, 1973).

An unusual feature of the barn owl's external ears is that they are asymmetrical (see Figure 1), which causes the left ear to be more sensitive to sounds originating from below and the right ear to be more sensitive to sounds emanating from above (Payne, 1971; Knudsen, 1980). As a result, barn owls experience an interaural intensity difference that depends upon the elevation of the source (interaural timing differences still vary with the azimuth of the source, as they do in other animals). Since barn owls use interaural intensity differences to determine the elevation of a stimulus, they exhibit a large elevational component in their sound localization errors in experiments involving ear occlusion (see Figures 3 and 4).

In the behavioral experiments described in this chapter, sound localization was evaluated by measuring the accuracy with which an owl

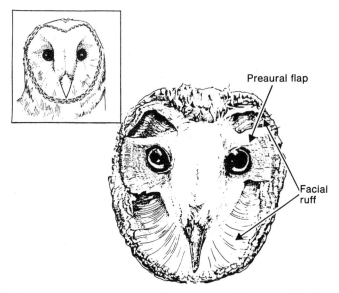

FIGURE 1. The sound collecting surface of the external ear is formed by the facial ruff. This illustration depicts the face of the owl as it would appear if the overlying facial disc feathers were removed. The owl's normal appearance is shown in the upper left. The openings to the ear canals are located immediately behind the preaural flaps.

turns its head to face a sound source. The experiments were conducted in a darkened, sound-attenuating chamber; acoustic stimuli consisted of repetitive noise bursts presented from random locations with a moveable speaker; head orientations were measured using an infrared beam and search coil system (Figure 2). The owl was given a food reward for responding with a quick head turn to the acoustic stimulus, whether the response was accurate or not. The owls received no feedback regarding the accuracy of their responses during the tests. Thus, their responses were shaped entirely by the experience they received in their home aviaries.

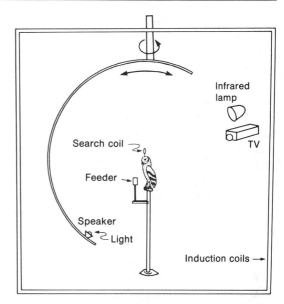

Definition of Terms

The discussion in this chapter requires the reader to distinguish the following concepts:

1. *Sensitive period:* a period in life when abnormal experience (including sensory deprivation) can cause abnormal development.
2. *Critical period:* a period in life when experience enables normal development.
3. *Accuracy:* the closeness of a judgment to the true value. For example, sound localization accuracy is the agreement between the judgment and the true location of the source.
4. *Precision:* the reproducibility of a judgment. For example, the precision of sound localization is the consistency or variability of the judgment.

The Role of Auditory Experience in Sound Localization

The ability of barn owls to localize sound sources reliably to within 1 to 2° in azimuth and elevation indicates that these animals associate particular values of acoustic cues with exactly the right locations in space. How does the nervous system form such exact asso-

FIGURE 2. Apparatus for measurement of sound localization. The owl is shown perched within an array of induction coils that generate magnetic fields. The amount of current induced in the search coil varies with the orientation of the search coil with respect to the magnetic fields. The owl is trained to orient to the target speaker and light. The targets move under remote control along the semicircular track, and the track rotates about a vertical axis. The owl is observed with an infrared closed-circuit video system during experiments. The apparatus is located in an anechoic chamber.

ciations? The answer appears to be that crude associations are established a priori by innate mechanisms and that these associations are then shaped and modified by early sensory experience (Clements and Kelly, 1978; Olmstead and Villablanca, 1980; Knudsen, 1984a, 1985; Knudsen and Knudsen, 1986a). The "experience" that guides the shaping of these associations is the agreement of auditory spatial information with visual spatial information (Knudsen and Knudsen, 1985a,b). A consistent disagreement between auditory and visual localization leads to a compensatory change in the auditory system's interpretation of localization cues. In the course of normal development, an auditory-visual mismatch could arise as the head and ears grow, because growth changes the correspondence of cue values with locations in space. Under experimental conditions, a more severe mismatch can be induced either by plugging one ear (monaural occlusion), which alters auditory localization cues, or by mounting deviating prisms over the eyes, which shifts the visual world to one side.

The influence of experience on sound localization was first demonstrated using chronic monaural occlusion (Knudsen et al., 1982, 1984a,b). An owl that is monaurally occluded mislocalizes sound sources toward the side of the open ear, i.e., to the left and down when the right is plugged, and to the right and up when the left ear is plugged (recall that the owl's ears are asymmetrical). If an owl is young (less than about 60 days old) when one ear is occluded, it learns to interpret correctly the abnormal cues induced by the earplug and gradually recovers accurate localization over a period of weeks (Figure 3). When the earplug is removed, the owl mislocalizes sound sources toward the opposite side, and the degree of subsequent adjustment depends on the age at the time of earplug removal (see later in this section). However, if the owl is older than 60 days when the ear is occluded, no such adjustment of sound localization occurs: the animal maintains a localization error for months while the ear is occluded, and when the earplug is removed, it localizes sounds immediately with normal, or near-normal, accuracy. Thus, monaural occlusion causes the auditory system to alter its interpretation of localization cues only during a restricted sensitive period that ends early in life.

When this sensitive period was first described, it was noted that it coincides closely with the period of growth of the head and ears (Knudsen et al. 1984a). This coincidence suggested the possibility that the continuously

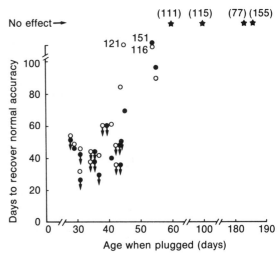

FIGURE 3. The sensitive period for adjusting sound localization accuracy revealed by monaural occlusion. The number of days required for a barn owl to regain normal localization accuracy (error less than 3°) is plotted against the age of the owl at the time the ear was plugged. Filled circles represent the day that normal accuracy in azimuth was attained; open circles signify the day that normal accuracy in elevation was attained. A downward-pointing arrow indicates that the owl exhibited normal accuracy in that dimension on the first day it was tested. A star signifies that normal accuracy was never achieved through the last day of testing, indicated by the number above (in fact, these birds made no adjustment of localization accuracy at all). All birds plugged before 60 days of age adjusted sound localization, whereas no bird plugged after 60 days of age adjusted sound localization. (From Knudsen et al., 1984a.)

changing cue values that accompany growth keep the sensitive period open; once growth stops and cues stabilize at adult values, the sensitive period ends. This experience-based hypothesis for the regulation of the sensitive period was subsequently tested and found to be false (Knudsen and Knudsen, 1986a). The experiment involved exposing owls to constantly changing cue values by alternating monaural occlusion (the earplug was switched to the opposite ear every other day) from the time the ear canals opened until past the normal end of the sensitive period (Figure 4). Alternating monaural occlusion was followed immediately by chronic monaural occlusion to test the ability of the owls to adjust to abnormal cue values. None of the owls did, indicating that the sensitive period had ended due to an age-dependent, rather than an experience-dependent, process.

A remarkable feature of these birds was that, despite being reared with abnormal cues that changed drastically every other day, the birds exhibited approximately normal local-ization accuracy once all earplugs were removed (see Figure 4) (Knudsen and Knudsen, 1986a). The associations upon which these localizations were based cannot have come about by experience, since the animals never experienced normal correspondences of cues with locations until the earplug was removed. Hence, they must have been established by innate mechanisms in anticipation of normal cue-location correspondences. The primary effect of alternating monaural occlusion was to decrease the precision of sound localization relative to that exhibited by animals raised under normal conditions, which suggests that

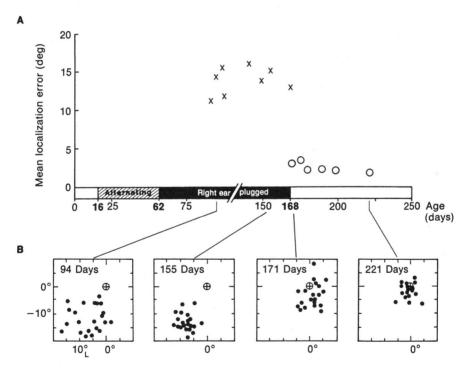

FIGURE 4. The effect of alternating monaural occlusion followed by chronic monaural occlusion on the development of sound localization accuracy by a barn owl. (A) Mean localization error is plotted against age. The auditory history of the bird is indicated along the abscissa. The period of alternating monaural occlusion, during which the earplug was switched every other day, is represented by diagonal hatching. An X indicates localization error measured with one ear plugged; an open circle indicates the measurement made with both ears open. (B) The raw data from particular days. Note that after the right earplug was removed, the owl's error was small, although its precision was poor. (From Knudsen and Knudsen, 1986a.)

the role of experience is to fine-tune a basically innate localization circuit.

Further evidence of the auditory system's innate predisposition to form cue-location associations as they correspond normally is found in the temporal discrepancy between the sensitive period for monaural occlusion, which ends at about 60 days of age, and the critical period for the development of sound localization accuracy, which ends at about 200 days of age (Knudsen et al., 1984b). The duration of the critical period was demonstrated by restoring normal hearing to animals of various ages that had been raised with one ear occluded since early in the sensitive period, and monitoring their ability to adjust sound localization accuracy (Figure 5). Animals younger than 150 days old when the earplug was removed adjusted within a few weeks. Beyond this age, rates of adjustment slowed, and birds remaining monaurally occluded after 200 days of age never recovered accurate localization following earplug removal. The

inability of owls more than 200 days old to recover accurate localization when exposed to normal cue values indicates that a critical period for adjusting cue-location associations had ended. The long period in development between 60 and 200 days of age, when animals can no longer adjust to the imposition of abnormal cue values yet still can adjust to normal cue values, indicates an asymmetry in the plasticity of the localization mechanism that favors associations based on normal cue-location correlations.

The Role of Visual Experience in Sound Localization

Sensitive and critical periods, similar to those demonstrated by monaural occlusion, can also be demonstrated by raising birds with deviating prisms (Figure 6) (Knudsen and Knudsen, 1985b). An owl that views the world through deviating prisms from the day its eyes open (at about ten days of age) adjusts sound

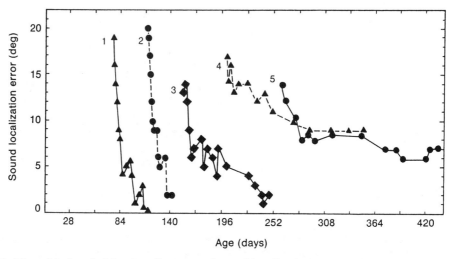

FIGURE 5. The critical period for the adjustment of sound localization accuracy in barn owls, as revealed by removal of unilateral earplugs. Data are shown for five representative owls. All animals had one ear occluded at 20 to 40 days of age, i.e., before the end of the sensitive period (see Figure 1). The rate of adjustment following earplug removal slows with age, and adjustment virtually ceases beyond 250 days of age. (After Knudsen et al., 1984b.)

FIGURE 6. A baby barn owl being raised with binocular deviating prisms.

ization circuit, although substantial, is restricted.

The plasticity of the system also decreases with age (see Figure 7) (Knudsen and Knudsen, 1986b). Animals that are younger than 20 days of age when the prisms are attached shift sound localization by up to 20°, whereas birds 60 to 80 days old shift localization by no more than 10°. The maximum shift that adult owls are capable of making has yet to be determined, although 100-day-old animals shift by less than 5° when subjected to 10° visual deviations. Preliminary data indicate a sensitive period characterized by a decreasing capacity to associate normal auditory cue values with abnormal locations in space. This highly plastic period ends by 100 days of age and perhaps much sooner.

In contrast, the ability of the auditory sys-

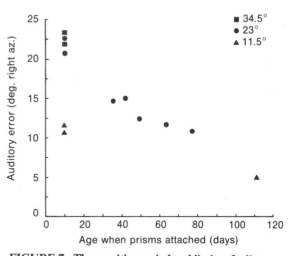

FIGURE 7. The sensitive period and limits of adjustment of sound localization as revealed by prism-rearing. Each data point represents the mean sound localization error exhibited by one barn owl after wearing rightward-deviating prisms for from 60 to 80 days. The different symbols represent different prismatic deviations as indicated. Notice that the localization error induced by wearing 34.5° prisms is no greater than that induced by 23° prisms.

localization to match the visual deviation caused by the prisms. For example, an animal raised with prisms that shift vision by 10° to the right responds to a sound by facing 10° to the right of the source. The maximum shift in sound localization that can be induced by prisms appears to be about 20°; animals reared with greater optical deviations adjust localization by up to 20°, then they stop adjusting (Figure 7) (Knudsen and Knudsen, 1986b). Thus, the plasticity of the sound localization to match the visual deviation tem to reassign cue values to *normal* locations persists until animals are much older (Figure 8) (Knudsen and Knudsen, 1986b). Prism-reared owls up to 200 days of age adjust sound localization within weeks after removal of deviating prisms. If the prisms remain in place until the owls are older than about 200 days, localization errors become permanent, indicating that the critical period for the development of sound localization accuracy has ended.

The effect of prism-rearing on sound localization demonstrates that vision guides the development of auditory localization: when confronted with a constant misalignment between visual and auditory spatial information, the brain uses visual space to adjust its interpretation of auditory localization cues. (This makes sense, since vision is a more precise and reliable source of spatial information.) Monaural occlusion and deviating prisms both introduce visual-auditory misalignments. However, the nature of the misalignment is quite different in the two cases. Monaural occlusion creates a misalignment by altering auditory cues: signal frequencies at one ear are differentially attenuated and time shifted so that the auditory system must reinterpret the spatial significance of interaural difference cues in a frequency-specific manner (Knudsen et al., 1984a). Moreover, because interaural intensity varies irregularly across space, a given attenuation causes shifts that are much larger for some regions of space than for others, i.e., auditory space is distorted as well as shifted (Knudsen, 1985). In contrast, deviating prisms cause a relatively coherent shift of visual space. In this case, the spatial significance of all auditory cues must be altered by a constant number of degrees in a particular direction. Considering the very different effects of monaural occlusion and the prismatic deviation, it is remarkable that the sensitive and critical periods revealed by these manipulations are so similar.

The corrective force guiding the adjustment of sound localization results from the brain's attempt to relate auditory and visual information originating from a common stimulus object. It is interesting to ponder which cues the brain uses to identify sensory information belonging to a common source.

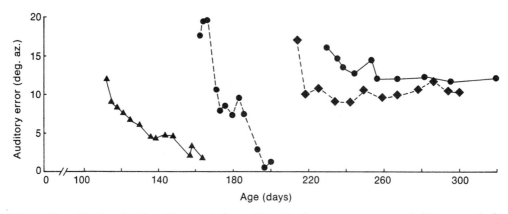

FIGURE 8. The critical period for adjustment of sound localization accuracy as revealed by removal of prisms following prism-rearing. These data are mean sound localization errors from four barn owls, each raised from 10 days of age wearing 23° rightward-deviating prisms. In each case, the data begin on the day the prisms were removed.

One important cue is likely to be synchrony across modalities: in this case, visual stimuli that move in synchrony with modulations of an acoustic signal. Whatever the cues are, the nervous system weights them far more heavily than it does auditory spatial cues, since large visual-auditory discrepancies (in excess of 30°) result in the adjustment of sound localization. One might expect that, when exposed to large intermodal spatial discrepancies, the nervous system would reject the possibility that the visual and auditory information originated from a common source, and would not change sound localization at all. This is not the case. Even when the visual-auditory spatial discrepancy is beyond the range of plasticity of the sound localization mechanism (see Figure 7), cues that identify sensory information with a common source dominate over cue-location associations that underlie sound localization.

Not only does visual spatial information guide the fine-tuning of sound localization, but vision seems to be the only source of information that provides this guidance under these conditions. Owls raised with vision blocked by opaque occluders develop sound localization that is significantly less precise than that exhibited by sighted owls (Knudsen and Knudsen, 1985b). In addition, animals that exhibit sound localization errors do not correct their errors when they are prevented from seeing (Knudsen and Knudsen, 1985b). Thus, potential sources of spatial information such as kinesthesia, somatic sensation, and auditory spatial cues resulting from head movements made in sound fields are not used to adjust sound localization.

The Spatial Tuning of Auditory Neurons in the Optic Tectum

The previously discussed behavioral experiments do not address the neurophysiological basis of sound localization, how the cir-cuitry is shaped by experience, and what restricts its plasticity by the end of the sensitive and critical periods. As a first step in addressing these issues, we have studied multimodal units in the owl's optic tectum, which form a neurophysiological representation of the auditory system's interpretation of localization cues. Units in the optic tectum respond to auditory stimuli and are sharply tuned for sound source location (Knudsen, 1982, 1984b). These same units also respond to visual stimuli, and the visual and auditory spatial tuning of most units are closely aligned. The alignment of visual and auditory spatial tuning means that tectal units are tuned for auditory cue values that result from sound sources located in their visual receptive fields. Thus the bimodal spatial tuning of these units indicates the correlation of auditory cue values with locations in visual space. Based on the results of experiments on sound localization behavior, one might expect the alignment of auditory with visual spatial tuning in the tectum to be altered by abnormal experience during the sensitive period and to be unaffected by experience after the critical period ends.

In experiments that tested these predictions, the auditory and visual spatial tuning of units in the optic tectum of the barn owl was measured in a darkened sound-attenuating chamber that was equipped with a remote-controlled, moveable loudspeaker. The speaker moved on an imaginary sphere centered on the head of the anesthetized owl. Because neither the eyes nor the ears of the owl move relative to the head, by fixing the head, the eyes and ears were held in stationary and normal positions. Auditory spatial tuning was tested with repetitive noise bursts presented from the speaker; visual spatial tuning was plotted with bars and spots of light projected onto a translucent hemisphere that was brought into the sound chamber for visual testing and then removed.

When units in the optic tectum are stimulated with broad-band sounds, they reveal a tuning for source location that is rarely observed among units in the classical auditory pathway (Knudsen, 1982, 1984b). The units are excited only when the source is located within a restricted receptive field and are inhibited by sounds from outside of this receptive field (Figure 9). Within the receptive field is a small "best area." Only when a sound source is located in a unit's best area does the unit respond maximally. Thus, a unit's best area represents the location from which sounds give rise to the optimal combination of localization cue values for the unit.

The auditory and visual spatial tuning of units varies systematically across the tectum, forming maps of auditory and visual space (Knudsen, 1982). Units at the rostral end of the tectum respond to frontal or slightly ipsilateral locations, units at the caudal end respond to far contralateral locations, units positioned dorsally respond to locations above the ani-

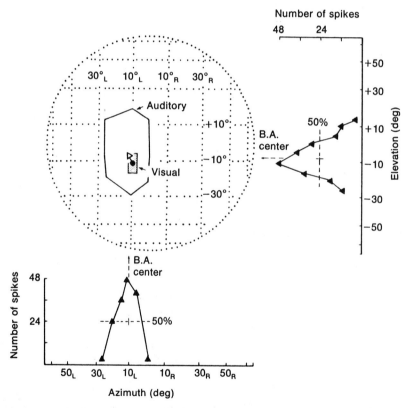

FIGURE 9. The auditory spatial tuning and visual receptive field of a single unit from the optic tectum of the barn owl are plotted on a dotted globe of space. Below and to the right of the globe are the responses of the unit to eight presentations at each location of a noise burst 20 dB above the unit's threshold. Locations from which the noise burst elicited 50% of the maximum response were used to define the limits of the best area. The geometric center of these limits was the best area (B.A.) center, indicated by the open triangle in the field plot. The filled circle indicates the geometric center of the visual receptive field (stippled region). (Data are from Knudsen, 1985.)

mal, and units positioned ventrally respond to locations below the animal. In the rostral half of the tectum, where frontal space is represented, the alignment of auditory best areas with visual receptive fields is quite good: the median alignment error is 2° in azimuth and 4° in elevation. Toward the caudal end, the accuracy and consistency of alignment decreases. Thus, in the rostral half of the tectum, the location of a unit's visual receptive field reliably predicts the location of its auditory best area.

To determine whether the merging of auditory with visual spatial information in the tectum is guided by experience in the same way that sound localization behavior is, we measured the alignment of auditory and visual spatial tuning of units in animals that had been raised with one ear occluded (Knudsen, 1983, 1985). The immediate effect of monaural occlusion is to shift auditory spatial tuning toward the side of the plugged ear or, if the effect of the occlusion is strong enough, even to render the unit unresponsive to sound stimuli. In these experiments, we tested whether the auditory system would adjust the tuning of tectal units for auditory localization cues in such a way that accurate alignment of auditory and visual spatial tuning is re-established.

Results from an animal that was monaurally occluded as an adult are shown in Figure 10. The owl had its left ear occluded at 250 days of age, and units were recorded beginning 335 days later (at 585 days of age). With the earplug in place, auditory best areas were consistently located to the left and below visual receptive fields. Moreover, units with visual receptive fields located to the left and down did not respond to sounds at all, which is expected if these units were tuned to interaural intensity differences favoring the left ear that could not occur because the left ear was plugged. Thus, the tuning of these units to auditory localization cues seemed not to have been altered by more than ten months of ex-

posure to abnormal cues. This interpretation was confirmed by data gathered after the earplug was removed. Auditory best areas were found to be aligned with visual receptive fields, and units with visual receptive fields located to the left and down responded strongly to acoustic stimuli. Thus, long-term exposure of an animal to monaural occlusion did not alter the tuning of units to auditory cues when the occlusion was introduced after the animal had reached adulthood.

In contrast, adaptive alterations in auditory spatial tuning did occur when animals were monaurally occluded at an early age. Figure 11 shows data from an owl that had its left ear plugged at 41 days of age, i.e., before the end of the sensitive period. When measured 294 days later (at 335 days of age), auditory best areas were aligned with visual receptive fields as long as the earplug was in place. After the earplug was removed, at 342 days of age, auditory best areas were located to the right and above visual receptive fields. The direction of the auditory-visual misalignment following earplug removal depended on which ear had been plugged during early life: in owls raised with the right ear plugged, auditory best areas were located to the left and below visual receptive fields. Apparently, the tuning of units to auditory localization cues was altered by experience in such a way that auditory and visual spatial tuning became realigned while the earplug was in place. However, this plasticity was evident only in animals that were monaurally occluded before the end of the sensitive period.

The change in auditory spatial tuning that resulted from early monaural occlusion was not permanent if the earplug was removed before the end of the critical period. The data in Figure 12 are from an owl that was raised with the right ear occluded. At 170 days of age, the earplug was removed and the spatial tuning of tectal units was measured. On the day the earplug was removed, the median audi-

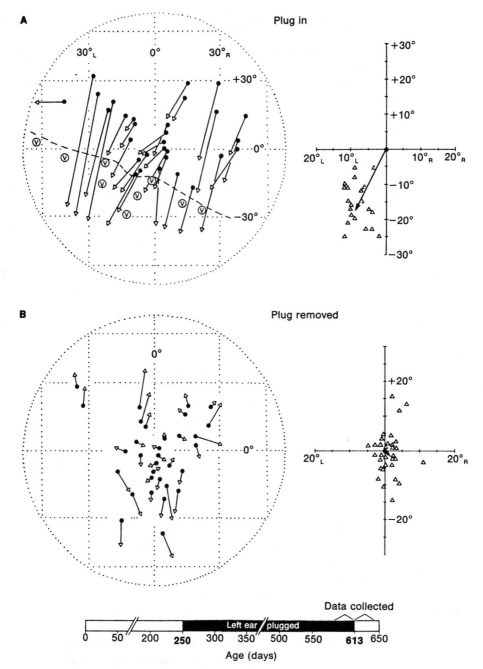

FIGURE 10. Alignment of auditory best areas with visual receptive fields recorded before (A) and after (B) ear-plug removal for an owl that was monaurally occluded as an adult. On the left, centers of best areas (open triangles) and centers of visual receptive fields (filled circles) for individual units are connected by lines. The circled Vs in (A) indicate the visual receptive fields of units in the intermediate tectal layers that did not respond to auditory stimuli. All units with visual receptive fields located below the dashed line (n = 11) could not be driven with acoustic stimuli. On the right, the centers of auditory best areas of all bimodal units located in the rostral tectum are plotted relative to the centers of their visual receptive fields. The vectors represent the median auditory-visual misalignments of these samples. The auditory history of the owl and the periods over which the data were collected are indicated at the bottom. (Data are from Knudsen, 1985.)

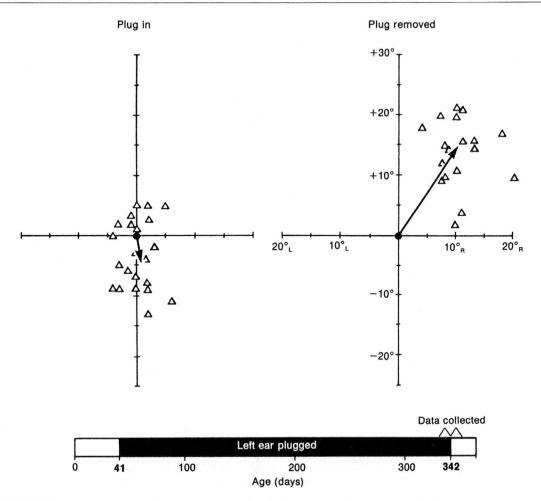

FIGURE 11. Alignment of auditory best areas with visual receptive fields recorded before and after earplug removal for an owl that was monaurally occluded at 41 days old. The centers of auditory best areas (open triangles) of all bimodal units located in the rostral tectum are plotted relative to the centers of their visual receptive fields. The vectors represent the median auditory-visual misalignment of these samples. The auditory history of this owl and the period over which the data were collected are indicated at the bottom. (Data are from Knudsen, 1985.)

tory-visual misalignment was left 6.5° and down 10.8° (*n* = 7). Six days later the median misalignment had decreased to left 4.0° and down 5.3° (*n* = 9), and after 19 days the auditory-visual misalignment was left 2.0° and up 2.0° (*n* = 9). Because auditory spatial tuning was changing over time, estimates of

auditory-visual misalignment in this animal had to be based on the data that could be collected in a single day. The unit-to-unit variability in misalignment was great, especially for units tuned to different locations, and the sample sizes were small. Thus, the most convincing evidence of a progressive

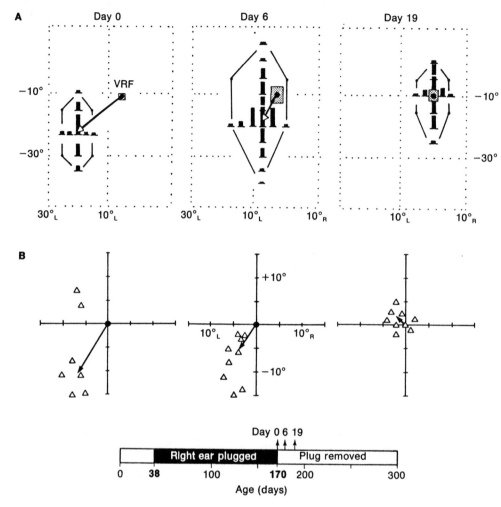

FIGURE 12. Progressive realignment of auditory spatial tuning with visual receptive fields over time in an owl that was unplugged at 170 days of age. **(A)** The auditory and visual receptive fields of three units are plotted. The units had similarly located visual receptive fields but they were recorded on successively later days following earplug removal, as indicated above each plot. The open triangles represent auditory best area centers; the filled circles represent visual receptive field centers. Hence, the lines indicate the direction and magnitude of the auditory-visual misalignments. The strength of the units' responses to eight noise bursts at 20 dB above unit threshold is represented by the heights of the bars, the bases of which indicate the locations of the stimuli. **(B)** The locations of auditory best area centers are plotted relative to the locations of visual receptive field (VRF) centers for all bimodal units in the rostral tectum recorded on the day indicated at the top in **(A)**. The vectors represent the median auditory-visual misalignments for these samples. The owl's auditory history and the days on which these measurements were made are indicated at the bottom. (Data are from Knudsen, 1985.)

realignment of auditory with visual spatial tuning was the decrease in the auditory-visual misalignment of units recorded from approximately the same site in the tectum in sequential experiments (see Figure 12A).

Auditory-visual misalignments in spatial tuning that appeared following the removal of an earplug seemed to be permanent in animals that were monaurally occluded until past the end of the behavioral critical period. The data in Figure 13 were recorded from an animal that was monaurally occluded from 43

to 202 days of age. After earplug removal, the owl was returned to the aviary, where it remained for an additional 126 days. Then it was trained for behavioral measurements of sound localization and was prepared for neurophysiological recording from the tectum. The neurophysiological data, gathered between four and five months after earplug removal, demonstrate a large alignment error between auditory best areas and visual receptive fields; the median error was left 10° and down 9°. The direction and magnitude of the alignment

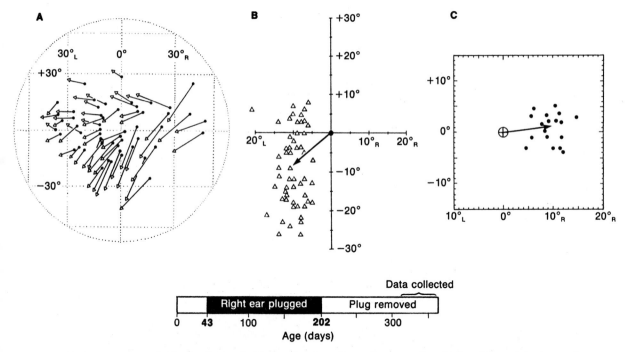

FIGURE 13. Physiological and behavioral effects of early prolonged monaural occlusion. These data were collected beginning 4 months after earplug removal from an owl that had its right ear occluded from 43 to 202 days of age, as indicated on the bottom. (A) Auditory best area centers (open triangles) and visual receptive field centers (filled circles) are plotted on coordinates of space. The lines connect data from individual units and indicate the direction and magnitude of the misalignments. Not all data are shown for units in the rostral tectum, because the overlap of lines and points would have obscured the data. (B) Best area centers of all bimodal units in the rostral tectum are plotted relative to the centers of their visual receptive fields. The vector indicates the median auditory-visual misalignment for this sample. (C) The orienting responses of this animal to noise bursts are plotted relative to the location of the speaker. The vector represents the mean localization error based on this sample. (Data are from Knudsen, 1985.)

errors varied across the tecta: the smallest errors were found among units with visual receptive fields directly in front or to the left of the animal, whereas much larger errors were found among units with visual receptive fields to the animal's right (on the side of the ear that had been occluded). This variation in the magnitude and direction of the alignment errors reflects the fact that auditory cues can vary irregularly and in a complicated pattern as a function of sound source location, and that an earplug alters different auditory cues differently. For example, the variation in elevation tuning, which presumably results from the tuning of units to interaural spectrum differences (Knudsen, 1980; Moiseff and Konishi, 1981), followed a pattern that probably reflects the spatial pattern of interaural spectral differences for that individual.

A comparison of the physiological data with the behavioral data collected from this owl reveals that auditory-visual misalignments measured physiologically in the tectum correspond roughly to sound localization errors measured behaviorally (see Figure 13). However, the variability in the behavioral responses was less than the variability in the alignment of auditory best areas with visual receptive fields. This difference between behavior and physiology is probably due to the nature of the behavioral test, which required the owl to orient its head in the direction of an ongoing acoustic stimulus. Because the owl's final head orientation was based on its judgment of a sound source located almost directly in front of its face, the appropriate comparison is between the behavioral responses and the auditory-visual misalignments of units with receptive fields located directly in front of the bird. With this restriction, the match between the physiological and behavioral data is good.

Summary

The results described in this chapter demonstrate interactions of innate and ex-periential influences on the development of sound localization in owls. Early in life, innate mechanisms form neural circuits that anticipate a normal correspondence of auditory cues with locations in space. Within a sensitive period, this connectivity can be modified substantially by experience-based forces. The shaping forces are directed by visual spatial information that the nervous system uses to calibrate its interpretation of auditory cues. Under normal circumstances, experience confirms innate associations and refines them, thereby improving the precision of sound localization. However, when a consistent mismatch between auditory and visual spatial information occurs, either because of altered auditory cues or altered vision, auditory cues are reinterpreted according to the spatial information provided by the visual system. This shaping force can be generated by extremely large discrepancies in auditory-visual localization, although adaptive modification of sound localization seems to be limited (for unknown reasons) to about 20°. Despite the auditory system's strong predisposition to form normal cue-location associations, once abnormal associations are established, they remain stable until they are actively reshaped by a new discrepancy between auditory and visual localization.

As development progresses, the extent to which abnormal experience can alter sound localization decreases. By the end of the sensitive period, shortly after the owl reaches adult size, the ability of the auditory system to form abnormal associations based on experience is relatively restricted. However, the critical period for developing accurate sound localization remains open: after the end of the sensitive period, but before the end of the critical period, adjustments of cue-location associations are constrained to those that are consistent with a normal correspondence of cues with locations, associations to which the auditory system is innately predisposed. Once animals reach about 200 days of age (approximately the age at which barn owls reach sexual maturity), the neural circuitry subserving sound localization has crystallized and there is little capacity for further experience-dependent modification. At this stage, the development of sound localization is complete.

The sensitive and critical periods that affect the development of sound localization behavior also affect the development of the auditory spatial tuning of units in the optic tectum. Auditory spatial tuning can be modified by monaural occlusion only if monaural occlusion is imposed before the end of the sensitive period, at about 60 days of age, and subsequent recovery of auditory-visual alignment following earplug removal occurs only if the earplug is removed before the end of the critical period, at about 200 days of age.

The striking parallels between the experience-dependent changes in sound localization behavior and those in the spatial tuning of auditory units in the tectum indicate that the same, or equivalent, changes affect both processes. However, the data do not address the question of where in the auditory pathway these changes take place, except that they must occur at or before the optic tectum. In fact, anatomical experiments have demonstrated that the alterations in spatial tuning occur at or before the level of the inferior colliculus (Knudsen, 1985). The data do show, however, that during sensitive and critical periods in development, sensory experience can modify the properties of auditory units in such a way as to customize them to the sensory environment of the individual.

Acknowledgments

The symposium at which these papers were originally presented was made possible through the generosity of the following:

The University of Michigan
    Center for Human Growth and Development
    Consortium for Developmental and Reproductive Biology
    Department of Anatomy and Cell Biology
    Developmental Biology Training Grant
    LS&A Enrichment Fund
    The Medical School
    Mental Health Research Institute
    Office of the Vice President for Research
Mager Scientific
The Upjohn Company

Chapter 3: D.M. FEKETE, P. FERRETTI, H. GORDON, AND J.P. BROCKES
We thank Frank Wilson and Sue Hall for technical assistance, and Brian B. Boycott for his helpful comments on the manuscript. We particularly thank Chris Kintner, who originally isolated the 22/18 antibody.

    D.M.F. is a Smith-Kline and French Fellow of the Life Sciences Research Foundation.

Chapter 4: D. CHIKARAISHI
I would like to thank my colleagues Murray Brilliant, Brenda Fung, Paul Patterson, and Irene Pech for allowing me to describe their experiments. I would also like to thank Barbara D'Angelo for her excellent help in preparing the manuscript.

    This work is supported by NIH grants GM33991 and NS22675.

Chapter 6: R.D. McKINNON, P. DANIELSON, M.A. BROW, M. GODBOUT, J.B. WATSON, AND J.G. SUTCLIFFE
We acknowledge our collaborators on these studies, Ian Brown, Thomas Shinnick, Mary Kiel, Robert Milner, and Joel Gottesfeld; and we thank Linda Elder for preparing the manuscript.

    These studies were supported in part by NIH grants GM32355 and NS22111 to

J.G.S.; a National Cancer Institute of Canada Research Fellowship to R.D.McK.; a Medical Research Council of Canada Fellowship to M.G.; and a National Huntington's Disease Association Fellowship to J.B.W.

This is Publication No. 4453-MB from The Research Institute of Scripps Clinic.

Chapter 8: M. WESTERFIELD AND J.S. EISEN
We wish to thank Paul Myers and James McMurray for help with some of the experiments, Charles Kimmel and Carla Shatz for comments on the manuscript, and Dave Brumbley and Harrison Howard for technical assistance.

This work is supported by NIH grants NS21132 and NS17963, the Muscular Dystrophy Association, and the Medical Research Foundation of Oregon. M.W. was an Alfred P. Sloan Fellow.

Chapter 9: L. LANDMESSER
I would like to thank Lisa Dahm and Hideaki Tanaka, who participated in some of the unpublished experiments described here, Liviu Cupceancu for his technical assistance, and Sara Putnam for typing the manuscript.

Supported by NIH grant NS19640.

Chapter 10: K. KALIL
I thank Cheryl Adams for skilled technical assistance and Carolyn Norris for many helpful discussions.

Research from my laboratory is supported by NSF grant BNS-8311517 and NIH grant NS14428, a Javits Neuroscience Investigator Award.

Chapter 11: M.E. HATTEN AND J.C. EDMONDSON
Carol Mason, Ronald Liem, and Michael Shelanski collaborated on some of the experiments reported here, and we are grateful to them for their comments on the manuscript.

Supported by NIH grants NS15429 and 21907 to M.E.H.

Chapter 12: J.W. LICHTMAN, M.M. RICH, AND R.S. WILKINSON
Our research is supported by grants from the NIH, MDA, and a Research Development Award from the McKnight Foundation to J.W.L.

We thank Sue Eads for secretarial assistance.

Chapter 13: E. FRANK, C. SMITH, AND B. MENDELSON
It is a pleasure to acknowledge a short but fruitful discussion with Douglas Fambrough on possible mechanisms of peripheral specification.

This work was supported by grants from the NIH and NSF to E.F.

Chapter 14: D. WEISBLAT
Support by Basil O'Connor Starter Research Grant 5-483 from the March of Dimes Birth Defects Foundation and by NSF grant PCM-8409785 is gratefully acknowledged. Most of the previously published work summarized here was performed in the laboratory of Gunther Stent.

I thank David Wake for helpful discussions about amphibians.

Chapter 15: J. WESTON, K.S. VOGEL, AND M.F. MARUSICH

We are grateful to Judith Eisen, Charles Kimmel, Kathleen Morrison-Graham, and Sherry Rogers for criticism of the manuscript and to Diane Nelson for technical assistance.

The research described in this chapter has been supported by PHS grant DE-04315, NSF grant PCM-82118899, and grants from the National Neurofibromatosis Foundation and the Dysautonomia Foundation, Inc.

Chapter 16: K. HERRUP AND M.W. VOGEL

The research performed by the authors and described in this chapter was supported by NIH grants NS20591, NS18381 (K.H.), and NS507679 (M.W.V.) and March of Dimes grant 1-763 (K.H.).

Chapter 18: B.S. MCEWEN, V.N. LUINE, AND C.T. FISCHETTE

Research in our laboratory related to this article is supported by NIH grants NS07080 (B.McE.) and HD12011 (V.N.L.) and by grant GAPS 8532 from the Rockefeller Foundation for research in reproductive biology.

We thank Inna Perlin for editorial assistance.

Chapter 19: W.T. GREENOUGH

Preparation of this chapter and research not yet published was supported by NIMH grants 35321 and 40631, ONR N00014-85-K-0587, the Retirement Research Foundation, the System Development Foundation, and the University of Illinois Senior Scholars Program.

I thank J. Black and J. M. Juraska for comments.

Chapter 20: E.I. KNUDSEN

Preparation of this chapter was supported by grants from the March of Dimes (1-863), the National Institutes of Health (R01 NS16099-06), and a Neuroscience Development Award from the McKnight Foundation.

Bibliography

THE NUMBERS THAT FOLLOW EACH ENTRY
IDENTIFY THE CHAPTER(S) IN WHICH THE REFERENCE IS CITED.

Abercrombie, M., Heaysman, J. M. and Pegrum, S. M. 1970. The locomotion of fibroblasts in culture. I. Movements of the leading edges. *Exp. Cell Res. 59*: 393-398. (11)

Aebersold, H., Creutzfeld, O. D., Kuhnt, U. and Sanides, D. 1981. Representation of the visual field in the optic tract and the optic chiasma of the cat. *Exp. Brain Res. 42*: 127-145. (8)

Akam, M. 1985. Developmental genes: Mediators of cell communication? *Nature 319*: 447-448. (14)

Allan, I. J., Newgreen, D. F. 1980. The origin and differentiation of enteric neurons of the intestine of the fowl embryo. *Amer. J. Anat. 157*: 137-154. (15)

Allen, L. S., Hines, M., Shryne, J. E. and Gorski, R. A. 1986. Two sexually dimorphic cell groups in the human brain. *Endocrinology* (suppl.) *118*: 633. (17)

Allen, R. D., Allen, N. S. and Travis, J. L. 1981. Video-enhanced contrast, differential interference contrast (AVEC-DIC) microscopy: A new method capable of analyzing microtubule-related motility in the reticulopodial network of *Allgromia laticollaris. Cell Motil. 1*: 291-302. (11)

Altman, J. and Bayer, S. A. 1978. Development of the diencephalon in the rat. I. Autoradiographic study of the time of origin and settling patterns of neurons in the hypothalamus. *J. Comp. Neurol. 182*: 945-972. (17)

Altman, J. and Das, G. D. 1964. Autoradiographic examination of the effects of enriched environment on the rate of glial multiplication in the adult rat brain. *Nature 204*: 1161-1163. (19)

Amara, S. G., Evans, R. M. and Rosenfeld, M. G. 1984. Calcitonin/calcitonin gene-related peptide transcription unit: Tissue-specific expression involves selective use of alternative polyadenylation sites. *Molec. Cell Biol. 4*: 2151-2160. (4)

Anderson, C. H. 1978. Time of neuron origin in the anterior hypothalamus of the rat. *Brain Res. 154*: 119-122. (17)

Anderson, D. J. and Axel, R. 1985. Molecular probes for the development and plasticity of neural crest derivatives. *Cell 42*: 649-662. (15)

Anderson, M. J. and Cohen, M. W. 1977. Nerve-induced and spontaneous redistribution of acetylcholine receptors on cultured muscle cells. *J. Physiol.* (Lond.) *268*: 757-773. (2)

Anzil, A. P., Beiser, A. and Wernig, A. 1984. Light and electron microscopic identification of nerve terminal sprouting and retraction in normal adult frog muscle. *J. Physiol.* (Lond.) *350*: 393-399. (12)

Arai, Y., Matsumoto, A. and Nishizuka, M. 1978. Synaptogenic action of estrogen on the hypothalamic arcuate nucleus (ARCN) of the developing brain and of the deafferented adult brain in female rats. In *Hormones and Brain Development,* Dorner, G. and Kawakami, M. (eds.), Elsevier/North Holland, Amsterdam, pp. 43-48. (17)

Argiro, V., Bunge, M. B. and Johnson, M. I. 1984. Correlation between growth cone form and movement and their dependence on neuronal age. *J. Neurosci. 4*: 3051-3062. (10)

Arimatsu, Y. and Seto, A. 1982. Ontogeny of sexual difference in alpha-bungarotoxin binding capacity in the mouse amygdala. *Brain Res. 234*: 27-39. (18)

Arimatsu, Y., Seto, A. and Amano, T. 1981. Sexual dimorphism in α-bungarotoxin binding capacity in the mouse amygdala. *Brain Res. 213*: 432- 437. (18)

Arnold, A. P. and Gorski, R. A. 1984. Gonadal steroid induction of structural sex differences in the CNS. In *Annual Review of Neuroscience,* Vol. 7, Cowan, W.M. (ed.), Annual Reviews, Inc., Palo Alto, pp. 413-442. *(17)*

Atlas, M. and Bond, V. P. 1965. The cell generation cycle of the 11-day mouse embryo. *J. Cell Biol. 26*: 19-24. (16)

Attardi, B., Geller, L. N. and Ohno, S. 1976. Androgen and estrogen receptors in brain cytosol from male, female and Tfm mice. *Endocrinology 98*: 864-874. (17)

Auffray, C. and Rougeon, F. 1980. Purification of mouse immunoglobin heavy- chain messenger RNAs from total myeloma tumor RNA. *Eur. J. Biochem. 107*: 303-314. (4)

Ayer-LeLievre, C. S. and Le Douarin, N. M. 1982. The early development of cranial sensory ganglia and the potentialities of their component cells studied in quail-chick chimeras. *Devel. Biol. 94*: 291-310. (15)

Bak, A. L. and Jørgensen, A. L. 1984. RNA polymerase III control regions in retrovirus LTR, Alu-type repetitive DNA, and Papovavirus. *J. Theor. Biol. 108*: 339-348. (6)

Baker, R. E., Corner, M. A. and Veltman, W. A. M. 1978. Topography of cutaneous mechanoceptive neurones in dorsal root ganglia of skin-grafted frogs. *J. Physiol.* (Lond.) *284*: 181-192. (13)

Ball, E. E., Ho, R. K. and Goodman, C. S. 1985. Development of neuromuscular specificity in the grasshopper embryo: Guidance of motoneuron growth cones by muscle pioneers. *J. Neurosci. 5*: 1808-1819. (8,9)

Ballivet, M., Nef, P., Stalder, R. and Fulpius, B. 1983. Genomic sequences encoding the alpha-subunit of acetylcholine receptor are conserved in evolution. *Cold Spring Harbor Symp. Quant. Biol. 48*: 83-87. (5)

Bantle, J. A. and Hahn, W. E. 1976. Complexity and characterization of polyadenylated RNA in mouse brain. *Cell 8*: 139-150. (4,6)

Barald, K. F. 1982. Monoclonal antibodies to embryonic neurons: Cell-specific markers for chick ciliary ganglion. In *Neuronal Development,* Spitzer, N. C. (ed.), Plenum, New York, pp. 101-119. (15)

Barbu, M., Ziller, C., Rong, P. M. and Le Douarin, N. M. 1986. Heterogeneity in migrating neural crest cells revealed by a monoclonal antibody. *J. Neurosci. 6*: 2215-2225. (15)

Bardin, C. W., Bullock, L., Schneider, G., Allison, J. E. and Stanley, A. J. 1970. Pseudohermaphrodite rat: End organ insensitivity to testosterone. *Science 167*: 1136-1137. (17)

Barta, A., Richards, R. I., Baxter, J. D. and Shine, J. 1981. Primary structure and evolution of the rat growth hormone gene. *Proc. Natl. Acad. Sci. USA 78*: 4867-4871. (6)

Bastiani, M. J. and Goodman, C. S. 1984. The first growth cones in the central nervous system of the grasshopper embryo. In *Cellular and Molecular Biology of Neuronal Development,* Black, I. B. (ed.). Plenum, New York. (8)

Bastiani, M. J. and Goodman, C. S. 1986. Guidance of neuronal growth cones in the grasshopper embryo. III. Recognition of specific glial pathways. *J. Neurosci. 6*: 3542-3551. (7)

Bastiani, M. J., Doe, C. Q., Helfand, S. L. and Goodman, C.S. 1985. Neuronal specificity and growth cone guidance in grasshopper and *Drosophila* embryos. *Trends Neurosci. 8*: 257-266. (7,8,9,10)

Bastiani, M. J., du Lac, S. and Goodman, C. S. 1986a. Guidance of neuronal growth cones in the grasshopper embryo. I. Recognition of a specific axonal pathway by the pCC neuron. *J. Neurosci. 6*: 3518-3531. (7)

Bastiani, M. J., du Lac, S. and Goodman, C. S. 1985a. The first neuronal growth cones in insect embryos. Model system for studying the development of neuronal specificity. In *Model Neural Networks and Behavior,* Selverston, A. I. (ed.). Plenum, New York. (8)

Bastiani, M. J., Harrelson, A. L., Snow, P. M. and Goodman, C. S. 1986b. Dynamic expression of surface glycoproteins during embryonic and post-embryonic development of the grasshopper. *Soc. Neurosci. Abst. 12*: 195. (7)

Bastiani, M. J., Harrelson, A. L., Snow, P. M. and Good-

man, C. S. 1987. Expression of fasciclin I and II glycoproteins on subsets of axon pathways during neuronal development in the grasshopper. *Cell 48*: 745-755. (7)

Bastiani, M. J., Raper, J. A. and Goodman, C. S. 1984. Pathfinding by neuronal growth cones in grasshopper embryos. III. Selective affinity of the G growth cone for the P cells within the A/P fascicle. *J. Neurosci. 4*: 2311-2328. (7)

Bate, C. M. 1976. Embryogenesis of an insect nervous system. I. A map of the thoracic and abdominal neuroblasts in *Locusta migratoria. J. Embryol. Exp. Morphol. 35*: 107-123. (8)

Bate, C. M. 1978. Development of sensory systems in anthropods. In *Development of Sensory Systems,* Jacobson, M. (ed.). Springer-Verlag, New York. (8)

Bate, C. M. and Grunewald, E. B. 1981. Embryogenesis of an insect nervous system. II. A second class of precursor cell and the origin of the intersegmental connectives. *J. Embryol. Exp. Morphol. 61*: 317-330. (8)

Baum, M. 1979. Differentiation of coital behavior in mammals: A comparative analysis. *Neurosci. Biobehav. Rev. 3*: 265-284. (18)

Bear, M. F. and Daniels, J. D. 1983. The plastic response to monocular deprivation persists in kitten visual cortex after chronic depletion of norepinephrine. *J. Neurosci. 3*: 407-416. (19)

Bear, M. F. and Singer, W. 1986. Modulation of visual cortical plasticity by acetylcholine and noradrenaline. *Nature 320*: 172-176. (19)

Beattie, C., Rodgers, C. and Soyka, L. 1972. Influence of ovariectomy and ovarian steroids on hypothalamic tyrosine hydroxylase activity in the rat. *Endocrinology 91*: 276-279. (18)

Beaulieu, C. and Colonnier, M. 1985. The effect of environmental complexity on the numerical density of neurons and on the size of their nuclei in the visual cortex of cat. *Soc. Neurosci. Abst. 11*: 225. (19)

Becker, J., Robinson, K. and Lorenz, K. 1982. Sex differences and estrous cycle variations in amphetamine-elicited rotational behavior. *Eur. J. Pharm. 80*: 65-72. (18)

Beckman, S. L., Chikaraishi, D. M., Deeb, S. S. and Sueoka, N. 1981. Sequence complexity of nuclear and cytoplasmic ribonucleic acids from clonal neurotumor cell lines and brain sections of the rat. *Biochemistry 20*: 2684-2692. (4)

Beer, M. and Becker, J. 1984. Influence of estrogen on striatal dopamine release. *Soc. Neurosci. Abst. 10*: 259.4. (18)

Benowitz, L. I. and Lewis, E. R. 1983. Increased transport of 44,000 to 49,000-dalton acidic proteins during regeneration of the goldfish optic nerve: A two-dimensional gel analysis. *J. Neurosci. 3*: 2153-2163. (10)

Benowitz, L. I., Shashoua, V. E. and Yoon, M. G. 1981. Specific changes in rapidly transported proteins during regeneration of goldfish optic nerve. *J. Neurosci. 1*: 301-307. (10)

Benowitz, L. I., Yoon, M. G. and Lewis, E. R. 1983. Transported proteins in the regenerating optic nerve: Regulation by interactions with the optic tectum. *Science* 222: 185-187. (10)

Bentley, D. and Caudy, M. 1983. Pioneer axons lose directed growth after selective killing of guidepost cells. *Nature 304*: 62-65. (8, 10)

Bentley, D. and Caudy, M. 1984. Navigational substrates for peripheral pioneer growth cones: Limb-axis polarity cues, limb-segment boundaries and guidepost neurons. *Cold Spring Harbor Symp. Quant. Biol. 48*: 573-585. (9)

Bentley, D. and Keshishian, H. 1982a. Pathfinding by peripheral pioneer neurons in grasshoppers. *Science 218*: 1082-1088. (8)

Bentley, D. and Keshishian, H. 1982b. Pioneer neurons and pathways in insect appendages. *Trends Neurosci. 5*: 354-358. (8)

Beresford, B. 1983. The contribution of somites to specific muscles in the chick wing. *J. Embryol. Exp. Morphol. 77*: 99-116. (9)

Berlot, J. and Goodman, C. S. 1984. Guidance of peripheral pioneer neurons in the grasshopper: Adhesive hierarchy of epithelial and neuronal surfaces. *Science 223*: 493-496. (8)

Bevan, S. and Steinbach, J. H. 1977. The distribution of alpha-bungarotoxin binding sites on mammalian skeletal muscle developing in vivo. *J. Physiol.* (Lond.) *267*: 195-213. (2)

Bhide, P. G. and Bedi, K. S. 1984. The effects of a lengthy period of environmental diversity on well-fed and previously undernourished rats. II. Synapse to neuron ratios. *J. Comp. Neurol. 227*: 305-310. (19)

Bibb, H. D. 1977. The production of ganglion hypertrophy in *Rana pipiens* larvae. *J. Exp. Zool. 200*: 265-276. (13)

Bibb, H. D. 1978. Neuronal death in the development of normal and hypertrophic spinal ganglia. *J. Exp. Zool. 206*: 65-72. (13)

Black, I. R. et al. 1984. Neurotransmitter plasticity at the molecular level. *Science 225*: 1266-1270. (15)

Black, J. E., Greenough, W. T., Anderson, B. J. and Isaacs, K. R. In press. Environment and the aging brain. *Canad. J. Psychol.* (19)

Black, J. E., Parnisari, R., Eichbaum, E. and Greenough, W. T. 1986. Morphological effects of housing environment and voluntary exercise on cerebral cortex and cerebellum of old rats. *Soc. Neurosci. Abst. 12*: 1579. (19)

Blackshaw, S. E., Nicholls, J. G. and Parnas, I. 1982. Expanded receptive fields of cutaneous mechanoreceptor cells following deletion of single neurons in the CNS of the leech. *J. Physiol. 326*: 261-268. (8)

Blair, S. S. 1982. Interactions between mesoderm and ectoderm in segment formation in the embryo of a glossiphoniid leech. *Devel. Biol. 89*: 389-396. (14)

Blair, S. S. 1983. Blastomere ablation and the developmental origin of identified monoamine containing neurons in the leech. *Devel. Biol. 95*: 65-72. (14)

Blair, S. S. and Palka, J. 1985a. Axon guidance in cultured wing discs and disc fragments of *Drosophila. Devel. Biol. 108*: 411-419. (8)

Blair, S. S. and Palka, J. 1985b. Axon guidance in the wing of *Drosophila. Trends Neurosci. 8*: 284-288. (8, 9)

Blair, S. S. and Weisblat, D. A. 1982. Ectodermal interactions during neurogenesis in the glossiphoniid leech *Helobdella triserialis. Devel. Biol. 91*: 64-72. (14)

Blair, S. S. and Weisblat, D. A. 1984. Cell interactions in the developing epidermis of the leech *Helobdella triserialis. Devel. Biol. 101*: 318-325. (14)

Blair, S. S., Murray, M. A. and Palka, J. 1985. Axon guidance in cultured epithelial fragments of *Drosophila* wing. *Nature 315*: 406-409. (8)

Blaustein, J., Ryer, H. and Feder, H. 1980. A sex difference in the progestin receptor system of guinea pig brain. *Neuroendocrinology 31*: 403- 409. (18)

Bleier, R., Byne, W. and Siggelkow, I. 1982. Cytoarchetectonic sexual dimorphisms of the medial preoptic and anterior hypothalamic areas in guinea pig, rat, hamster and mouse. *J. Comp. Neurol. 212*: 118-130. (17)

Blight, A. R. 1978. Golgi staining of primary and secondary motoneurons in the developing spinal cord of an amphibian. *J. Comp. Neurol. 180*: 679-690. *(8)*

Bone, Q. 1978. Locomotor muscle. In *Fish Physiology,* Hoar, W. S. and Randall, D. J. (eds.). Academic Press, New York. (8)

Booth, J. E. 1977. Sexual behavior of male rats injected with the antioestrogen MER-25 during infancy. *Physiol. Behav. 19*: 35-39. (17)

Borges, S. and Berry, M. 1978. The effects of dark rearing on the development of the visual cortex of the rat. *J. Comp. Neurol. 180*: 277-300. (19)

Boss, V. C. and Schmidt, J. T. 1984. Activity and the formation of ocular dominance patches in dually innervated tectum of goldfish. *J. Neurosci. 4*: 2891-2905. (9, 13)

Boulter, J. and Patrick, J. 1977. Purification of an acetylcholine receptor from a nonfusing muscle cell line. *Biochemistry 16*: 4900. (5)

Boulter, J. B., Luyten, W., Evans, K., Mason, P., Ballivet, M., et al. 1985. Isolation of a clone coding for the alpha-subunit of a mouse acetylcholine receptor. *J. Neurosci. 5*: 2545-2552. (5)

Boulter, J., Evans, K., Goldman, D., Martin, G., Treco, D., et al. 1986a. Isolation of a cDNA clone coding for a possible neural nicotinic acetylcholine receptor alpha-subunit. *Nature 319*: 368-374. (5)

Boulter, J., Goldman, D., Evans, K., Martin, G., Stengelin, S., Heinemann, S. et al. 1986b. Isolation, sequence and preparation of a cDNA clone coding for the gamma subunit of mouse muscle nicotinic acetylcholine receptor. *J. Neurosci. Res. 16*: 37-50. (5)

Boulter, J., Evans, K., Mason, P., Martin, G., Heinemann, S. et al. In preparation. Isolation of a clone coding for the precursor to the beta-subunit of mouse muscle nicotinic acetylcholine receptor. (5)

Bovolenta, P. and Mason, C. A. In press. Growth cone morphology varies with position in the developing mouse visual pathway from retina to first targets. *J. Neurosci.* (10)

Boydston, W. R. and Sohal, G. S. 1979. Grafting of additional periphery reduces the embryonic loss of neurons. *Brain Res. 178*: 403-410. (16)

Bradley, P. and Bery, M. 1978. The Purkinje cell dendrite tree in mutant mouse cerebellum. A quantitative Golgi study of *weaver* and *staggerer* mice. *Brain Res. 142*: 135-141. (16)

Breedlove, S. M. and Arnold, A. P. 1981. Sexually dimorphic motor nucleus in the rat lumbar spinal cord: Response to adult hormone manipulation, absence in androgen-insensitive rats. *Brain Res. 225*: 297-307. (17)

Breedlove, S. and Arnold, A. 1983. Hormonal control of a developing neuromuscular system. *J. Neurosci. 3*: 417-423. (18)

Brilliant, M. H., Sueoka, N. and Chikaraishi, D. M. 1984. Cloning of DNA corresponding to rare transcripts of rat brain: Evidence of transcriptional and post-transcriptional control and for the existence of poly A$^-$ mRNA. *Mol. Cell Biol. 4* 2187-2197. (4)

Britten, R. J. and Davidson, E. H. 1969. Gene regulation for higher cells: A theory. *Science 165*: 349-358. (6)

Brockes, J. P. 1984. Mitogenic growth factors and nerve dependence of limb regeneration. *Science 225*: 1280-1287. (3, 15)

Brockes, J. P. and Hall, Z. W. 1975. Acetylcholine receptors in normal and denervated rat diaphragm muscle. II. Comparison of junctional and extrajunctional receptors. *Biochemistry 14*: 2100-2106. (2)

Brockes, J. P. and Kintner, C. R. 1986. Glial growth factor and nerve-dependent proliferation in the regeneration blastema of urodele amphibians. *Cell 45*: 301-306. (3)

Brow, M. A., Danielson, P., McKinnon, R. D., Arriza, J., Brown, I. R. et al. In preparation. Structural characterization of rat brain Pol III transcripts BC1 and BC2. (6)

Brown, I. R. 1978. Postnatal appearance of a short DNA repeat length in neurons of the cerebral cortex. *Biochem. Biophys. Res. Commun. 84*: 285-292. (6)

Brown, I. R. and Sutcliffe, J. G. 1987. Atypical nucleosome spacing of rat neuronal identifier elements in non-neuronal chromatin. *Nucleic Acids Res. 15*: 3563-3571. (6)

Brown, I. R. and Sutcliffe, J. G. In preparation. Peculiar nucleosome structure of rat identifier elements. (6)

Brown, I. R., Danielson, P., Rush, S., Godbout, M. and Sutcliffe, J. G. In press. Rat and mouse identifier sequences are preferentially but not exclusively located in cortical neuronal genes expressed postnatally. *J. Neurosci. Res.* (6)

Buller, R. E. and O'Malley, B. M. 1976. The biology and mechanism of steroid hormone receptor interaction with the eukaryotic nucleus. *Biochem. Pharm. 25*: 1-12. (18)

Bunge, M. B. 1986. The axonal cytoskeleton: Its role in generating and maintaining cell form. *Trends Neurosci. 9*: 477-482. (III)

Burden, S. J. 1977. Development of the neuromuscular junction in the chick embryo: The number, distribution, and stability of acetylcholine receptors. *Devel. Biol. 57*: 317-329. (2)

Burden, S. J. 1985. The subsynaptic 43-kd protein is concentrated at developing nerve-muscle synapses in vivo. *Proc. Natl. Acad. Sci. USA 82*: 8270-8273. (2)

Burden, S. J., DePalma, R. L. and Gottesman, G. S. 1983. Crosslinking of proteins in acetylcholine receptor-rich membranes: Association between the beta subunit and the 43 kd subsynaptic protein. *Cell 35*: 687-692. (2)

Burden, S. J., Sargent, P. B. and McMahan, U. J. 1979. Acetylcholine receptors in regenerating muscle accumulate at original synaptic sites in the absence of nerve. *J. Cell Biol. 82*: 412-425. (2)

Burke, R. E. and Rudomín, P. 1977. Spinal neurons and synapses. In *Handbook of Physiology. The Nervous System,* American Physiological Society, Bethesda, MD. pp. 877-944. (13)

Burnstine, T. H., Beck, D. A. and Greenough, W. T. In preparation. Hypertrophic changes in spine density and dendritic branching in auditory cortex of neonatally blinded and dark reared mice. (19)

Caddy, K. W. T. and Briscoe, T. J. 1979. Structural and quantitative studies on the normal C3H and *lurcher* mutant mouse. *Phil. Trans. Roy. Soc. Lond. B. 287.* (16)

Camel, J. E., Withers, G. S. and Greenough, W. T. 1986. Persistence of visual cortex dendritic alterations induced by postweaning exposure to a "superenriched" environment in rats. *Behav. Neurosci. 100*: 810-813. (19)

Cameron, J. A., Hilgers, A. R. and Hinterberger, T. J. 1986. Evidence that reserve cells are a source of regenerated adult newt muscle in vitro. *Nature 321*: 607-610. (3)

Capano, C. P., Gioio, A. E., Giuditta, A. and Kaplan, B. B. 1986. Complexity of nuclear and polysomal RNA from squid optic lobe and gill. *J. Neurochem. 46*: 1517-1521. (4)

Carr, V. M. 1984. Dorsal root ganglia development in chicks following partial ablation of the neural crest. *J. Neurosci. 4*: 2434-2444. (13)

Caudy, M. and Bentley, D. 1986. Pioneer growth cone steering along a series of neuronal and non-neuronal cues of different affinities. *J. Neurosci. 6*: 1781-1795. (8, III, 10)

Chan, S. Y., Murakami, K. and Routtenberg, A. 1986. Phosphoprotein Fi: Purification and characterization of a brain kinase C substrate related to *plasticity. J. Neurosci. 6*: 3618-3627. (10)

Chang, F.-L. F. and Greenough, W. T. 1982. Lateralized effects of monocular training on dendritic branching in adult split-brain rats. *Brain Res. 232*: 283-292. (19)

Chang, F.-L. F. and Greenough, W. T. 1984. Transient and enduring morphological correlates of synaptic activity and efficacy change in the rat hippocampal slice. *Brain Res. 309*: 35-46. (19)

Chang, S., Ho, R. and Goodman, C. S. 1983. Selective groups of neuronal and mesodermal cells recognized early in grasshopper embryogenesis by a monoclonal antibody. *Dev. Bio. Res. 9*: 297-304. (7)

Changeux, J.-P. and Danchin, A. 1976. Selective stabilization of developing synapses as a mechanism for the specification of neuronal networks. *Nature 264*: 705-712. (5, 16)

Chaudhari, N. and Hahn, W. E. 1983. Genetic expression in the developing brain. *Science 220*: 924-928. (4, 6)

Chikaraishi, D. M. 1979. Complexity of cytoplasmic polyadenylated and nonpolyadenylated rat brain RNAs. *Biochemistry 18*: 3249-3256. (4)

Chikaraishi, D. M., Deeb, S. and Sueoka, N. 1978. Sequence complexity of nuclear RNAs in adult rat tissues. *Cell 13*: 111-120. (4)

Chow, I. and Cohen, M. W. 1983. Developmental changes in the distribution of acetylcholine receptors in the myotomes of *Xenopus laevis. J. Physiol.* (Lond.) *339*: 553-571. (2)

Christensen, L. W. and Gorski, R. A. 1978. Independent masculinization of neuroendocrine systems by intracerebral implants of testosterone or estradiol in the neonatal female rat. *Brain Res. 146*: 325-340. (17)

Chu-Wang, I. W. and Oppenheim, R. W. 1978. Cell death of motoneurons in the chick embryo spinal cord. I. A light and electron microscope study of naturally occurring and induced cell loss during development. *J. Comp. Neurol. 177*: 38-58. (16)

Chuong, C.-M. and Edelman, G. M. 1985a. Expression of cell adhesion molecules in embryonic induction: 1. Morphogenesis of nestling feathers. *J. Cell Biol. 101*: 1009-1026. (1)

Chuong, C.-M. and Edelman, G. M. 1985b. Expression of cell adhesion molecules in embryonic induction: II. Morphogenesis of adult feathers. *J. Cell Biol. 101*: 1027-1043. (1)

Chuong, C.-M., Crossin, K. L. and Edelman, G. M. 1987. Sequential expression and differential function of multiple adhesion molecules during the formation of cerebellar cortical layers. *J. Cell Biol. 104*: 331-342. (1)

Chuong, C.-M., McClain, D. A., Streti, P. and Edelman, G. M. 1982. Neural cell adhesion molecules in rodent brains isolated by monoclonal antibodies with cross-species reactivity. *Proc. Natl. Acad. Sci. USA 79*: 4234-4238. (1)

Ciment, G. 1983. Neurogenesis in the branchial arch mesenchyme of avian embryos. In *Developing and Regenerating Vertebrate Nervous Systems*, A. R. Liss, New York, pp. 159-165. (15)

Ciment, G. and Weston, J. A. 1982. Early appearance in neural crest and crest-derived cells of an antigenic determinant present in avian neurons. *Devel. Biol. 93*: 355-367. (15)

Ciment, G. and Weston, J. A. 1983. Enteric neurogenesis by neural crest-derived branchial arch mesenchymal cells. *Nature 305*: 424-427. (15)

Ciment, G. and Weston, J. A. 1985. Segregation of developmental abilities in neural-crest derived cells: Identification of partially restricted intermediate cell types in the branchial arches of avian embryos. *Devel. Biol. 111*: 73-83. (15)

Ciment, G., Glimelius, B., Nelson, D. M. and Weston, J. A. 1986a. Reversal of a developmental restriction in neural crest-derived cells of avian embryos by a phorbal ester drug. *Devel. Biol. 118*: 392-398. (15)

Ciment, G., Ressler, A., Letourneau, P. and Weston, J. A. 1986b. Identification of an intermediate filament-associated protein, NAPA-73, which binds to different filament types at different stages of neurogenesis. *J. Cell Biol. 102*: 245-251. (15)

Clark, A., Nock, B., Feder, H. and Roy, E. 1985. Alpha-1-nonadrenergic receptor blockade decreases nuclear estrogen receptor binding in guinea pig hypothalamus and preoptic area. *Brain Res. 330*: 197-199. (18)

Clarke, P. B. S., Schwartz, R. D., Paul, S. M., Pert, C. B. and Pert, A. 1985. Nicotinic binding in rat brain: Autoradiographic comparison of [^{3}H]acetylcholine, [^{3}H]alicotine, and [^{125}I]α-bungarotoxin. *J. Neurosci. 5*: 1307-1315. (5)

Clarke, P. G. H. and Cowan, W. M. 1976. The development of the isthmo-optic tract in the chick with special reference to the occurrence and correction of developmental errors in the location and connections of isthmo-optic neurons. *J. Comp. Neurol. 167*: 143-164. (16)

Claudio, T., Ballivet, M., Patrick, J. and Heinemann, S. 1983. Nucleotide and deduced amino acid sequences of *Torpedo californica* acetylcholine receptor gamma-subunit. *Proc. Natl. Acad. Sci. USA 80*: 1111-1115. (5)

Clements, M. and Kelly, J. B. 1978. Auditory spatial responses of young guinea pigs (*Cavia procellus*) during and after ear blocking. *J. Comp. Physiol. Psychol. 92*: 34-44. (20)

Clough, R. W. and Rodriquez-Sierra, J. F. 1982. Puberty associated neural synaptic changes in female rats administered estrogen. *Soc. Neurosci. Abst. 8*: 196. (17)

Cohen, A. M. and Konigsberg, I. R. 1975. A clonal approach to the problem of neural crest determination. *Devel. Biol. 46*: 262-280. (15)

Coleman, P. D., Kaplan, B. B., Osterburg, H. H. and Finch, C. E. 1980. Brain Poly (A) RNA during aging: Stability of yield and sequence complexity in two rat strains. *J. Neurochem. 34*: 335-345. (4)

Commins, D. and Yahr, P. 1984. Adult testosterone levels influence the morphology of a sexually dimorphic area in the Mongolian gerbil brain. *J. Comp. Neurol. 224*: 132-140. (17)

Conklin, E. G. 1905. The organization and cell lineage of the ascidian egg. *J. Acad. Nat. Sci.* (Philadelphia) *13*: 1-119. (V)

Connor, J. R., Diamond, M .C. and Johnson, R. E. 1980. Occipital cortical morphology of the rat: Alterations with age and environment. *Exp. Neurol. 68*: 158-170. (19)

Constantine-Paton, M. and Capranica, R. P. 1976. Axonal guidance of developing optic nerves in the frog. I. Anatomy of the projection from transplanted eye primordia. *J. Comp. Neurol. 170*: 17-32. (10)

Conti-Tronconi, B. M. and Raftery, M. A. 1982. The nicotinic cholinergic receptor: Correlation of molecular structure with functional properties. *Annu. Rev. Biochem. 51*: 491-530. (5)

Conti-Tronconi, B. M., Dunn, S. M. J., Barnard, E. A., Dolly, J. O., Lai, F. A. et al. 1985. Brain and muscle nicotinic acetylcholine receptors are different but homologous proteins. *Proc. Natl. Acad. Sci. USA 82*: 5208- 5212. (5)

Cooke, C. T., Nolan, T. M., Dyson, S. E. and Jones, D. G. 1974. Pentobarbitol-induced configurational changes at the synapse. *Brain Res. 76*: 330-335. (19)

Coulombe, J. and Bronner-Fraser, M. 1986. Cholinergic neurones acquire adrenergic neurotransmitters when transplanted into an embryo. *Nature 324*: 569-572. (15)

Covault, J. and Sanes, J. R. 1985. Neural cell-adhesion molecule (N-CAM) accumulates in denervated and paralyzed skeletal-muscles. *Proc. Natl. Acad. Sci. USA 82*: 4544-4548. (1)

Cowan, W. M. 1978. Aspects of neural development. In *International Review of Physiology and Neurophysiology III*, Porter, R. (ed.), University Park Press, Baltimore, pp. 149-191. (17)

Cowan, W. M. and Wenger, E. 1968. Degeneration in the nucleus of origin of the preganglionic fibers to the chick ciliary ganglion following early removal of the optic vesicle. *J. Exp. Zool. 168*: 105-124. (16)

Crepel, F. and Mariani, J. 1975. Anatomical, physiological, and biochemical studies of the cerebellum from mutant mice. I. Electrophysiological analysis of cerebellar neurons in the *staggerer* mouse. *Brain. Res. 98*: 135-147. (16)

Crepel, F., Delhaye-Bouchard, N., Goustivino, J. M. and Sampaio, I. 1980. Multiple innervation of cerebellar Purkinje cells by climbing fibers in *staggerer* mutant mouse. *Nature 283*: 483-485. (16)

Crick, F. H. C. and Lawrence, P. A. 1975. Compartments and polyclones in insect development. *Science 189*: 340-347. (14)

Crossin, K. L., Chuong, C.-M. and Edelman, G. M. 1985. Expression sequences of cell adhesion molecules. *Proc. Natl. Acad. Sci. USA* 82: 6942-6946. (1)

Crossin, K. L., Hoffman, S., Grumet, M., Thiery, J.-P. and Edelman, G. M. 1986. Site-restricted expression of cytotactin during development of the chicken embryo. *J. Cell Biol. 102*: 1917-1930. (1)

Crossin, K. L., Richardson, G. P., Chuong, C.-M. and Edelman, G. M. In press. Modulation of adhesion molecules during induction and differentiation of the auditory placode. In *Functions of the Auditory System*, Edelman, G. M., Gall, W. E., Cowan, W. M. (eds.). John Wiley & Sons, New York. (1)

Cruce, W. L. R. 1974. Supraspinal input to hindlimb motoneurons in lumbar spinal cord of the frog, *Rana catesbeiana*. *J. Neurophysiol. 37*: 691-704. (13)

D'Eustachio, P., Owens, G., Edelman, G. M. and Cun-

ningham, B. A. 1985. Chromosomal location of the gene encoding to the neural cell adhesion molecule (N-CAM) in the mouse. *Proc. Natl. Acad. Sci. USA 82*: 7631-7635. (1)

Daniels, G. R. and Deininger, P. L. 1985. Repeat sequence families derived from mammalian tRNA genes. *Nature 317*: 819-822. (6)

Daniloff, J. K., Chuong, C.-M., Levi, G. and Edelman, G. M. 1986a. Differential distribution of cell adhesion molecules during histogenesis of the chick nervous system. *J. Neurosci. 6*: 739-758. (1)

Daniloff, J. K., Levi, G., Grumet, M., Rieger, F. and Edelman, G. M. 1986b. Altered expression of neuronal cell adhesion molecules induced by nerve injury and repair. *J. Cell Biol. 103*: 929-945. (1)

David, S. and Aguayo, A. J. 1981. Axonal elongation into peripheral nervous system bridges after central nervous system injury in adult rats. *Science 214*: 931-933. (10)

Davies, A. M. 1986. The survival and growth of embryonic proprioceptive neurons is promoted by a factor present in skeletal muscle. *Devel. Biol. 115*: 56-67. (15)

Davis, M. R. and Constantine-Paton, M. 1983a. Hyperplasia in the spinal sensory system of the frog. I. Plasticity in the most caudal dorsal root ganglion. *J. Comp. Neurol. 221*: 444-452. (13)

Davis, M. R. and Constantine-Paton, M. 1983b. Hyperplasia in the spinal sensory system of the frog. II. Central and peripheral connectivity patterns. *J. Comp. Neurol. 221*: 453-465. (13)

Davis, P. and Barfield, R. 1979. Activation of masculine sexual behavior by intracranial estradiol benzoate implants in male rats. *Neuroendocrinology 28*: 217-227. (18)

Daw, N. W., Robertson, T. W., Rader, R. K., Videen, T. O. and Coscia, C. J. 1984. Substantial reduction of cortical norepinephrine by lesions of adrenergic pathway does not prevent effects of monocular deprivation. *J. Neurosci. 4*: 1354-1360. (19)

Daw, N. W., Videen, T. O., Parkinson, D. and Rader, R. K. 1985. DSP-4 depletes noradrenaline in kitten visual cortex without altering the effects of monocular deprivation. *J. Neurosci. 5*: 1925-1933. (19)

de Ferra, F., Engh, H., Hudson, L., Kamholz, J., Puckett, C. et al. 1985. Alternative splicing accounts for the four forms of myelin basic protein. *Cell 43*: 721-727. (4)

DeVoogd, T. and Nottebohm, F. 1981. Gonadal hormones induce dendritic growth in the adult avian brain. *Brain 214*: 202-204. (17)

DeVoogd, T., Nixdorf, B. and Nottebohm, F. 1982. Recruitment of additional synapses into a brain network takes extra brain space. *Soc. Neurosci. Abst. 8*: 140. (17)

DeVries, G. J., Buijs, R. M. and van Leeuwen, F. W. 1984b. Sex differences in vasopressin and other neurotransmitter systems in the brain. In *Progress in Brain Research, Vol. 61: Sex Differences in the Brain*, DeVries, G. J. DeBruin, J. P. C., Uylings, H. B. M. and Corner, M. A. (eds.), Elsevier, Amsterdam, pp. 185-203. (17, 18)

DeVries, G. J., DeBruin, J. P. C., Uylings, H. B. M. and Corner, M. A. (eds.). 1984a. *Progress in Brain Research, Vol. 61: Sex Differences in the Brain.* Elsevier, Amsterdam. (17)

Deeb, S. S. 1983. Sequence complexity of nuclear RNA in brain sections of the sheep. *Cell and Molec. Biol. 29:* 113-119. (4)

Dennis, M. J. 1981. Development of the neuromuscular junction: Inductive interactions between cells. *Annu. Rev. Neurosci. 4:* 43-68. (2)

Derby, M. A. 1982. Environmental factors affecting neural crest differentiation by crest cells exposed to cell-free (deoxycholate-extracted) dermal mesenchymal matrix. *Cell Tissue Res. 225:* 379-386. (15)

Devillers-Thiery, A., Giraudat, J., Bentaboulet, M. and Changeux, J-P. 1983. Complete mRNA coding sequence of the acetylcholine binding alpha-subunit from *Torpedo marmorata* acetylcholine receptor: A model for the transmembrane organization of the polypeptide chain. *Proc. Natl. Acad. Sci. USA 80:* 2067-2071. (5)

Dewey, M. J., Gervais, A. G. and Mintz, B. 1976. Brain and ganglion development from two genotypic classes of cells in allophenic mice. *Devel. Biol. 50:* 68-81. (16)

Dhar, R., Ellie, R. W., Shih, T. Y., Oroszlan, S., Shapiro, B. et al. 1982. Nucleotide sequence of the p21 tranforming protein of Harvey Murine Sarcoma Virus. *Science 217:* 934-937. (6)

Diamond, J. 1971. The mauthner cell. In *Fish Physiology V,* Hoar, W. S. and Randall, D. J. (eds.). Academic Press, New York. (8)

Diamond, J. and Miledi, R. 1962. A study of foetal and newborn rat muscle fibres. *J. Physiol.* (Lond.) *162:* 393-408. (2)

Diamond, M. C., Law, F., Rhodes, H., Lindner, B., Rosenzweig, M. R. et al. 1966. Increases in cortical depth and glia numbers in rats subjected to enriched environment. *J. Comp. Neurol. 128:* 117-125. (19)

Diamond, M. C., Lindner, B., Johnson, R., Bennett, E.L. and Rosenzweig, M. R. 1975. Differences in occipital cortical synapses from environmentally enriched, impoverished, and standard colony rats. *J. Neurosci. Res. 1:* 109-119. (19)

Dippold, W. G., Lloyd, K. O., Li, L. T. C., Ikeda, H., Oettgen, H. F. and Old, L. J. 1980. Cell surface antigens of human malignant melanoma: Definition of six antigenic systems with mouse monoclonal antibodies. *Proc. Natl. Acad. Sci. USA 77:* 6114-6118. (15)

Doe, C. Q. and Goodman, C. S. 1985a. Early events in insect neurogenesis. I. Development and segmental differences in the pattern of neuronal precursor cells. *Devel. Biol. 111:* 193-205. (14)

Doe, C. Q. and Goodman, C. S. 1985b. Early events in insect neurogenesis. II. The role of cell interactions and cell lineage in the determination of neuronal precursor cells. *Devel. Biol. 111:* 206-219. (14)

Doe, C. Q., Bastiani, M. J. and Goodman, C. S. 1986. Guidance of neuronal growth cones in the grasshopper embryo. IV. Temporal delay experiments. *J. Neurosci. 6:* 3552-3563. (7)

Dohanich, G., Wicher, J., Weaver, D. and Clemens, L. 1982. Alteration of muscarinic binding in specific brain areas following estrogren treatment. *Brain Res. 241:* 347-350. (18)

Döhler, K. D., Coquelin, A., Davis, F., Hines, M., Shryne, J. E. et al. 1982a. Differentiation of the sexually dimorphic nucleus in the preoptic area of the rat brain is determined by the prenatal hormone environment. *Neurosci. Lett. 33:* 295-298. (17)

Döhler, K. D., Coquelin, A., Davis, F., Hines, M., Shryne, J. E. and Gorski, R. A. 1984. Pre- and postnatal influence of testosterone propionate and diethylstilbestrol on differentiation of the sexually dimorphic nucleus of the preoptic area in male and female rats. *Brain Res. 302:* 291-295. (17)

Döhler, K. D., Hines, M., Coquelin, A., Davis, F., Shryne, J. E. and Gorski, R. A. 1982b. Pre- and postnatal influence of diethylstilbestrol on differentiation of the sexually dimorphic nucleus in the preoptic area of the female rat brain. *Neuroendocrinol. Lett. 4*(6): 361-365. (17)

Donlon, P. and Stenson, R. 1976. Neuroleptic-induced extrapyramidal symptoms. *Dis. Nerv. Syst. 37:* 629-635. (18)

Dörner, G. and Staudt, J. 1968. Structural changes in the preoptic anterior hypothalamic area of the male rat, following neonatal castration and androgen substitution. *Neuroendocrinology 3:* 136-140. (17)

Doughty, C. and McDonald, P. G. 1974. Hormonal control of sexual differentiation of the hypothalamus in the neonatal female rat. *Differentiation 2:* 275-285. (17)

Doupe, A. J., Landis, S. C. and Patterson, P. H. 1985. Environmental influences in the development of neural crest derivatives: Glucocorticoids, growth factors and chromaffin cell plasticity. *J. Neurosci. 5:* 2119-2142. (15)

Dubin, M. W., Stark, L. A. and Archer, S. M. 1986. A role for action-potential activity in the development of neuronal connections in the kitten retinogeniculate pathway. *J. Neurosci. 6:* 1021-1036. (13)

du Lac, S., Bastiani, M. J. and Goodman, C. G. 1986. Guidance of neuronal growth cone in the grasshopper embryo. II. Recognition of a specific axonal pathway by the aCC neuron. *J. Neurosci. 6:* 3532-3541. (7)

Dupin, E. 1984. Cell division in the ciliary ganglion of quail embryos in situ and after back-transplantation into the neural crest migration pathways of chick embryos. *Devel. Biol. 105:* 288-299. (15)

Dyson, S. E. and Jones, D. G. 1980. Quantitation of terminal parameters and their interrelationships in maturing central synapses: A perspective for experimental studies. *Brain Res. 183:* 43-59. (19)

Easter, S. S., Jr., Purves, D., Rakic, P. and Spitzer, N. C. 1985. The changing view of neural specificity. *Science 230:* 507-511. (8, 9, IV)

Easter, S. S., Rusoff, A. C. and Kish, P. E. 1981. The growth and organization of the optic nerve and tract in juvenile and adult goldfish. *J. Neurosci. 1*: 793-811. (8)

Eaton, R. C. and Hackett, J. T. 1984. The role of the Mauthner cell in fast-starts involving escape in teleost fishes. In *Neural Mechanisms of Startle Behavior,* Eaton, R. C. (ed.). Plenum Press, New York. (8)

Eayrs, J. T. and Ireland, K. F. 1950. The effect of total darkness on the growth of the newborn albino rat. *J. Endocrinol. 6*: 386-397. (19)

Eccles, J. C., Eccles, R. M. and Lundberg, A. 1957. The convergence of monosynaptic excitatory afferents onto many different species of alpha motoneurones. *J. Physiol.* (Lond.) *137*: 22-50. (13)

Eccles, J. C., Eccles, R. M. and Magni, F. 1960. Monosynaptic excitatory action on motoneurones regenerated to antagonistic muscles. *J. Physiol.* (Lond.) *154*: 68-88. (13)

Eccles, J. C., Eccles, R. M. and Shealy, C. N. 1962a. An investigation into the effect of degenerating primary afferent fibres on the monosynaptic innervation of motoneurones. *J. Neurophysiol. 25*: 544-558. (13)

Eccles, J. C., Eccles, R. M., Shealy, C. N. and Willis, W. D. 1962b. Experiments utilizing monosynaptic excitatory action of motoneurones for testing hypotheses relating to specificity of neuronal connections. *J. Neurophysiol. 25*: 559-580. (13)

Edelman, G. M. 1984a. Modulation of cell adhesion during induction, histogenesis, and perinatal development of the nervous system. *Annu. Rev. Neurosci. 7*: 339-377. (1)

Edelman, G. M. 1984b. Cell adhesion and morphogenesis: The regulator hypothesis. *Proc. Natl. Acad. Sci. USA 81*: 1460-1464. (1)

Edelman, G. M. 1984c. Cell surface modulation and marker multiplicity in neural patterning. *Trends Neurosci. 7*: 78-84. (1)

Edelman, G. M. 1985. Expression of cell adhesion molecules during embryogenesis and regeneration. *Exp. Cell Res. 161*: 1-16. (1)

Edelman, G. M. 1986a. Epigenetic rules for expression of cell adhesion molecules during morphogenesis. In *Junctional Complexes of Epithelial Cells.* CIBA Foundation Symposium No. 125. John Wiley & Sons, London. (1)

Edelman, G. M. 1986b. Cell adhesion molecules in the regulation of animal form and tissue pattern. *Annu. Rev. Cell Biol. 2*: 81-116. (1)

Edelman, G. M. and Chuong, C.-M. 1982. Embryonic to adult conversion of neural cell-adhesion molecules in normal and *staggerer* mice. *Proc. Natl. Acad. Sci. USA 79*: 7036-7040. (1)

Edelman, G. M., Gallin, W. J., Delouvee, A., Cunningham, B. A. and Thiery, J.-P. 1983. Early epochal maps of two different cell adhesion molecules. *Proc. Natl. Acad. Sci. USA 80*: 4384-4388. (1)

Edmondson, J. C. and Hatten, M. E. 1987. Glial-guided neuronal migration in vitro: A high-resolution time-lapse video microscopic study. *J. Neurosci. 7*: 1928-1934. (11)

Edmondson, J. C. and Hatten, M. E. In press. Astrotactin, a novel neuronal cell surface antigen that mediates neuron-astroglial interactions in cerebellar microcultures. *J. Cell Biol.* (11)

Edwards, H. P., Barry, W. F. and Wyspianski, J. O. 1969. Effect of differential rearing on photic evoked potentials and brightness discrimination in the albino bat. *Devel. Psychobiol. 2*: 133-138. (19)

Edwards, J. 1977. Pathfinding by insect sensory neurons. In *Identified Neurons and Behavior of Arthropods,* Hoyle, G. (ed.). Plenum Press, New York. (8)

Edwards, J., Chen, S. W. and Berns, M. W. 1981. Cercal sensory development following laser microlesions of embryonic apical cells in *Acheta domesticus. J. Neurosci. 1*: 250-258. (8)

Ehrhardt, A. A., Meyer-Bahlburg, H. F., Rosen, L. R., Feldman, J. F. Verdiano, N. P. et al. 1985. Sexual orientation after prenatal exposure to exogenous estrogen. *Arch. Sexual Behav. 14*: 57-75. (18)

Eide, A.-L., Jansen, J. K. S. and Ribchester, R. R. 1982. The effect of lesions in the neural crest on the formation of synaptic connexions in the embryonic chick spinal cord. *J. Physiol.* (Lond.) *324*: 453-478. (13)

Eisen, J. S., Myers, P. Z. and Westerfield, M. 1986. Pathway selection by growth cones of identified motoneurones in live zebra fish embryos. *Nature 320*: 269-271. (8, 13)

Eisenbarth, G. S., Walsh, F. S. and Nirenberg, M. 1979. A monoclonal antibody to a plasma membrane antigen of neurons. *Proc. Natl. Acad. Sci. USA 76*: 4913-4917. (15)

Elliot, J., Blanchard, S. G., Wu, W., Miller, J., Strader, J. et al. 1980. Purification of *Torpedo californica* postsynaptic membranes and fractionation of their constituent proteins. *Biochem. J. 185*: 667-677. (2)

Ellis, L., Wallis, I., Abreu, E. and Pfenninger, K. H. 1985. Nerve growth cones isolated from fetal rat brain. IV. Preparation of a membrane subfraction and identification of a membrane glycoprotein expressed on sprouting neurons. *J. Cell Biol. 101*: 1977-1989. (1)

Erickson, C. 1986. Morphogenesis of neural crest cells. In *Cellular Basis of Morphogenesis. Developmental Biology: A Comprehensive Synthesis,* Browder, L. (ed.). Plenum Press, N.Y., pp. 481-543. (15)

Erickson, C., Tosney, K. and Weston, J. A. 1980. Analysis of migratory behavior of neural crest and fibroblastic cells in embryonic tissues. *Devel. Biol. 77*: 142-156. (15)

Ermini, M. and Kuenzle, C. C. 1978. The chromatin repeat length of cortical neurons shorten during early postnatal development. *FEBS Lett. 90*: 167-172. (6)

Ernst, S. G., Britten, R. J. and Davidson, E. H. 1979. Distinct single-copy sequence sets in sea urchin nuclear RNAs. *Proc. Natl. Acad. Sci. USA 76*: 2209-2212. (4)

Etgen, A. 1985. Effects of body weight, adrenal status and estrogen priming on hypothalamic progestin receptors in male and female rats. *J. Neurosci. 5:* 2439-2442. (18)

Fallon, J. R. 1985. Preferential outgrowth of CNS neurites on astrocytes and Schwann cells as compared with nonglial cells in vitro. *J. Cell Biol. 100:* 198-207. (11)

Fambrough, D. M. 1979. Control of acetylcholine receptors in skeletal muscle. *Physiol. Rev. 59:* 165-227. (2)

Farel, P. B. and Bemelmans, S. E. 1985. Specificity of motoneurons projection patterns during development of the bullfrog tadpole (*Rana catesbeiana*). *J. Comp. Neurol. 238:* 122-134. (8, 9)

Feder, H. H. and Whalen, R. E. 1964. Feminine behavior in neonatally castrated and estrogen-treated male rats. *Science 147:* 306-307. (17)

Fekete, D. M. and Brockes, J. P. 1987. A monoclonal antibody detects a difference in the cellular composition of developing and regenerating limbs of newts. *Development 99:* 589-602. (3)

Fekete, D. M. and Brockes, J. P. In press. The regeneration blastemas of aneurogenic limbs are antigenically distinct from those of normally-innervated limbs in newt larvae. *Soc. Neurosci. Abst.* (3)

Felten, D. L., Hallman, H. and Jonsson, G. 1982. Evidence for a neurotrophic role of noradrenaline neurons in the postnatal development of rat cerebral cortex. *J. Neurocytol. 11:* 119-135. (19)

Ferchmin, P. A., Bennett, E. L. and Rosenzweig, M. R. 1975. Direct contact with enriched environments is required to alter cerebral weights in rats. *J. Comp. Physiol. Psychol. 88:* 360-367. (19)

Fernandez, J. 1980. Embryonic development of the glossiphoniid leech *Theromyzon rude*: Characterization of developmental stages. *Devel. Biol. 76:* 245-262. (14)

Fernandez, J. and Stent, G. S. 1980. Embryonic development of the glossiphoniid leech *Theromyzon rude*: Structure and development of the germinal bands. *Devel. Biol. 78:* 407-434. (14)

Fifkova, E. 1968. Changes in the visual cortex of rats after unilateral deprivation. *Nature 220:* 379-381. (19)

Fifkova, E. and Van Harreveld, A. 1977. Long-lasting morphological changes in dendritic spines of dentate granular cells following stimulation of the entorhinal area. *J. Neurocytol. 6:* 211-230. (19)

Finer-Moore, J. and Stroud, R. M. 1984. Amphipathic analysis and possible conformation of the ion channel in an acetylcholine receptor. *Proc. Natl. Acad. Sci. USA 81:* 155-159. (5)

Fischbach, G. D. and Cohen, S. A. 1973. The distribution of acetylcholine sensitivity over uninnervated and innervated muscle fibers grown in cell culture. *Devel. Biol. 31:* 147-162. (2)

Fischette, C., Biegon, A. and McEwen, B. S. 1983. Sex differences in serotonin-1 receptor binding in rat brain. *Science 222:* 333-335. (18)

Fischette, C., Biegon, A. and McEwen, B. S. 1984. Sex steroid modulation of the serotonin behavioral syndrome. *Life Sci. 35:* 1197-1206. (18)

Fixen, W., Sternberg, P. Ellis, H. and Horvitz, R. 1985. Genes that affect cell fates during the development of *Caenorhabditis elegans. Cold Spring Harbor Symp. Quant. Biol. 50:* 99-104. (14)

Floeter, M. K. and Greenough, W. T. 1979. Cerebellar plasticity: Modification of Purkinje cell structure by differential rearing in monkeys. *Science 206:* 227-229. (19)

Flood, D. G., Buell, S. J., DeFiore, C. H., Horwitz, G. J. and Coleman, P. D. 1985. Age-related dendritic growth in dentate gyrus of human brain is followed by regression in the "oldest old." *Brain Res. 345:* 366-368. (19)

Foote, S. L., Bloom, F. E. and Aston-Jones, G. 1983. Nucleus locus ceruleus: New evidence of anatomical and physiological specificity. *Physiol. Rev. 63:* 844-914. (19)

Forehand, C. J. and Farel, P. B. 1982. Spinal cord development in anuran larvae: I. Primary and secondary neurons. *J. Comp. Neurol. 209:* 386-394. (8)

Forgays, D. G. and Forgays, J. W. 1952. The nature of the effect of free-environmental experience in the rat. *J. Comp. Physiol. Psychol. 45:* 322-328. (19)

Fraille, I. G., Pfaff, D. W. and McEwen, B. S. 1987. Progestin receptors with and without estrogen induction in male and female hamster brains. *Neuroendocrinology 45:* 487-491. (18)

Frank, E. and Fischbach, G. D. 1979. Early events in neuromuscular junction formation in vitro. Induction of acetylcholine receptor clusters in the postsynaptic membrane and morphology of newly-formed synapses. *J. Cell Biol. 83:* 143-158. (2)

Frank, E. and Jackson, P. C. 1986. Normal electrical activity is not required for the formation of specific sensory-motor synapses. *Brain Res. 378:* 147-151. (13)

Frank, E. and Westerfield, M. 1982a. Synaptic organization of sensory and motor neurones innervating triceps brachii muscles in the bullfrog. *J. Physiol. (Lond.) 324:* 479-494. (13)

Frank, E. and Westerfield, M. 1982b. The formation of appropriate central and peripheral connexions by foreign sensory neurones of the bullfrog. *J. Physiol. (Lond.) 324:* 495-505. (13)

Frank, E. and Westerfield, M. 1983. Development of sensory-motor synapses in the spinal cord of the frog. *J. Physiol. (Lond.) 343:* 593-610. (13)

Frank, E., Harris, W. A. and Kennedy, M. B. 1980. Lysophosphatidyl choline facilitates labelling of CNS projections with horseradish peroxidase. *J. Neurosci. Methods 2:* 183-189. (13)

Frankfurt, M., Fuchs, E. and Wuttke, W. 1984. Sex differences in α-aminobutyric acid and glutamate concentrations in discrete nuclei. *Neurosci. Letts. 50:* 245-250. (18)

Fraser, S. E., Murray, B. A., Chuong, C.-M. and Edelman, G. M. 1984. Alteration of the retinotectal map in *Xenopus* by antibodies to neural cell adhesion molecules. *Proc. Natl. Acad. Sci. USA 81:* 4222-4226. (1)

Friedlander, D. R., Grumet, M. and Edelman, G. M. 1986. Nerve growth factor enhances expression of neuron-glia cell adhesion molecule in PC12 cells. *J. Cell Biol.* *102*: 413-419. (1)

Froehner, S. C. 1984. Peripheral proteins of postsynaptic membranes from *Torpedo* electric organ identified with monoclonal antibodies. *J. Cell Biol.* *99*: 88-96. (2)

Fuchs, J. L., Bajjalieh, S. M., Hoffman, C. A. and Greenough, W. T. 1983. Regional brain 2-deoxyglucose uptake during performance of a learned reaching task. *Soc. Neurosci. Abst.* *9*: 54. (19)

Fukada, K. 1985. Purification and partial characterization of a cholinergic neuronal differentiation factor. *Proc. Natl. Acad. Sci. USA 82*: 8795-8799. (4)

Furshpan, E. J., MacLeish, P. R., O'Lague, P. H. and Potter, D. D. 1976. Chemical transmission between rat sympathetic neurons and cardiac myocytes developing in microcultures: Evidence for cholinergic adrenergic and dual- function neurons. *Proc. Natl. Acad. Sci. USA 73*: 4225-4229. (15)

Gallien, L. and Durocher, M. 1957. Table chronologies du developpement chez *Pleurodeles waltlii* Michah. *Bull. Biol.,* facs., *2*: 97-114. (3)

Gallin, W. J., Chuong, C.-M., Finkel, L. H. and Edelman, G. M. 1986. Antibodies to L-CAM perturb inductive interactions and alter feather patterns and structure. *Proc. Natl. Acad. Sci. USA 83*: 8235-8239. (1)

Gear, A. R. L. 1974. Rhodamine GG, a potent inhibitor of mitochondrial oxidative phosphorylation. *J. Biol. Chem.* *249*: 3628-3637. (12)

George, F. and Ojeda, S. 1982. Changes in aromatase activity in the rat brain during embryonic, neonatal and infantile development. *Endocrinology 111*: 522-529. (18)

Ghysen, A. and Janson, R. 1980. Sensory pathways in Drosophila. In *Development and Neurobiology of Drosophilia,* edited by Siddiqi, Babu, Hall and Hall. New York: Plenum Press, pp. 247-265. (7)

Gietzen, D., Hope, W. and Woolley, D. 1983. Dopaminergic agonists increase 3H-estradiol binding in hypothalamus of female rats, but not of males. *Life Sci. 33*: 2221-2228. (18)

Gimlich, R. L. and Braun, J. 1985. Improved fluorescent compounds for tracing cell lineage. *Devel. Biol. 109*: 509-514. (14)

Girdlestone, J. and Weston, J. A. 1985. Identification of early neuronal subpopulations in avian neural crest cell cultures. *Devel. Biol. 109*: 274-287. (15)

Glick, S. and Hinds, P. 1984. Sex differences in sensitization to cocaine-induced rotation. *Eur. J. Pharm. 99*: 119-121. (18)

Glimelius, B. and Pentar, J. E. 1981. Analysis of developmentally homogeneous neural crest cue populations in vitro. IV. Cell proliferation and synthesis of glycosaminoglycans. *Cell Diff. 10*: 173-182. (15)

Glimelius, B. and Weston, J. A. 1981. Analysis of developmentally homogeneous neural crest cell populations in vitro. III. Role of culture environment in cluster formation and differentiation. *Cell Diff. 10*: 57-67. (15)

Globus, A., Rosenzweig, M. R., Bennett, E. L. and Diamond, M. C. 1973. Effects of differential experience on dendritic spine counts in rat cerebral cortex. *J. Comp. Physiol. Psychol. 82*: 175-181. (19)

Globus, M. and Vethamany-Globus, S. 1977. Transfilter mitogenic effect of dorsal root ganglia on cultured regeneration blastemata in the newt *Notophthalmus viridescens. Devel. Biol. 56*: 316-328. (3)

Globus, M., Vethamany-Globus, S., Kesik, A. and Milton, G. 1983. Roles of neural peptide substance P and calcium in blastema cell proliferation in the newt *Notophthalmus viridescens.* In *Limb Regeneration and Development,* Kelly, R. O., Goetinck, P. F. and MacCabe, J. A. (eds.). Alan R. Liss Inc. New York. (3)

Glover, J. C. and Mason, A. 1986. Morphogenesis of an identified leech neuron: Segmental specification of axonal outgrowth. *Devel. Biol. 115*: 256- 260. (8)

Gold, P. E. 1984. Memory modulation: Neurobiological contexts. In *Neurobiology of Learning and Memory,* Lynch, G.J., McGaugh, L. and Weinberger, N.M. (eds.), The Guilford Press, New York. pp. 374-382. (19)

Goldman, D., Simmons, D., Swanson, L., Patrick, J. and Heinemann, S. 1986. Mapping brain areas expressing RNA homologous to two different acetycholine receptor alpha subunit cDNAs. *Proc. Natl. Acad. Sci. USA 83*: 4076-4080. (5)

Goldowitz, D. and Mullen, R. J. 1982. Granule cell as a site of gene action in the *weaver* mouse cerebellum: Evidence from heterozygous mutant chimeras. *J. Neurosci. 2*: 1474-1485. (11, 16)

Goodman, C. S., Bastiani, M. J., Doe, C. Q., du Lac, S. Helfand, S. L. et al. 1984. Cell recognition during neuronal development. *Science 225*: 1271- 1279. (7, 13)

Goodman, C. S., Raper, J. A., Ho, R. K. and Chang, S. 1982. Pathfinding of neuronal growth cones in grasshopper embryos. In *Developmental Order: Its Origin and Regulation,* Subtelny, S. and Green, P. B., (eds.), New York: Alan R. Liss, pp. 275-316. (III, 7, 8)

Gordon, A. S. and Milfay, D. 1986. v_1M_r 43,000 component of postsynaptic membranes, is a protein kinase. *Proc. Natl. Acad. Sci. USA 83*: 4172-4174. (2)

Gordon, J. H., Nance, D. M., Wallis, C. J. and Gorski, R. A. 1979. Effects of septal lesions and chronic estrogen treatment on dopamine, GABA and lordosis behavior in male rats. *Brain Res. Bull. 4*: 85-89. (17)

Gorio, A., Carmignoto, G., Finesso, M., Polato, P. and Nunzi, M. G. 1983. Muscle reinnervation. II. Sprouting, synapse formation and repression. *Neurosci. 8*: 403-416. (12)

Gorski, R. A. 1971. Gonadal hormones and the perinatal development of neuroendocrine function. In *Frontiers in Neuroendocrinology, 1971,* Martini, L. and Ganong, W. F. (eds.), Oxford University Press, New York, pp. 237- 290. (17)

Gorski, R. A. 1974. The neuroendocrine regulation of sexual behavior. In *Advances in Psychobiology*, Vol. II, Newton, G. and Riesen, A. H. (eds.), John Wiley & Sons, New York, pp. 1-58. (17)

Gorski, R. A. 1984. Sexual differentiation of brain structure in rodents. In *Sexual Differentiation: Basic and Clinical Aspects*, Serio, M., Zanisi, M., Motta, M. and Martini, L. (eds.), Raven Press, New York, pp. 65-77. (17)

Gorski, R. A. 1985a. Sexual differentiation of the brain: Possible mechanisms and implications. *Can. J. Physiol. Pharmacol. 63*: 577-594. (17)

Gorski, R. A. 1985b. Sexual dimorphisms of the brain. *J. Animal Sci. 61* (suppl. 3): 38-61. (17)

Gorski, R. A. 1985c. Gonadal hormones as putative neurotrophic substances. In *Synaptic Plasticity*, Cotman, W.C., ed., Guilford Press, New York, pp. 287-310. (17)

Gorski, R. A. In press. Sex differences in the rodent brain: Their nature and origin. In *Masculinity/Feminity: Basic Perspectives*, Reinisch, J.M. Rosenblum, L.A. and Sanders, S.A., eds., Oxford University Press, New York. (17)

Gorski, R. A., Csernus, V. J. and Jacobson, C. D. 1981. Sexual dimorphism in the preoptic area. In *Advances in Physiological Sciences*, Vol. 15: *Reproduction and Development*, Flerko, B., Setalo, G. and Tima, L., eds. Pergamon Press and Akadamiai Kiado, Budapest, pp. 121-130. (17)

Gorski, R. A., Gordon, J. H., Shryne, J. E. and Southam, A. M. 1978. Evidence for a morphological sex difference within the medial preoptic area of the rat brain. *Brain Res. 148*: 333-346. (17, 18)

Gorski, R. A., Harlan, R. E., Jacobson, C. D., Shryne, J. E. and Southam, A. M. 1980. Evidence for the existence of a sexually dimorphic nucleus in the preoptic area of the rat. *J. Comp. Neurol. 193*: 529-539. (17)

Gospodarowicz, D. and Mescher, A. L. 1980. Fibroblast growth factor and the control of vertebrate regeneration and repair. *Ann. N.Y. Acad. Sci. 339*: 151-174. (3)

Gottesfeld, Z. and Jacobowitz, D. M. 1979. Cholinergic projections from the septal-diagonal band area to the habenular nuclei. *Brain Res. 176*: 391-394. (5)

Goy, R. W. and McEwen, B. S. 1980. *Sexual differentiation of the brain*. MIT Press, Cambridge, MA. (17)

Green, E. J., Greenough, W. T. and Schlumpf, B. E. 1983. Effects of complex or isolated environments on cortical dendrites of middle-aged rats. *Brain Res. 246*: 233-240. (19)

Green, M. C. 1981. *Genetic Variants and Strains of the Laboratory Mouse*. Gustav Fischer Verlag, New York. (16)

Greenberg, J. R. 1976. Isolation of L-cell messenger RNA which lacks poly (adenylate). *Biochem. 13*: 3677-3682. (4)

Greene, L. A. and Tischler, A. S. 1976. Establishment of a noradrenergic clonal line of rat adrenal pheochromocytoma cells which respond to nerve growth factor. *Proc. Natl. Acad. Sci. USA 73*: 2424-2428. (5, 11)

Greenough, W. T. 1976. Enduring brain effects of differential experience and training. In *Neural Mechanisms of Learning and Memory*, Rosenzweig, M.R. and Bennett, E.L. (eds.), MIT Press, Cambridge. pp. 255-278. (19)

Greenough, W. T. 1984. Structural correlates of information storage in the mammalian brain: A review and hypothesis. *Trends Neurosci. 7*: 229-233. (4, 19)

Greenough, W. T. and Chang, F.-L. C. 1985. Synaptic structural correlates of information storage in mammalian nervous systems. In *Synaptic Plasticity and Remodeling*, Cotman, C.W. (ed.), Guilford Press, New York. pp. 335-372. (19)

Greenough, W. T. and Volkmar, F. R. 1973. Pattern of dendritic branching in occipital cortex of rats reared in complex environments. *Exp. Neurol. 40*: 491-504. (19)

Greenough, W. T., Carter, C. S., Steerman, C. and DeBoogd, T. J. 1977. Sex differences in dendritic patterns in hamster preoptic area. *Brain Res. 126*: 63-72. (17)

Greenough, W. T., Hwang, H.-M., Gorman, C. 1985a. Evidence for active synapse formation or altered postsynaptic metabolism in visual cortex of rats reared in complex environments. *Proc. Natl. Acad. Sci. USA 82*: 4549-4552. (19)

Greenough, W. T., Juraska, J. M. and Volkmar, F. R. 1979. Maze training effects on dendritic branching in occipital cortex of adult rats. *Behav. Neur. Biol. 26*: 287-297. (19)

Greenough, W. T., Larson, J. and Withers, G. 1985b. Effects of unilateral and bilateral training in a reaching task on dendritic branching of neurons in the rat motor-sensory forelimb cortex. *Behav. Neur. Biol. 44*: 301-314. (19)

Greenough, W. T., McDonald, J. W., Parnisari, R. M. and Camel, J. E. 1986. Environmental conditions modulate degeneration and new dendrite growth in cerebellum of senescent rats. *Brain Res. 380*: 136-143. (19)

Greenough, W. T., West, R. W. and DeVoogd, T. J. 1978. Subsynaptic plate perforations: Changes with age and experience in the rat. *Science 202*: 1096-1098. (19)

Greenwald, I., Sternberg, P. W. and Horvitz, H. R. 1983. Lin-12 specifies cell fates in *C. elegans. Cell 34*: 435-444. (14)

Griffin, K. J. P., Fekete, D. M. and Carlson, B. M. In preparation. A monoclonal antibody stains myogenic cells in regenerating newt muscle. (3)

Grinnell, A. D., Herrera, A. and Wolowske, B. 1983. Ultrastructural correlates of differences in synaptic effectiveness at frog neuromuscular junctions. *Soc. Neurosci. Abst. 9*: 1026. (19)

Grouse, L. D., Schrier, B. K., Bennett, E. L., Rosenzweig, M. R. and Nelson, P. G. 1978. Sequence diversity studies of rat brain RNA: Effects of environmental complexity on rat brain RNA diversity. *J. Neurochem. 30*: 191-203. (4)

Grouse, L. D., Schrier, B. K. and Nelson, P. G. 1979. Effect of visual experience on gene expression during the development of stimulus specificity in cat brain. *Exp. Neurol. 64*: 354-364. (4)

Grumet, M. and Edelman, G. M. 1984. Heterotypic binding between neuronal membrane vesicles and glial cells is mediated by a specific cell adhesion *molecule*. *J. Cell Biol. 98*: 1746-1756. (11)

Grumet, M., Hoffman, S., Crossin, K. L. and Edelman, G. M. 1985. Cytoactin, a extracellular matrix protein of neural and non-neural tissues that mediates glia-neuron interaction. *Proc. Natl. Acad. Sci. USA 82*: 8075-8079. (1)

Guillery, R. W. 1982. The optic chiasm of the vertebrate brain. In *Contributions to Sensory Physiology*, Neff, W. D. (ed.). Academic Press, New York. (8)

Güldner, F. H. 1982. Sexual dimorphisms of axo-spine synapses and postsynaptic density material in the suprachiasmatic nucleus of the rat. *Neurosci. Lett. 28*: 145-150. (17)

Gunderson, R. W. and Barrett, J. N. 1980. Characterization of the turning response with dorsal root neurites toward nerve growth factor. *J. Cell Biol. 87*: 546-554. (III)

Gutierrez-Hartman, A. I., Lieberburg, D., Gardner, D., Baxter, J. D. and Cathala, G. G. 1984. Transcription of two classes of rat growth hormone gene-associated repetitive DNA: Differences in activity and effects of tandem repeat structure. *Nucleic Acids Res. 12*: 7153-7173. (6)

Guy, H. R. 1984. A structural model of the acetylcholine receptor channel based on partition energy and helix packing calculations. *Biophys. J. 45*: 249-261. (5)

Hahn, W. E., Chaudhari, N., Beck, L., Wilber, K. and Peffley, D. 1983. Genetic expression and postnatal development of the brain: Some characteristics of nonpolyadenylated mRNAs. *Cold Spring Harbor Symp. Quant. Biol. 47*: 465-475. (4)

Hamburger, V. 1939. The development and innervation of transplanted limb primordia of chick embryos. *J. Exp. Zool. 80*: 347-389. (9)

Hamburger, V. 1958. Regression versus peripheral control of differentiation in motor hypolasia. *Amer. J. Anat. 102*: 365-409. (16)

Hamburger, V. 1975. Cell death in the development of the lateral motor column of the chick embryo. *J. Comp. Neurol. 160*: 535-546. (16)

Hamburger, V. and Hamilton, H. C. 1951. A series of normal stages in the development of the chick embryo. *J. Morphol. 88*: 49-92. (13)

Hamburger, V. and Oppenheim, R. 1982. Naturally occurring neuronal death in vertebrates. *Neurosci. Comment 1*: 39-55. (16, 17)

Hamburger, V., Brunso-Bechtold, J. K. and Yip, J. W. 1981. Neuronal death in the spinal ganglia of the chick embryo and its reduction by nerve growth factor. *J. Neurosci. 1*: 60-71. (14)

Hamilton, S. L., McLaughlin, M. and Karlin, A. 1979. Formation of disulfide-linked oligomers of acetylcholine receptor in membrane from *Torpedo* electric tissue. *Biochemistry 18*: 155-163. (2)

Hammer, R. 1984. The sexually dimorphic region of the preoptic area in rats contains denser opiate receptor binding sites in females. *Brain Res. 308*: 172-176. (18)

Han, J. H., Rall, L. and Rutter, W. J. 1986. Selective expression of rat pancreatic genes during embryonic development. *Proc. Natl. Acad. Sci. USA 83*: 110-114. (1)

Handa, R., Reid, D. and Resko, J. 1986. Androgen receptors in brain and pituitary of female rats: Cyclic changes and comparisons with the male. *Biol. Reprod. 34*: 293-303. (18)

Hanken, J. 1984. Miniaturization and its effects on cranial morphology in plethodontid salamanders, genus *Thorius* (Amphibia: Plethodontidae). I. Osteological variation. *Biol. J. Linnean Soc. 23*: 55-75. (14)

Harkmark, W. 1956. The influence of the cerebellum on the development and maintenance of the inferior olive and the pons: An experimental investigation on chick embryos. *J. Exp. Zool. 131*: 333-371. (16)

Harrelson, A. L., Bastiani, M. J., Snow, P. M. and Goodman, C. S. 1986. Cell surface glycoproteins expressed on subsets of axon pathways during embryonic development of the grasshopper. *Soc. Neurosci. Abst. 12*: 195. (7)

Harris, R. M. and Woolsey, T. A. 1981. Dendritic plasticity in mouse barrel cortex following postnatal vibrissa follicle damage. *J. Comp. Neurol. 196*: 357-376. (19)

Harris, W. A. 1980. The effects of eliminating impulse activity on the development of the retinotectal projection in salamanders. *J. Comp. Neurol. 194*: 303-317. (13)

Harris, W. A. 1984. Axonal pathfinding in the absence of normal pathways and impulse activity. *J. Neurosci. 4*: 1153-1162. (10)

Harris, W. A. 1986. Homing behavior of axons in the embryonic vertebrate brain. *Nature 320*: 266-269. (8, 10)

Harris, W. A., Holt, C. E., Smith, T. A. and Gallenson, N. 1985. Growth cones of developing retinal cells in vivo, on culture surfaces and in collagen matrices. *J. Neurosci. Res. 13*: 101-122. (8, 10)

Harrison, R. G. 1924. Neuroblast versus sheath cell in the development of peripheral nerves. *J. Comp. Neurol. 37*: 123-205. (3)

Hart, B. 1967. Testosterone regulation of sexual reflexes in spinal male rats. *Science 155*: 1283-1284. (18)

Hatten, M. E. 1984. Embryonic cerebellar astroglia in vitro. *Devel. Brain Res. 13*: 309-313. (11)

Hatten, M. E. 1985. Neuronal regulation of astroglial morphology and proliferation in vitro. *J. Cell Biol. 100*: 384-396. (11)

Hatten, M. E. 1987. Neuronal regulation of astroglia proliferation is membrane-mediated. *J. Cell Biol. 104*: 1353-1360. (11)

Hatten, M. E. and Liem, R. K. H. 1981. Astroglial cells provide a template for the positioning of developing cerebellar neurons in vitro. *J. Cell Biol. 90*: 622-630. (11)

Hatten, M. E., Liem, R. K. H. and Mason, C. A. 1984. Two forms of cerebellar glial cells interact differently with neurons in vitro. *J. Cell Biol.* 98: 193-204. (11)

Hatten, M. E., Liem, R. K. H. and Mason, C. A. 1986. *Weaver* mouse cerebellar granule neurons fail to migrate on wild-type astroglial processes in vitro. *J. Neurosci.* 6: 2676-2683. (11)

Hatton, J. D. and Ellisman, M. H. 1982. A restructuring of hypothalamic synapses is associated with motherhood. *J. Neurosci.* 2: 704-707. (19)

Haverkamp, L. 1986. Anatomical and physiological development of the *Xenopus* embryonic motor system in the absence of neural activity. *J. Neurosci.* 6: 1338-1348. (9)

Hay, E. D. 1974. Cellular basis of regeneration. In *Concepts of Development,* Lash, J. W. and Whittaker, J. R. (eds.). Sinauer Associates, Sunderland, MA. (3)

Hayashi, M., Edgar, D. and Thoenen, H. 1985. Nerve growth factor changes the relative levels of neuropeptides in developing sensory and sympathetic ganglia of the chick embryo. *Devel. Biol.* 108: 49-55. (15)

Heacock, A. M. and Agranoff, B. W. 1982. Protein synthesis and transport in the regenerating goldfish visual system. *J. Neurochem. Res.* 7: 771-788. (10)

Hebb, D. O. 1949. *The Organization of Behavior.* New York: John Wiley & Sons. (19)

Hemperly, J. J., Murray, B. A., Edelman, G. M. and Cunningham, B. A. 1986. Sequence of a cDNA clone encoding the polysialic acid-rich and cytoplasmic domains of the neural cell adhesion molecule N-CAM. *Proc. Natl. Acad. Sci. USA 83:* 3037-3041. (1)

Herget, T., Reich, M., Stuber, K. and Starzinski-Powitz, A. 1986. Regulated expression of repetitive sequences including the identifier sequence during myotube formation in culture. *EMBO J.* 5: 659-664. (6)

Herkenham, M. and Nauta, W. J. H. 1977. Afferent connections of the habenular nuclei in the rat. A horseradish peroxidase study, with a note on the fiber-of-passage problem. *J. Comp. Neurol.* 173: 123-146. (5)

Herrup, K. 1983. Role of *staggerer* gene in determining cell number in cerebellar cortex. I. Granule cell death is an indirect consequence of *staggerer* gene action. *Devel. Brain. Res.* 11: 267-274. (16)

Herrup, K. 1986a. Cell lineage relationships in the development of the mammalian CNS. III. Role of cell lineage in regulation of Purkinje cell number. *Devel. Biol.* 115: 148-154. (16)

Herrup, K. In press. Roles of cell lineage in the developing mammalian brain. In *Current Topics in Developmental Biology,* Volume 18, *Neural Development,* Hunt, R. K. (ed.). Academic Press, New York. (16)

Herrup, K. and Mullen, R. J. 1979a. Regional variation and absence of large neurons in the cerebellum of the *staggerer* mouse. *Brain Res.* 172: 1-12. (16)

Herrup, K. and Mullen, R. J. 1979b. *Staggerer* chimeras: Intrinsic nature of Purkinje cell defects and implication for normal cerebellar development. *Brain Res.* 178: 443-457. (16)

Herrup, K. and Sunter, K. 1986. Cell lineage dependent and independent control of Purkinje cell number in the mammalian CNS: Further quantitative studies of *lurcher* chimeric mice. *Devel. Biol.* 117: 417-427. (16)

Herrup, K. and Sunter, K. 1987. Numerical matching during development: Quantitative analysis of granule cell death in *staggerer* mouse chimeras. *J. Neurosci.* 7: 829-836. (16)

Herrup, K., Diglio, T. J. and Letsou, A. 1984a. Cell lineage relationships in the development of the mammalian CNS. I. The facial nerve nucleus. *Devel. Biol.* 103: 329-336. (16)

Herrup, K., Wetts, R. and Biglio, T. J. 1984b. Cell lineage relationships in the development of the mammalian CNS. II. Bilateral independence of CNS clones. *J. Neurogenetics* 1: 275-288. (16)

Heuser, J. E. and Reese, T. S. 1973. Evidence for recycling of synaptic vesicle membrane during transmitter release at the frog neuromuscular junction. *J. Cell Biol.* 57: 314-344. (12)

Hiemke, C., Bruder, D., Poetz, B. and Ghraf, R. 1985. Sex-specific effects of estradiol on hypothalamic noradrenaline turnover in gonadectomized rats. *Exp. Brain Res.* 59: 68-72. (18)

Hier, D. B. 1979. Sex differences in hemispheric specialization: Hypothesis for the excess of dyslexia in boys. *Bull. Orton Soc.* 29: 74-83. (18)

Hille, B. 1984. *Ionic Channels of Excitable Membranes.* Sinauer Associates, Sunderland, MA. (5)

Hines, M., Davis, F. C., Coquelin, A., Goy, R. W. and Gorski, R. A. 1985. Sexually dimorphic regions in the medial preoptic area and the bed nucleus of the stria terminalis of the guinea pig brain: A description and an investigation of their relationship to gonadal steroids in adulthood. *J. Neurosci.* 5: 40-47. (17)

Hirn, M., Pierres, M., Deagostini-Bazin, Hirsch, M. and Goridis, C. 1981. Monoclonal antibody against cell surface glycoprotein of neurons. *Brain Res.* 214: 433-439. (11)

Hitti, Y. S. and Deeb, S. S. 1984. Complexity of polysomal RNA in sheep brain sections and other organs. *Cell and Molec. Biol.* 30: 169-174. (4)

Ho, R. K. and Goodman, C. S. 1982. Peripheral pathways are pioneered by an array of central and peripheral neurones in grasshopper embryos. *Nature 297:* 404-406. (8)

Ho, R. K. and Weisblat, D. A. 1987. A provisional epithelium in leech embryo: Cellular origins and influence on developmental equivalence group. *Devel. Biol.* 120: 520-534. (14)

Ho, R. K., Ball, E. E. and Goodman, C. S. 1983. Muscle pioneers: Large mesodermal cells that erect a scaffold for developing muscles and motoneurones in grasshopper embryos. *Nature 301:* 66-69. (8)

Hoffman, S., Friedlander, D. F., Chuong, C.-M., Grumet, M., and Edelman, G. M. 1986. Differential contributions of Ng-CAM to N-CAM to cell adhesion in different neural regions. *J. Cell Biol.* 103: 145-158. (1)

Hokfelt, T., Johansson, O. and Goldstein, M. 1984. Chemical anatomy of the brain. *Science 225:* 1326-1334. (15)

Hollyday, M. 1980. Organizational motor pools in the chick lumbar lateral motor column. *J. Comp. Neurol. 194*: 143-170. (9)

Hollyday, M. 1981. Rules of motor innervation in chick embryos with supernumerary limbs. *J. Comp. Neurol. 202*: 439-465. (9)

Hollyday, M. 1983. Development of motor innervation of chick limbs. In *Limb Development and Regeneration, Part A.* Fallon, J. and Caplan, A. (eds.). Alan R. Liss, New York. pp. 183-193. (9)

Hollyday, M. and Hamburger, V. 1976. Reduction of the naturally occurring motor neuron loss by enlargement of the periphery. *J. Comp. Neurol. 170*: 311-320. (16)

Holton, B. and Weston, J. A. 1982. Analysis of glial cell differentiation in peripheral nervous tissue. *Devel. Biol. 89*: 64-81. (15)

Hong, J., Yoshikawa, K. and Lamartiniere, C. 1982. Sex-related differences in the rat pituitary [Met5]-enkephalin level-altered by gonadectomy. *Brain Res. 251*: 380-383. (18)

Honig, M. and Hume, R. 1986. Fluorescent carbocyanine dyes allow living neurons of identified origin to be studied in long-term cultures. *J. Cell Biol. 103*: 171-187. (9)

Hopkins, W. G., Brown, M. C. and Keynes, R. J. 1985. Postnatal growth of motor nerve terminals in muscles of the mouse. *J. Neurocytol. 14*: 525-540. (12)

Horder, T. J. and Martin, K. A. C. 1978. Morphogenesis as an alternative to chemospecificity in the formation of nerve connections. In *Society for Experimental Biology, Biology Symposium, Cell-Cell Recognition,* Vol. 32, Curtis, A. S. G. (ed.)., Cambridge University Press, Cambridge, England. (10)

Horvitz, H. R., Ellis, H. M. and Sternberg, P. M. 1982. Programmed cell death in nematode development. *Neurosci. Commun. 1*: 56-65. (14)

Hosley, M. A., Hughes, S. E., Morton, L. L. and Oakley, B. In press. A sensitive period for the neural induction of taste buds. *J. Neuroscience.* (I)

Hosley, M. A., Hughes, S. E. and Oakley, B. In press. Neural induction of taste buds. *J. Comp. Neurol.* (I)

Hubel, D. H. and Wiesel, T. N. 1965. Binocular interaction in striate cortex of kittens reared with artificial squint. *J. Neurophysiol. 28*: 1041-1059. (13)

Hubel, D. H. and Wiesel, T. N. 1970. The period of susceptibility to the physiological effects of unilateral eye closure in kittens. *J. Physiol.* (Lond.) *206*: 416-436. (13)

Hubel, D. H., Wiesel, T. N. and LeVay, S. 1977. Plasticity of ocular dominance columns in monkey striate cortex. *Philos. Trans. R. Soc. Lond. B. 278*: 377-409. (8, 13)

Hudson, R. C. L. 1969. Polyneuronal innervation of the fast muscles of the marine teleost *Cottus scorpius* L. *J. Exp. Biol. 50*: 47-67. (8)

Huganir, R. L., Delcour, A. H., Greengard, P. and Hess, G. P. 1986. Phosphorylation of the nicotinic acetylcholine receptor regulates its rate of desensitization. *Nature 321*: 774-776. (2)

Hunt, S. P. and Schmidt, J. 1978. The electron microscopic autoradiographic localization of α-bungarotoxin binding sites within the central nervous system of the rat. *Brain Res. 142*: 152-159. (5)

Hunter, T. and Garrels, J. I. 1977. Characterization of the mRNAs for α-, β-, and γ-actin. *Cell 12*: 767-781. (4)

Hwang, H. M. and Greenough, W. T. 1984. Spine formation and synaptogenesis in rat visual cortex: A serial section developmental study. *Soc. Neurosci. Abst. 10*: 579. (19)

Hwang, H.-M. and Greenough, W. T. 1986. Synaptic plasticity in adult rat occipital cortex following short-term, long-term, and reversal of differential housing environment complexitiy. *Soc. Neurosci. Abst. 12*: 1284. (19)

Hymovitch, B. 1952. The effects of experimental variations on problem solving in the rat. *J. Comp. Physiol. Psychol. 45*: 313-321. (19)

Iacovitti, L., Joh, T. H., Park, D. H. and Bunge, R. R. 1981. Dual expression of neurotransmitter synthesis in cultural autonomic neurons. *J. Neurosci. 1*: 685-690. (4)

Ifft, J. D. 1972. An autoradiographic study of the time of final division of neurons in rat hypothalamic nuclei. *J. Comp. Neurol. 144*: 193-204. (17)

Imada, M. and Sueoka, N. 1978. Clonal sublines of rat neurotumor RT4 and cell differentiation. I. Isolation and characterization of cell lines and cell type conversion. *Devel. Biol. 66*: 97-108. (4)

Innocenti, G. M., Clarke, S. and Kraftsik, R. 1986. Interchange of callosal and association projections in the developing visual cortex. *J. Neurosci. 6*: 1384-1409. (10)

Inouye, M. and Murakami, U. 1980. Temporal and spatial patterns of Purkinje cell formation in the mouse cerebellum. *J. Comp. Neurol. 194*: 499-504. (16)

Inouye, M. and Oda, S.-I. 1980. Strain specific variations in the folial pattern of the mouse cerebellum. *J. Comp. Neurol. 190*: 357-362. (16)

Iten, L. E. and Bryant, S. V. 1973. Forelimb regeneration from different levels of amputation in the newt, *Notophthalmus viridescens*: Length rate and stages. *Wilhelm Roux Arch. Entwicklungsmech. Org. 173*: 263-282. (3)

Jackson, P. C. and Frank, E. 1987. Development of synaptic connections between muscle sensory afferents and motor neurons: Anatomical evidence that postsynaptic dendrites grow into a pre-formed sensory neuropil. *J. Comp. Neurol. 255*: 538-547. (13)

Jacob, M. H., Berg, D. K. and Lindstrom, J. M. 1984. Shared antigenic determinant between the electrophorus acetylcholine receptor and a synaptic component on chicken ciliary ganglion neurons. *Proc. Natl. Acad. Sci. USA 81*: 3223-3227. (5)

Jacob, M., Choo, Q. L. and Thomas, C. 1982. Vimentin and 70-kD neurofilament protein coexist in embryonic neurones from spinal ganglia. *J. Neurochem. 38*: 969-977. (3)

Jacobson, C.-O. 1959. The localization of the presumptive cerebral regions in the neural plate of the axolotl larva. *J. Embryol. Exp. Morphol.* 7: 1- 21. (16)

Jacobson, C. D. and Gorski, R. A. 1981. Neurogenesis of the sexually dimorphic nucleus of the preoptic area of the rat. *J. Comp. Neurol. 196*: 512-529. (17)

Jacobson, C. D., Csernus, V. J., Shryne, J. E. and Gorski, R. A. 1981. The influence of gonadectomy, androgen exposure, or a gonadal graft in the neonatal rat on the volume of the sexually dimorphic nucleus of the preoptic area. *J. Neurosci. 1*: 1142-1147. (17)

Jacobson, C. D., Davis, F. C. and Gorski, R. A. 1985. Formation of the sexually dimorphic nucleus of the preoptic area: Neuronal growth, migration and changes in cell number. *Dev. Brain Res. 21*: 7-18. (17)

Jacobson, C. D., Shryne, J. E., Shapiro, F. and Gorski, R. A. 1980. Ontogeny of the sexually dimorphic nucleus of the preoptic area. *J. Comp. Neurol. 193*: 541-548. (17)

Jacobson, M. 1978. *Developmental Neurobiology,* 2nd ed. Plenum Press, New York. (13)

Jacobson, R. D., Virâg, I. and Skene, J. H. P. 1986. A protein associated with axon growth, GAP-43, is widely distributed and developmentally regulated in rat CNS. *J. Neurosci. 6*: 1843-1855. (10)

Jan, L. Y. and Jan, Y. N. 1982. Antibodies to horseradish peroxidase as specific neuronal markers in *Drosophila* and in grasshopper embryos. *Proc. Natl. Acad. Sci. USA 79*: 2700-2704. (7)

Jhaveri, S. and Frank, E. 1983. Central projections of the brachial nerve in bullfrogs: Muscle and cutaneous afferents project to different regions of the spinal cord. *J. Comp. Neurol. 221*: 304-312. (13)

Johnson, M. I., Ross, C. D. and Bunge, R. P. 1980. Morphological and biochemical studies on the development of cholinergic properties in cultured sympathetic neurons. II. Dependence on postnatal age. *J. Cell Biol. 84*: 692-704. (4)

Johnson, M., Ross, D., Meyers, M., Rees, R., Bunge, R. et al. 1976. Synaptic vesicle cytochemistry changes when cultured neurons develop cholinergic interactions. *Nature 262*: 308-310. (4)

Jones, K., Chikaraishi, D. and Harrington, C. In press. In situ hybridization detection of estradiol-induced changes in ribosomal RNA levels in rat brain. *Mol. Brain Res.* (18)

Jones, K., Pfaff, D. W. and McEwen, B. S. 1985. Early estrogen-induced nuclear changes in rat hypothalamic ventromedial neurons: An ultrastructural and morphometric analysis. *J. Comp. Neurol. 239*: 255-266. (18)

Jonklaas, J. and Buggy, J. 1985. Angiotensin-estrogen central interaction: Localization and mechanism. *Brain Res. 326*: 239-249. (18)

Jost, A. 1970. Hormonal factors in the sex differentiation of the mammalian foetus. *Phil. Trans. Roy. Soc. Lond. B. 259*: 119-130. (18)

Juraska, J. M. 1984. Sex differences in dendritic response to differential experience in the rat visual cortex. *Brain Res. 295*: 27-34. (19)

Juraska, J. M. 1986. Sex differences in developmental plasticity in behavior and the brain. In *Developmental Neuro/Psychobiology,* Greenough, W.T. and Juraska, J.M. (eds.). Academic Press, New York. pp. 409-422. (19)

Juraska, J. M., Greenough, W. T., Elliot, G., Mack, K. and Berkowitz, R. 1980. Plasticity in adult rat visual cortex: An examination of several cell populations after differential rearing. *Behav. Neur. Biol. 29*: 157-167. (19)

Kahn, C. R. and Sieber-Blum, M. 1983. Cultured quail neural crest cells attain competence for terminal differentiation into melanocytes before competence for terminal differentiation into adrenergic neurons. *Devel. Biol. 95*: 232-238. (15)

Kalil, K. 1984. Development and regrowth of the rodent pyramidal tract. *Trends Neurosci. 7*: 394-398. (10)

Kalil, K. 1985. Development and plasticity of the sensorimotor cortex and pyramidal tract. In *Development, Organization, and Processing in Somatosensory Pathways,* Willis, W. D. and Rowe, M. (eds.), Alan R. Liss, New York. (10)

Kalil, K. and Norris, C. 1985. Rearrangement of axon fascicles in the decussation of the pyramidal tract. *Soc. Neurosci. Abst. 11*: 584. (10)

Kalil, K. and Norris, C. 1986. Axon fasciculation and growth cone morphology in the developing corpus callosum. *Soc. Neurosci. Abst. 12*: 503. (10)

Kalil, K. and Reh, T. 1979. Regrowth of severed axons in the neonatal central nervous system: Establishment of normal connections. *Science 205*: 1158-1161. (10)

Kalil, K. and Reh, T. 1982. A light and electron microscopic study of regrowing pyramidal tract fibers. *J. Comp. Neurol. 211*: 265-275. (10)

Kalil, K. and Skene, J. H. P. 1986. Elevated synthesis of an axonally transported protein correlates with axon outgrowth in normal and injured pyramidal tracts. *J. Neurosci. 6*: 2563-2570. (10)

Kalra, S. P. and Kalra, P. S. 1983. Neural regulation of luteinizing hormone secretion in the rat. *Endocr. Rev. 4*: 311-351. (17, 18)

Kamberi, I. and Kobayashi, Y. 1970. Monoamine oxidase activity in the hypothalamus and various other brain areas and in some endocrine glands of the rat during the estrous cycle. *J. Neurochem. 17*: 261-268. (18)

Kao, P. N., Dwork, A. J., Kaldany, R. J., Silver, M. L., Wideman, J. et al. 1984. Identification of two alpha-subunit half-cystines specifically labeled by an affinity reagent for the acetylcholine binding site. *J. Biol. Chem. 259*: 1162-1165. (5)

Kapfhammer, J. P., Grunewald, B. E. and Raper, J. A. 1986. The selective inhibition of growth cone extension by specific neurites in culture. *J. Neurosci. 9*: 2527-2534. (III)

Kaplan, B. B. and Gioio, A. E. 1986. Diversity of gene expression in goldfish brain. *Comp. Biochem. Physiol. 83B*: 305-308. (4)

Karlin, A. 1980. Molecular properties of nicotinic acetyl-choline receptors. In *The Cell Surface and Neuronal Function,* Cotman, C., Post, G. and Nicholson, G. (eds.). Elsevier/North Holland, New York. pp. 191-260. (2, 5)

Karlin, A. and Cowburn, D. A. 1973. The affinity-labeling of partially purified acetylcholine receptor from electric tissue of electrophorus. *Proc. Natl. Acad. Sci. USA* 70: 3636-3640. (5)

Karlin, A., DiPaola, M., Kao, P. N. and Lobel, P. 1986. Functional sites and transient states of the nicotinic acetylcholine receptor. In *Proteins of Excitable Membrane,* Hille, B. and Fambrough, D. M. (eds.). John Wiley & Sons, New York. (5)

Kasamatsu, T. and Pettigrew, J. D. 1976. Depletion of brain catecholamines: Failure of ocular dominance shift after monocular occlusion in kittens. *Science* 194: 206-209. (19)

Kasamatsu, T. and Pettigrew, J. D. 1979. Preservation of binocularity after monocular deprivation in the striate cortex of kittens treated with 6- hydroxydopamine. *J. Comp. Neurol. 185*: 139-162. (19)

Kato, J. 1985. Progesterone receptors in brain and hypophysis. *Curr. Topics Neuroendo. 5*: 31-81. (18)

Katz, B. 1966. *Nerve, Muscle and Synapse.* McGraw-Hill, New York. (5)

Katz, M. J. and Lasek, R. J. 1978. Eyes transplanted to tadpole tails send axons rostrally in two spinal cord tracts. *Science 199*: 202-204. (III)

Katz, M. J. and Lasek, R. J. 1979. Substrate pathways which guide growing axons in *Xenopus* embryos. *J. Comp. Neurol. 183*: 817-832. (III)

Kauffman, S. L. 1968. Lengthening of the generation cycle during embryonic differentiation of the mouse neural tube. *Exp. Cell. Res. 49*: 420-424. (16)

Keifer, J. and Kalil, K. 1985. Functional role of the hamster pyramidal tract during locomotion. *Soc. Neurosci. Abst. 11*: 1030. (10)

Keshishian, H. and Bentley, D. 1983. Embryogenesis of peripheral nerve pathways in grasshopper legs. I. The initial nerve pathway to the CNS. II. The major nerve routes. III. Development without pioneer neurons. *Devel. Biol. 96*: 89-124. (8)

Kikuyama, S. 1966. Influence of thyroid hormone on the induction of persistent estrus by androgen in the rat. *Sci. Papers Coll. Gen. Educ. Univ. Tokyo 16*: 265-270. (18)

Kikuyama, S. 1969. Alteration by neonatal hypothroidism of the critical period for the induction of persistent estrus in the rat. *Endocr. Japan 16*: 269-273. (18)

Killackey, H. P. and Leshin, S. 1975. The organization of specific thalamocortical projections to the posteromedial barrel subfield of the rat somatic sensory cortex. *Brain Res 86*: 469-472. (19)

Kimble, J. 1981. Lineage alterations after ablation of cells in the somatic gonad of *Caenorhabditis elegans. Devel. Biol. 87*: 286-300. (14)

Kimble, J., Sulston, J. and White, J. 1979. Cell lineage, stem cells and cell determination. INSERM symposium No. 10, Le Douarin, N. (ed.). Elsevier, Amsterdam. pp. 59-68. (14)

Kimmel, C. B. and Law, R. D. 1985. Cell lineage of zebrafish blastomeres. *Devel. Biol. 108*: 78-101. (8)

Kintner, C. R. and Brockes, J. P. 1984. Monoclonal antibodies identify blastemal cells derived from dedifferentiating muscle in newt limb regeneration. *Nature 308*: 67-69. (3)

Kintner, C. R. and Brockes, J. P. 1985. Monoclonal antibodies to cells of a regenerating limb. *J. Embryol. Exp. Morph. 89*: 37-55. (3)

Kintner, C. R., Lemke, G. E. and Brockes, J. P. 1985. Glial growth factor and the neuronal control of cell division in amphibian limb regeneration. In *Molecular Bases of Neural Development,* Edelman, G. M., Gall, W. E. and Cowan, W. M. (eds.). John Wiley & Sons, New York. (3)

Kleene, K. C. and Humphreys, T. 1977. Similarity of hnRNA sequences in blastula and pluteus stage sea urchin embryos. *Cell 12*: 143-155. (4)

Knudsen, E. I. 1980. Sound localization in birds. In *Comparative Studies of Hearing in Vertebrates,* Popper and Fay (eds.). Springer-Verlag, New York. pp. 287-322. (20)

Knudsen, E. I. 1982. Auditory and visual maps of space in the optic tectum of the owl. *J. Neurosci. 2*: 1177-1194. (20)

Knudsen, E. I. 1983. Early auditory experience aligns the auditory map of space in the optic tectum of the barn owl. *Science 222*: 939-942. (20)

Knudsen, E. I. 1984a. The role of auditory experience in the development and maintenance of sound localization. *Trends Neurosci. 7*: 326-330. (20)

Knudsen, E. I. 1984b. Auditory properties of space-tuned units in owl's optic tectum. *J. Neurophysiol. 52*: 709-723. (20)

Knudsen, E. I. 1985. Experience alters spatial tuning of auditory units in the optic tectum during a sensitive period in the barn owl. *J. Neurosci. 5*: 3094-3109. (20)

Knudsen, E. I. and Knudsen, P. F. 1985a. Vision guides the adjustment of auditory localization in young barn owls. *Science 230*: 545-548. (13, 20)

Knudsen, E. I. and Knudsen, P. F. 1985b. Vision guides the development of auditory localization in barn owls. *Soc. Neurosci. Abst. 11*: 735. (20)

Knudsen, E. I. and Knudsen, P. F. 1986a. The sensitive period for auditory localization in barn owls is limited by age, not by experience. *J. Neurosci. 6*: 1918-1924. (20)

Knudsen, E. I. and Knudsen, P. F. 1986b. Sensitive and critical periods for the visual control of sound localization. *Soc. Neurosci. Abst. 12*: 1052. (20)

Knudsen, E. I. and Konishi, M. 1979. Mechanisms of sound localization by the barn owl (*Tyto alba*). *J. Comp. Physiol. 133*: 13-21. (13)

Knudsen, E. I., Blasdel, G. C., and Konishi, M. 1979. Sound localization by the barn owl measured with the search coil technique. *J. Comp. Physiol. 133*: 1-11. (20)

Knudsen, E. I., Esterly, S. D. and Knudsen, P. F. 1984a. Monaural occlusion alters sound localization during a sensitive period in the barn owl. *J. Neurosci. 4*: 1001-1011. (13, 20)

Knudsen, E. I., Knudsen, P. F., and Esterly, S. D. 1982. Early auditory experience modifies sound localization in barn owls. *Nature 295*: 238-240. (20)

Knudsen, E. I., Knudsen, P. F., and Esterly, S. D. 1984b. A critical period for the recovery of sound localization accuracy following monaural occlusion in the barn owl. *J. Neurosci. 4*: 1012-1020. (20)

Ko, P. K., Anderson, M. J. and Cohen, M. W. 1977. Denervated skeletal muscle fibers develop discrete patches of high acetylcholine receptor density. *Science 196*: 540-542. (12)

Kollros, J. J. 1943. Experimental studies on the development of the corneal reflex in amphibia. III. The influence of the periphery on the reflex center. *J. Exp. Zool. 92*: 121-142. (13)

Konishi, M. 1973. How the owl tracks its prey. *Amer. Sci. 61*: 414-424. (20)

Korenbrot, C. D., Paup, D. and Gorski, R. A. 1975. Effects of testosterone or dihydrotestosterone propionate on plasma FSH and LH levels in neonatal rats and on sexual differentiation of the brain. *Endocrinology 97*: 709-717. (17)

Kramer, A. P. and Goldman, J. R. 1981. The nervous system of the glossiphoniid leech *Haementeria ghilianii*. I. Identification of neurons. *J. Comp. Physiol. 144*: 435-448. (8)

Kramer, A. P. and Kuwada, J. Y. 1983. Formation of the receptive fields of leech mechanosensory neurons during embryonic development. *J. Neurosci. 3*: 2474-2486. (8)

Kramer, A. P. and Stent, G. S. 1985. Developmental arborization of sensory neurons in the leech *Haementeria ghilianii*. II. Experimentally induced variations in the branching pattern. *J. Neurosci. 5*: 768-775. (8)

Kramer, A. P. and Weisblat, D. A. 1985. Developmental neural kinship groups in the leech. *J. Neurosci. 5*: 388-407. (14)

Kramer, A. P., Goldman, J. R. and Stent, G. S. 1985. Developmental arborization of sensory neurons in the leech *Haementeria ghilianii*. I. Origin of natural variations in the branching pattern. *J. Neurosci. 5*: 759-767. (8)

Kranzler, J., Jones, E., MacLusky, N., Mamoto, H. and Naftolin, F. 1984. An exchange assay for the measurement of cell nuclear estrogen receptors in microdissected brain regions. *J. Neurochem.* 895-898. (18)

Krech, D., Rosenzweig, M. R. and Bennett, E. L. 1960. Effects of environmental complexity and training on brain chemistry. *J. Comp. Physiol. Psychol. 53*: 509-519. (19)

Krey, L., Lieberburg, I., MacLusky, N., Davis, P. and Robbins, R. 1982. Testosterone increases cell nuclear estrogen receptor levels in the brain of the Stanley-Gumbreck pseudohermaphrodite male rat: Implications for testosterone modulation of neuroendocrine activity. *Endocrinology 110*: 2168-2175. (18)

Kuwada, J. 1982. Primary axon outgrowth in embryonic leech neurons. In *Neuronal Development: Cellular Approaches in Invertebrates,* Goodman, C. and Pearson, K. (eds.). NRP Bulletin, MIT Press, Cambridge. (8)

Kuwada, J. 1986. Cell recognition by neuronal growth cones in a simple vertebrate embryo. *Science 233*: 740-746. (7, 10, 13)

Kuwada, J. and Goodman, C. S. 1985. Neuronal determination during embryonic development of the grasshopper nervous system. *Devel. Biol. 110*: 114-126. (14)

Kuwada, J. and Kramer, A. P. 1983. Embryonic development of the leech nervous system: Primary axon outgrowth of identified neurons. *J. Neurosci. 3*: 2098-2111. (8)

LaPolla, R. J., Mayne, K. and Davidson, N. 1984. Isolation and characterization of a cDNA clone for the complete protein coding region of the delta-subunit of the mouse acetylcholine receptor. *Proc. Natl. Acad. Sci. USA 81*: 7970-7974. (5)

Lamb, A. H. 1976. The projection patterns of the ventral horn to the hind limb during development. *Devel. Biol. 54*: 82-99. (8, 9)

Lamb, A. H. 1977. Neuronal death in the development of the somatotopic projections of the ventral horn in *Xenopus. Brain Res. 134*: 145-150. (8)

Lamb, A. H. 1979. Evidence that some developing limb motoneurones die for reasons other than peripheral competition. *Devel. Biol. 71*: 8-21. (8, 9)

Lamb, A. H. 1980. Motoneurone counts in *Xenopus* frogs reared with one bilaterally-innervated hindlimb. *Nature 284*: 347-350. (16)

Lamb, A. H. 1981. Target dependency of developing motoneurons in *Xenopus laevis. J. Comp. Neurol. 203*: 157-171. (16)

Lampidis, T. J., Salet, C., Moreno, G. and Che, L. B. 1984. Effects of the mitochondrial probe rhodamine 123 and related analogs on the function and viability of pulsating myocardial cells in culture. *Ag. and Act. 14*: 751- 757. (12)

Lance-Jones, C. 1982. Motoneuron cell death in the developing lumbar spinal cord of the mouse. *Devel. Brain Res. 4*: 473-479. (16)

Lance-Jones, C. and Lagenauer, C. In preparation. Development of specific nerve-muscle projections in the chick limb: Early motoneuron-somite relationships. (9)

Lance-Jones, C. and Landmesser, L. 1980. Motoneuron projection patterns in the chick hind limb following early partial spinal cord reversals. *J. Physiol. 302*: 581-602. (8, 9, 16)

Lance-Jones, C. and Landmesser, L. 1981a. Pathway selection by chick lumbosacral motoneurons during normal development. *Proc. Roy. Soc. Lond. B. 214*: 1-18. (8, 9, 10, 13)

Lance-Jones, C. and Landmesser, L. 1981b. Pathway selection by embryonic chick motoneurons in an experimentally altered environment. *Proc. Roy. Soc. Lond. B. 214*: 19-52. (9)

Landis, D. 1971. Cerebellar cortical development in the *staggerer* mutant mouse. *Proc. Amer. Soc. Cell Biol. 159.* (16)

Landis, D. M. D. and Sidman, R. L. 1978. Electron microscope analysis of postnatal histogenesis in the cerebellar cortex of *staggerer* mutant mice. *J. Comp. Neurol. 179*: 831-863. (16)

Landis, S. C. 1983. Factors which influence the transmitter functions of sympathetic ganglion cells. In *Autonomic Ganglia*, Elfvin, L.-G. (ed.). John Wiley & Sons. (4)

Landmesser, L. 1978a. The distribution of motoneurons supplying chick hindlimb muscles. *J. Physiol. 284*: 371-389. (8, 9)

Landmesser, L. 1978b. The development of motor projection patterns in the chick hind limb. *J. Physiol. 284*: 391-414. (8, 9)

Landmesser, L. 1980. The generation of neuromuscular specificity. *Annu. Rev. Neurosci. 3*: 279-301. (8)

Landmesser, L. 1981. Pathway selection by embryonic neurons. In *Studies in Developmental Neurobiology*, Cowan, W. M. (ed.). Oxford, New York. (8)

Landmesser, L. 1984. The development of specific motor pathways in the chick embryo. *Trends Neurosci. 7*: 336-339. (9, III)

Landmesser, L. 1986. Axonal guidance and the formation of neuronal circuits. *Trends Neurosci. 9*: 489-492. (III)

Landmesser, L. and Morris, D. G. 1975. The development of functional innervation in the hind limb of the chick embryo. *J. Physiol. 249*: 301-326. (8, 9, 16)

Landmesser, L. and O'Donovan, M. 1984a. Activation patterns of embryonic chick hindlimb muscles recorded in ovo and in an isolated spinal cord preparation. *J. Physiol. 347*: 189-204. (9)

Landmesser, L. and O'Donovan, M. 1984b. The activation patterns of embryonic chick motoneurones projecting to inappropriate muscles. *J. Physiol. 347*: 205-224. (9, 13)

Landmesser, L. and Pilar, G. 1974. Synapse formation during embryogenesis on ganglion cells lacking a periphery. *J. Physiol. 241*: 715-736. (16)

Landmesser, L. and Pilar, G. 1976. Fate of ganglionic synapses and ganglion cell axons during normal and induced cell death. *J. Cell Biol. 68*: 357-374. (16)

Landmesser, L. and Szente, M. 1986. Activation patterns of embryonic chick hind-limb muscles following blockade of activity and motorneurone cell death. *J. Physiol. 380*: 157-174. (9)

Langley, J. N. 1895. Note on regeneration of preganglionic fibres of the sympathetic. *J. Physiol. 18*: 280-284. (IV)

Lanser, M. E. and Fallon, J. F. 1984. Development of the lateral motor column in the *limbless* mutant chick embryo. *J. Neurosci. 4*: 2043-2050. (16)

Larramendi, L. M. H. 1969. Analysis of synaptogenesis in the cerebellum of the mouse. In *Neurobiology of Cerebellum. Evolution and Development*, Llinas, R. (ed.). American Medical Association, Chicago. pp. 803-843. (16)

Lawrence, C. B., McDonnell, D. P. and Ramsey, W. J. 1985. Analysis of repetitive sequence elements containing tRNA-like sequences. *Nucleic Acids Res. 13*: 4239-4251. (6)

Le Douarin, N. M. 1982. *The Neural Crest*. Cambridge University Press, New York. (13, V, 15)

Le Douarin, N. M, and McLaren, A. (eds.) 1984. *Chimeras in Developmental Biology*. Academic Press, New York. (16)

Le Douarin, N. M. and Teillet, M.-A. 1974. Experimental analysis of the migration and differentiation of neuroblasts of the autonomic nervous system and of neurectodermal mesenchymal derivatives. *Devel. Biol. 41*: 162-184. (15)

Le Douarin, N. M., Teillet, M. A., Ziller, C. and Smith, J. 1978. Adrenergic differentition of cells of the cholinergic ciliary and Remak ganglia in avian embryos following *in* vitro transplantation. *Proc. Nat. Acad. Sci. USA 75*: 2030-2034. (15)

Le Lievre, C. S. and Le Douarin, N. M. 1975. Mesenchymal derivatives of the neural crest: Analysis of chimaeric quail and chick embryos. *J. Embryol. Exp. Morphol. 34*: 125-154. (15)

Le Lievre, C. S., Schweitzer, G. G., Ziller, C. M. and Le Douarin, N. M. 1980. Restrictions of developmental capabilities in neural crest cell derivatives as tested by in vivo transplantation experiments. *Devel. Biol. 77*: 362-378. (15)

LeVay, S. and Stryker, M. P. 1979. The development of ocular dominance columns in the cat. In *Aspects of Developmental Neurobiology* (Society for Neuroscience Symposium), Ferrendelli, J. A. (ed.), Society for Neuroscience, Bethesda, MD, pp. 83-98. (13)

LeVay, S., Hubel, D. H. and Wiesel, T. N. 1977. The development of ocular dominance columns in normal and visually deprived monkeys. *J. Comp. Neurol. 191*: 1-51. (13)

LeVay, S., Stryker, M. P. and Shatz, C. J. 1978. Ocular dominance columns and their development in layer IV of the cat's visual cortex: A quantitative study. *J. Comp. Neurol. 179*: 223-244. (8)

LeVay, S., Wiesel, T. N. and Hubel, D. N. 1980. The development of ocular dominance columns in normal and visually deprived monkeys. *J. Comp. Neurol. 191*: 1-51. (5, 19)

Lee, K., Oliver, M., Schottler, F. and Lynch, G. 1981. Electron microscopic studies of brain slices: The effects of high-frequency stimulation on dendritic ultrastructure. In *Electrophysiology of Isolated Mammalian CNS Preparations*, Kerkut, G.A. and Wheal, H.V. (eds.). Academic Press: New York. pp. 189-211. (19)

Lee, K. S., Schottler, F. Oliver, M. and Lynch, G. 1980. Brief bursts of high-frequency stimulation produce two types of structural change in rat hippocampus. *J. Neurophysiol. 44*: 247-258. (19)

Lemischka, I. and Sharp, P. 1982. The sequences of an expressed rat alpha-tubulin gene and a pseudogene with an inserted repetitive element. *Nature 300*: 330-335. (6)

Letinsky, M. S., Fischbeck, K. H. and McMahan, U. J. 1976. Precision of reinnervation of original post-synaptic sites in frog muscle after a nerve crush. *J. Neurocytol.* 5: 691-718. (12)

Letourneau, P. C. 1975. Cell-to-substratum adhesion and guidance of axonal elongation. *Devel. Biol.* 44: 92-101. (III)

Letourneau, P. C. 1975. Possible roles for cell-to-substratum adhesion in neuronal morphogenesis. *Devel. Biol.* 44: 77-91. (10)

Letourneau, P. 1982. Nerve fiber growth and its regulation by extrinsic factors. In *Neuronal Development,* Spitzer, N. C. (ed.). Plenum Press, New York. (8)

Letourneau, P. C. 1983. Axonal growth and guidance. *Trends Neurosci.* 6: 451-455. (10)

Letourneau, P. C. 1986. Branching of sensory and sympathetic neurites in vitro is inhibited by treatment with taxol. *J. Neurosci.* 6: 1912-1917. (10)

Lev, Z., Thomas, T. L., Lee, A. S., Angerer, R. C., Britten, R. J. et al. 1980. Developmental expression of two cloned sequences coding for rare sea urchin embryo messages. *Devel. Biol.* 76: 322-340. (4)

Levy, W. B. 1985. Associative changes in the synapse: LTP in the hippocampus. In *Synaptic Modification, Neuron Selectivity and Nervous System Organization,* Levy, W. B., Anderson, J. and Lehmkuble, S. (eds.). Lawrence Erlbaum, Hillsdale, N. J. (5)

Lewis, E. J., Tank, A. W., Weiner, N. and Chikaraishi, D. M. 1983. Regulation of tyrosine hydroxylase mRNA by glucocorticoids and cAMP in rat pheochromocytoma cell line: Isolation of a cDNA clone for tyrosine hydroxylase mRNA. *J. Biol. Chem.* 258: 14632-14637. (4)

Lewis, J., Al-Ghaith, L., Swanson, G. and Khan, A. 1983. The control of axon outgrowth in the developing chick wing. In *Limb Development and Regeneration,* Part A. Eds., Fallon, J. and Caplan, A. Alan R. Liss, New York. (9)

Lewis, P. D., Patel, A. J., Johnson, A. L. and Balazs, R. 1976. Effect of thyroid deficiency on cell acquisition in the postnatal rat brain: A quantitative histological study. *Brain Res.* 104: 49-62. (16)

Libertun, C., Timiras, P. and Kragt, C. 1973. Sexual differences in the hypothalamic cholinergic system before and after puberty: Inductive effect of testosterone. *Neuroendocrinology* 12: 73-85. (18)

Lichtman, J. W. and Frank, E. 1984. Physiological evidence for specificity of synaptic connections between individual sensory and motor neurons in the brachial spinal cord of the bullfrog. *J. Neurosci.* 4: 1745-1753. (13)

Lichtman, J. W. and Purves, D. 1981. Regulation of the number of axons that innervate target cells. In *Development of the Nervous System.* Garrod, D. R. and Feldman, J. D. (eds.). Cambridge Univ. Press, Cambridge. pp. 233-243. (12)

Lichtman, J. W., Jhaveri, S. and Frank, E. 1984. Anatomical basis of specific connections between sensory axons and motor neurons in the bullfrog's brachial spinal cord. *J. Neurosci.* 4: 1754-1763. (13)

Lichtman, J. W., Magrassi, L. and Purves, D. 1986. Repeated visualization of neuromuscular junctions in the living mouse. *Soc. Neurosci. Abst. 12:* 390. (12)

Lichtman, J. W., Magrassi, L. and Purves, D. 1987. Visualization of motor nerve terminals over time in living mice. *J. Neurosci.* 7: 1215-1222. (12)

Lichtman, J. W., Wilkinson, R. S. and Rich, M. M. 1985. Multiple innervation of tonic endplates revealed by activity-dependent uptake of fluorescent probes. *Nature* 314: 357-359. (12)

Lidov, H. G. W. and Molliver, M. E. 1982. The structure of cerebral cortex in the rat following prenatal administration of 6-hydroxydopamine. *Devel. Brain Res.* 3: 81-108. (19)

Lieberburg, I. and McEwen, B. S. 1977. Brain cell nuclear retention of testosterone metabolites, 5-α-dihydrotestosterone and estradiol-17-β in adult rats. *Endocrinology 100:* 588-597. (18)

Lieberburg, I., Krey, L. and McEwen, B. S. 1979. Sex differences in serum testosterone and in exchangeable brain cell nuclear estradiol during the neonatal period in rats. *Brain Res.* 178: 207-212. (18)

Lieberburg, I., MacLusky, N. and McEwen, B. S. 1980. Cytoplasmic and nuclear estradiol-17-β binding in male and female rat brain: Regional distribution, temporal aspects and metabolism. *Brain Res.* 193: 487-503. (18)

Liem, R. K. H., Yen, S-H., Salomon, G. D. and Shelanski, M. L. 1978. Intermediate filaments in nervous tissue. *J. Cell Biol.* 28: 637-645. (11)

Lindner, J., Rathjen, F. G. and Schachner, M. 1985. L1 mono- and polyclonal antibodies modify cell migration in early postnatal mouse cerebellum. *Nature* 305: 427-430. (11)

Loeb, E. P., Chang, F.-L. F. and Greenough, W. T. 1987. Effects of neonatal 6-hydroxydopamine treatment upon morphological organization of the posteromedial barrel subfield in mouse somatosensory cortex. *Brain Res.* 403: 113-128. (19)

Lone, Y., Simon, M., Kahn, A. and Marie, J. 1986. Sequences complementary to the brain-specific identifier sequences exist in L-type pyruvate kinase mRNA (a liver specific messenger) and in transcripts especially abundant in muscle. *J. Biol. Chem.* 261: 1499-1502. (6)

Lookingland, K., Wise, P. and Barraclough, C. 1982. Failure of the hypothalamic noradrenergic system to function in adult androgen-sterilized rats. *Biol. Reprod.* 27: 268-281. (18)

Lopresti, V., Macagno, E. R. and Levinthal, C. 1973. Structure and development in neuronal connections in isogenic organisms: Cellular interactions in the development of the optic lamina of *Daphnia. Proc. Nat. Acad. Sci. USA 70:* 433-437. (8)

Loring, R. H., Chiappinelli, V. A., Zigmond, R. E. and Cohen, J. B. 1983. Characterization of a snake venom neurotoxin which blocks nicotinic transmission in autonomic ganglia. *Soc. Neurosci. Abst. 9:* 1143. (5)

Loring, J., Glimelius, B., Erickson, C. and Weston, J. A. 1981. Analysis of developmentally homogeneious neural crest populations in vitro. I. Formation, morphology and differentiative behavior. *Devel. Biol. 82*: 86-94. (15)

Loring, J., Glimelius, B. and Weston, J. A. 1982. Extracellular matrix materials influence quail neural crest cell differentiation in vitro. *Devel. Biol. 90*: 165-174. (15)

Loy, R. and Milner, T. A. 1980. Sexual dimorphism in extent of axonal sprouting in rat hippocampus. *Science 208*: 1282-1284. (17)

Lucki, I., Nobler, M. F. and Frazer, A. 1984. Differential action of serotonin antagonists on two behavioral models of serotonin receptor activation in the rat. *J. Pharm. Exp. Ther. 228*: 133-139. (18)

Luine, V. N. 1985. Estradiol increases choline acetyltransferase activity in specific basal forebrain nuclei and projection areas of female rats. *Exp. Neurol. 89*: 484-490. (18)

Luine, V. and McEwen, B. S. 1983. Sex differences in cholinergic enzymes of diagonal band nuclei in the rat preoptic area. *Neuroendocrinology 36*: 475- 482. (18)

Luine, V. N. and Rhodes, J. 1983. Gonadal hormone regulation of MAO and other enzymes in hypothalamic areas. *Neuroendocrinology 36*: 235-241. (18)

Luine, V., Khylchevskaya, R. and McEwen, B. S. 1975. Effect of gonadal steroids on activities of monoamines oxidase and choline acetylase in rat brain. *Brain Res. 86*: 293-306. (18)

Luine, V., Park, D., Joh, T., Reis, D. and McEwen, B. S. 1980. Immunochemical demonstration of increased choline acetyltransferase concentration in rat preoptic area after estradiol administration. *Brain Res. 191*: 273-277. (18)

Luine, V. N., Renner, K., Frankfurt, M. and Azmitia, E. 1984. Intrahypothalamic raphe transplants reverse facilitation of sexual behavior and restore 5HT in 5,7 DHT-treated rats. *Science 236*: 1436-1439. (18)

Luine, V., Renner, K. and McEwen, B. S. 1986. Sex-dependent differences in estrogen regulation of choline acetyltransferase are altered by neonatal treatments. *Endocrinology 119*: 1-5. (18)

Luria, S. E. and Delbrück, M. 1943. Mutations of bacteria from virus sensitivity to virus resistance. *Genetics 28*: 491. (15)

Lüscher, H. R., Ruenzel, P. and Henneman, E. 1980. Topographic distribution of terminals of Ia and group II fibers in spinal cord, as revealed by postsynaptic population potentials. *J. Neurophysiol. 43*: 968-985. (13)

Luyten, W. 1986. A model for the acetylcholine binding site of the acetylcholine receptor. *J. Neurosci. Res. 16*: 51-74. (5)

Lynch, G., Gall, C. and Dunwiddie, T. V. 1978. Neuroplasticity in the hippocampal formation. *Progr. Brain Res. 48*: 113-128. (13)

MacLusky, N. and McEwen, B. S. 1980. Progestin receptors in rat brain: Distribution and properties of cytoplasmic progestin binding sites. *Endocrinology 106*: 192-202. (18)

MacLusky, N., Lieberburg, I. and McEwen, B. S. 1979. The development of estrogen receptor systems in the rat brain: Perinatal development. *Brain Res. 178*: 129-142. (18)

Macagno, E. R. 1978. Mechanism for the formation of synaptic projections in the arthropod visual system. *Nature 275*: 318-320. (13)

Maden, M. 1977. The role of Schwann cells in paradoxical regeneration in the axolotl. *J. Embryol. Exp. Morph. 41*: 1-13. (3)

Magrassi, L., Purves, D. and Lichtman, J. W. 1987. Fluorescent probes that stain living nerve terminals. *J. Neurosci. 7*: 1207-1214. (12)

Mai, M. S. and Allison, W. S. 1983. Inhibition of an oligomycin-sensitive ATPase by cationic dyes, some of which are atypical uncouplers of intact mitochondria. *Arch. Biochem. Biophys. 221*: 467-476. (12)

Mariani, J. and Changeux, J.-P. 1980. Multiple innervation of Purkinje climbing fibers in the cerebellum of the adult *staggerer* mutant mouse. *J. Neurobiol. 11*: 41-50. (16)

Mark, R. F. 1969. Matching muscles and motoneurons. A review of some experiments on motor nerve regeneration. *Brain Res. 14*: 245-254. (13)

Marshall, L. M., Sanes, J. R. and McMahan, U. J. 1977. Reinnervation of original synaptic sites on muscle fiber basement membrane after disruption of the muscle cells. *Proc. Natl. Acad. Sci. USA 74*: 3073-3077. (2, 12)

Marusich, M., Pourmehr, K. and Weston, J. A. 1986a. A monoclonal antibody (SN1) identifies a subpopulation of avian sensory neurons whose distribution is correlated with axial level. *Devel. Biol. 118*: 494-504. (15)

Marusich, M., Pourmehr, K. and Weston, J. A. 1986b. The development of an identified subpopulation of avian sensory neurons is regulated by interaction with the periphery. *Devel. Biol. 118*: 505-510. (15)

Mason, C. 1985. How do growth cones grow? *Trends Neurosci. 8*: 304-306. (8)

Mason, C. A., Edmondson, J. C. and Hatten, M. E. In preparation. Astroglial process outgrowth and the development of neuron-glial association in vitro. (11)

Mastronarde, D. N., Thibeault, M. A. and Dubin, M. W. 1984. Non-uniform postnatal growth of the cat retina. *J. Comp. Neurol. 228*: 598-608. (8)

Matsumoto, A. and Arai, Y. 1979. Synaptogenic effect of estrogen on the hypothalamic arcuate nucleus of the adult female rat. *Cell Tiss. Res. 198*: 427-433. (18)

Matsumoto, A. and Arai, Y. 1980. Sexual dimorphism in "wiring" pattern in the hypothalamic arcuate nucleus and its modification by neonatal hormonal environment. *Brain Res. 190*: 238-242. (17)

Matsumoto, A. and Arai, Y. 1983. Sex difference in volume of the ventromedial nucleus of the hypothalamus in the rat. *Endocrinol. (Japan) 30*: 277-280. (17)

Matsumoto, A. and Arai, Y. 1986. Development of sexual dimorphism in synaptic organization in the ventromedial nucleus of the hypothalamus in rats. *Neurosci. Lett. 68*: 165-168. (17)

Matsumoto, A. and Arai, Y. 1986. Male-female difference in synaptic organization of the ventromedial nucleus of the hypothalamus in the rat. *Neuroendocrinology 42*: 232-236. (18)

Maurer, R. 1974. [^{3}H]estradiol binding macromolecules in the hypothalamus and anterior pituitary of normal female, androgenized female and male rats. *Brain Res. 67*: 175-177. (18)

Mauron, A., Nef, P., Oneyser, C., Stalder, R., Alloid, C. et al. 1985. Structure of chicken genes encoding the nicotinic acetylcholine receptor subunits and their variants. *Soc. Neurosci. Abst.* 171. (5)

Maxwell, G. 1976. Cell cycle changes during neural crest cell differentiation in vitro. *Devel. Biol. 49*: 66-79. (15)

Maxwell, G. and Sietz, P. D. 1983. Expression of the capacity to release [^{3}H]norepinephrine by neural crest cultures. *J. Neurosci. 3*: 1860-1867. (15)

Maxwell, G., Sietz, P. D. and Jean, S. 1983. Expression of somatostatin immunoreactivity in neural crest cultures. *Soc. Neurosci. Abst. 9*: 897. (15)

McArdle, J. J. 1975. Complex end-plate potentials at the regenerating neuromuscular junction in the rat. *Exp. Neurol. 49*: 629-638. (12)

McCarthy, M. P., Earnest, J. P., Young, E. G., Choe, S. and Stroud, R. M. 1986. The molecular neurobiology of the acetylcholine receptor. *Annu. Rev. Neurosci. 9*: 383-413. (5)

McEwen, B. S. 1983. Gonadal steroid influences on brain development and sexual differentiation. In *Reproductive Physiology IV*, Greep, R. O. (ed.). University Park Press, Baltimore. pp. 99-145. (18)

McEwen, B. S. 1984. Gonadal hormone receptors in developing and adult brain: Relationship to the regulatory phenotype. In *Fetal Neuroendocrinology*, Ellendorff, F., Gluckman, P. and Parvizi, N. (eds.). Perinatology Press. pp. 149-159. (18)

McEwen, B. S. and Parsons, B. 1982. Gonadal steroid action on the brain: Neurochemistry and neuropharmacology. *Annu. Rev. Pharm. Toxicol. 22*: 555-598. (18)

McEwen, B. S., Biegon, A., Davis, P., Krey, L. C., Luine, V. N. et al. 1982. Steroid hormones: Humoral signals which alter brain cell properties and functions. *Rec. Prog. Horm. Res. 38*, Academic Press, N. Y. pp. 41-92. (18)

McEwen, B. S., Biegon, A., Fischette, C., Luine, V. N., Parsons, B. et al. 1984. Toward a neurochemical basis of steroid hormone action. In *Frontiers in Neuroendocrinology*, Martini, L. and Ganong, W. (eds.). Raven Press, N. Y. 8: 153-176. (18)

McEwen, B. S., Biegon, A., Fischette, C., Luine, V. N., Parsons, B. et al. 1984. In *Sexual Differentiation: Basic and Clinical Aspects*, Serio, M. (ed.). Raven Press, N. Y. pp. 93-98. (18)

McEwen, B. S., Lieberburg, I., Chaptal, C. and Krey, L. C. 1977a. Aromatization: Important for sexual differenciation of the neonatal rat brain. *Horm. Behav. 9*: 249-263. (17)

McEwen, B. S., Lieberburg, I., Maclusky, N. and Plapinger, L. 1977b. Do estrogen receptors play a role in the sexual differentiation of the rat brain? *J. Steroid Biochem. 8*: 593-598. (17)

McEwen, B. S., Plapinger, L., Chaptal, C., Geralch, J. and Wallach, G. 1975. Role of fetoneonatal estrogen binding proteins in the association of estrogen with neonatal brain cell nuclear receptors. *Brain Res. 96*: 400-406. (18)

McGrath, P. A. and Bennett, M. R. 1979. The development of synaptic connections between different segmental motoneurones and striated muscles in axolotl limb. *Devel. Biol. 69*: 133-145. (8)

McGuire, J. C., Greene, L. A. and Furano, A. V. 1978. NGF stimulates incorporation of fucose or glucosamine into an external glycoprotein in cultured rat PC12 pheochromocytoma cells. *Cell 115*: 357-365. (1)

McKinnon, R. D., Danielson, P., Brow, M. A., Bloom, F. E. and Sutcliffe, J. G. In press. Neuronal-specific Pol III transcripts are expressed in cultured cells. *Mol. Cell. Biol.* (6)

McKinnon, R. D., Shinnick, T. M. and Sutcliffe, J. G. 1986. The neuronal identifier element is a *cis*-acting positive regulator of gene expression. *Proc. Natl. Acad. Sci. USA 83*: 3751-3755. (6)

McLaren, A. 1976. *Mammalian Chimeras*. Cambridge Univ. Press, London. (16)

McLennan. 1982. Size of motorneuron pool may be related to number of myotubes in developing muscle. *Devel. Biol. 92*: 263-265. (16)

Meaney, M., Aitken, D., Jensen, L., McGinnis, M. and McEwen, B. S. 1985. Nuclear and cytosolic androgen receptor levels in the limbic brain of the neonatal male and female rats. *Dev. Brain Res. 23*: 179-185. (18)

Meaney, M., Stewart, J., Poulin, P. and McEwen, B. S. 1983. Sexual differentiation of social play in rat pups is mediated by the neonatal androgen receptor system. *Neuroendocrinology 37*: 85-90. (18)

Meiri, K. F., Pfenninger, K. H. and Willard, M. B. 1986. Growth-associated protein, GAP-43, a polypeptide that is induced when neurons extend axons, is a component of growth cones and corresponds to pp46, a major polypeptide of a subcellular fraction enriched in growth cones. *Proc. Natl. Acad. Sci. USA 83*: 3537-3541. (10)

Meisel, R. and Pfaff, D. W. 1985. Brain region specificity in estradiol effects on neuronal ultrastructure in rats. *Mol. Cell. Endocrinol. 40*: 159-166. (18)

Melton, D. A., Krieg, P. A., Rebagliati, M. R., Maniatis, T., Zinn, K. et al. 1984. Efficient in vitro synthesis of biologically active RNA and RNA hybridization probes from plasmids containing a bacteriophage SP6 promoter. *Nucleic Acids Res. 12*: 7035-7056. (4)

Mendell, L. M. and Henneman, E. 1971. Terminals of single Ia fibers: Location, density and distribution within a pool of 300 homogeneous motoneurons. *J. Neurophysiol. 34*: 171-187. (13)

Mendell, L. M. and Hollyday, M. 1976. Spinal reflexes in anurans with an altered periphery. In *Frog Neurobiology*, Llinas, R. and Precht, W. (eds.), Springer-Verlag, Berlin. pp. 793-810. (13)

Mendell, L. M. and Scott, J. G. 1975. The effect of peripheral nerve cross- union on connections of single Ia fibers to motoneurons. *Exp. Brain Res. 22*: 221-234. (13)

Mendelson, B. 1985. Soma position is correlated with time of development in three types of identified reticulospinal neurons. *Devel. Biol. 112*: 489-493. (8)

Merlie, J. P. and Sebbane, R. 1981. Acetylcholine receptor subunits transit a precursor pool before acquiring α-bungarotoxin binding activity. *J. Biol. Chem. 256*: 3605-3608. (5)

Merzenich, M. M., Nelson, R. J., Stryker, M. P., Cynader, M. S., Schoppman, A. et al. 1984. Somatosensory cortical map changes following digit amputation in adult monkeys. *J. Comp. Neurol. 224*: 591-605. (5)

Mescher, A. L. 1976. Effects on adult newt limb regeneration of partial and complete skin flaps over the amputation surface. *J. Exp. Zool. 195*: 117-128. (3)

Mescher, A. L. and Loh, J.-J. 1981. Newt forelimb regeneration blastemas in vitro: Cellular response to explantation and effects of various growth- promoting substances. *J. Exp. Zool. 216*: 235-245. (3)

Meyer, R. L. 1982. Tetrodotoxin blocks the formation of ocular dominance columns in goldfish. *Science 218*: 589-591. (9)

Meyer-Bahlburg, H. F., Ehrhardt, A. A., Rosen, L. R., Feldman, J. F. Veridano, N. P. et al. 1984. Psychosexual milestones in women prenatally exposed to diethylstilbestrol. *Horm. Behav. 18*: 359-366. (18)

Miale, I. L. and Sidman, R. L. 1961. An autoradiographic analysis of histogenesis in the mouse cerebellum. *Exp. Neurol. 4*: 277-296. (16)

Middleton, P., Jaramillo, F. and Schuetze, S. M. 1986. Forskolin increases the rate of acetylcholine receptor desensitization of rat soleus endplates. *Proc. Natl. Acad. Sci. USA 83*: 4967-4971. (2)

Milcarek, C., Price, R. and Penman, S. 1974. The metabolism of a poly (A)⁻ mRNA fraction in Hela cells. *Cell 3*: 1-10. (4)

Miledi, R. 1960. The acetylcholine sensitivity of frog muscle fibers after complete or partial denervation. *J. Physiol.* (Lond.) *151*: 1-23. (2)

Milner, R. J., Bloom, F. E., Lai, C., Lerner, R. A. and Sutcliffe, J. G. 1984. Brain-specific genes have identifier sequences in their introns. *Proc. Natl. Acad. Sci. USA 81*: 713-717. (6)

Milner, R. J. and Sutcliffe, J. G. 1983. Gene expression in rat brain. *Nucleic Acids Res. 11*: 5497-5520. (4, 6)

Milner, T.A. and Loy, R. 1982. Hormonal regulation of axonal sprouting in the hippocampus. *Brain Res. 243*: 180-185. (17)

Miner, N. 1956. Integumental specification of sensory fibers in the development of cutaneous local sign. *J. Comp. Neurol. 105*: 161-170. (IV, 13)

Mintz, B. 1962. Formation of genotypically mosaic mouse embryos. *Am. Zool. 2*: 432. (16)

Mintz, B. 1965. Genetic mosaicism in adult mice of quadraparental lineage. *Science 148*: 1232-1233. (16)

Mintz, B. 1974. Gene control of mammalian differentiation. *Annu. Rev. Genet. 8*: 411-470. (16)

Mintz, B. and Sanyal, S. 1970. Clonal origin of the mouse visual retina mapped from genetically mosaic eyes. *Genetics 64*: 543-544. (16)

Mirmiran, M. and Uylings, H. B. M. 1983. The environmental enrichment effect upon cortical growth is neutralized by concomitant pharmacological suppression of active sleep in female rats. *Brain Res. 261*: 331-334. (19)

Mobley, W. C., Rutkowski, J. L., Teenekoon, G. I., Gemski, J., Buchanan, K. and Johnston, M. V. 1986. Nerve growth factor increases choline acetyltransferase activity in developing basal forebrain neurons. *Mol. Brain Res. 1*: 53-62. (VI)

Moguilewsky, M. and Raynaud, J. 1979. The relevance of hypothalamic and hypophyseal progestin receptor regulation in the induction and inhibition of sexual behavior in the female rat. *Endocrinology 105*: 516-522. (18)

Moiseff, A. and Konishi, M. 1981. Neuronal and behavioral sensitivity to binaural time differences in the owl. *J. Neurosci. 1*: 40-48. (20)

Mollgaard, M., Diamond, M. C., Bennett, E. L., Rosenzweig, M. R. and Lindner, B. 1971. Qualitative synaptic changes with different experience in rat brain. *Internat. J. Neurosci. 2*: 113-128. (19)

Morgan, T. H. 1901. *Regeneration*. Macmillan, London. (3)

Morrison, M. R., Brodeur, R., Pardue, S., Baskin, F., Hall, C. L. et al. 1979. Differences in the distribution of poly (A) size classes in individual messenger RNAs from neuroblastoma cells. *J. Biol. Chem. 254*: 7675-7683. (4)

Morton, D. G. and Sprague, K. U. 1984. In vitro transcription of a silkworm 5S RNA gene requires an upstream signal. *Proc. Natl. Acad. Sci. USA 81*: 5519-5522. (6)

Mos, L. P. 1976. Light rearing effects on factors of mouse emotionality and endocrine organ weight. *Physiol. Psychol. 4*: 503-510. (19)

Mullen, R. J. 1977. Site of pcd gene action and Purkinje cell mosaicism in cerebella of chimeric mice. *Nature 270*: 245-247. (16)

Mullen, R. J. 1978. Mosaicism in the central nervous system of mouse chimeras. In *The Clonal Basis of Development* (36th Symposium of the Society for Developmental Biology). Academic Press, New York. pp. 83-107. (16)

Mullen, R. J. and Herrup, K. 1979. Chimeric analysis of mouse cerebellar mutants. In *Neurogenetics: Genetic Approaches to the Nervous System*, Breakefield, X. O. (ed.). Elsevier/North Holland, New York. pp. 173-196. (16)

Mullen, R. J. and LaVail, M. M. 1976. Inherited retinal dystrophy: Primary defect in pigment epithelium determined with experimental rat chimeras. *Science 192*: 799-801. (16)

Mullen, R. J. and Whitten, W. K. 1971. Relationships of genotypes and degree of chimerism in coat color to sex ratios of gametogenesis of chimeric mice. *J. Exp. Zool. 178*: 165-176. (16)

Muller, K. J., Nicholls, J. G. and Stent, G. S. (eds.). 1981. *Neurobiology of the Leech.* Cold Spring Harbor Lab, New York. (14)

Muneoka, K. and Bryant, S. V. 1982. Evidence that patterning mechanisms in developing and regenerating limbs are the same. *Nature 298*: 369-371. (3)

Muneoka, K. and Bryant, S. V. 1984. Cellular contribution to supernumerary limbs resulting from the interaction between developing and regenerating tissues in the axolotl. *Devel. Biol. 105*: 179-187. (3)

Murphey, R. K. 1985. Competition and chemoaffinity in insect sensory systems. *Trends Neurosci. 8*: 120-125. (8)

Murphey, R. K. and Lemere, C. A. 1984. Competition controls the growth of an identified axonal arborization. *Science 224*: 1352-1355. (8)

Murphey, R. K., Jacklet, A. and Schuster, L. 1980. A topographic map of sensory cell terminal arborizations in the cricket CNS: Correlation with birthday and position in a sensory array. *J. Comp. Neurol. 191*: 53-64. (8)

Murray, B. A., Hemperly, J. J., Prediger, E. A., Edelman, G. M. and Cunningham, B. A. 1986a. Alternatively spliced mRNAs code for different polypeptide chains of the chicken neural cell adhesion molecule (N-CAM). *J. Cell Biol. 102*: 189-193. (1)

Murray, B. A., Owens, G. C., Prediger, E. A., Crossin, K. L. and Cunningham, B. A. 1986b. Cell surface modulation of the neural cell adhesion molecule resulting from alternative mRNA splicing in a tissue-specific developmental sequence. *J. Cell Biol. 103*: 1431-1439. (1)

Murray, J. G. and Thompson, J. W. 1957. The occurrence of collateral sprouting in the sympathetic nervous system of the cat. *J. Physiol.* (Lond.) *135*: 133-162. (13)

Myers, P. A. 1985. Spinal motoneurons of the larval zebrafish. *J. Comp. Neurol. 236*: 555-561. (8)

Myers, P., Eisen, J. and Westerfield, M. 1986. Development and axonal outgrowth of identified motoneurons in the zebrafish. *J. Neurosci. 8*: 2278- 2289. (8, 9)

Nabeshima, T., Yamaguchi, K., Furukawa, H. and Kemeyama, T. 1984. Role of sex hormones in sex-dependent differences in phenycyclidine-induced stereotyped behaviors in rats. *Eur. J. Pharm. 105*: 197-206. (18)

Naess, O., Haug, E., Attramadal, A., Aakvaag, A., Hansson, V. and F. French. 1976. Androgen receptors in the anterior pituitary and central nervous system of the androgen "insensitive" (Tfm) rat: Correlation between receptor binding and effects of androgens on gonadotropic secretion. *Endocrinology 99*: 1295-1303. (17)

Naftolin, F., Ryan, K. J., Davies, I. J., Reddy, V. V., Flores, F., et al. 1975. The formation of estrogens by central neuroendocrine tissues. *Recent Prog. Horm. Res. 31*: 295-319. (17)

Nakamura, H., Ayer-Le Lievre, C. S. 1982. Mesoectodermal capabilities of the trunk neural crest of birds. *J. Embryol. Exp. Morph. 70*: 1-18. (15)

Nakatomi, Y., Fujishima, M., Ogata, J., Tamaki, K., Ishitsuka, T. et al. 1983. Sex differences in barbiturate-protection in experimental cerebral ischemia in spontaneously hypertensive rats (SHR). *Brain Res. 270*: 146-148. (18)

Namenwirth, M. 1974. The inheritance of cell differentiation during limb regeneration in the axolotl. *Devel. Biol. 41*: 42-56. (3)

Nance, D. M., Shryne, J. and Gorski, R. A. 1975. Facilitation of female sexual behavior in male rats by septal lesions: An interaction with estrogen. *Horm. Behav. 6*: 289-299. (17)

Narayanan, C. H. and Narayanan, Y. 1978. Neuronal adjustments in developing nuclear centers of the chick embryo following transplantation of an additional optic primordium. *J.E.E.M. 44*: 57-70. (16)

Nardi, J. B. 1983. Neuronal pathfinding in developing wings of the moth *Maduca sexta*. *Devel. Biol. 96*: 163-174. (8)

Nawa, H., Hirose, T., Takashima, H., Inayama, S. and Nakanishi, S. 1983. Nucleotide sequences of cloned cDNAs for two types of bovine brain substance P precursor. *Nature 306*: 32-36. (4)

Nelson, S. G. and Mendell, L. M. 1978. Projection of single knee flexor Ia fibers to homonymous and heteronymous motoneurons. *J. Neurophysiol. 41*: 778-787. (13)

Nemer, M., Graham, M. and Dubroff, L. M. 1974. Coexistence of non-histone messenger RNA species lacking and containing polyadenylic acid in sea urchin embryos. *J. Mol. Biol. 89*: 435-454. (4)

Neubig, R. R., Krodel, E. K., Boyd, N. D. and Cohen, J. B. 1979. Acetylcholine and local anesthetic binding to *Torpedo* nicotinic post-synaptic membranes after removal of non-receptor peptides. *Proc. Natl. Acad. Sci. USA 76*: 690-694. (2)

Newgreen, D. and Jones, R. 1975. Differentiation *in vitro* of sympathetic cells from chick embryo sensory ganglia. *J. Emb. Exp. Morphol. 33*: 43-56. (15)

Nichols, D. H. and Weston, J. A. 1977. Melanogenesis in cultures of peripheral nervous tissue. I. The origin and prospective fate of cells giving rise to melanocytes. *Devel. Biol. 60*: 217-225. (15)

Nichols, D. H., Kaplan, R. and Weston, J. A. 1977. Melanogenesis in cultures of peripheral nervous tissue. II. Environmental factors determining the fate of pigment-forming cells. *Devel. Biol. 60*: 226-237. (15)

Nishizuka, M. and Arai, Y. 1981. Organizational action of estrogen on synaptic pattern in the amygdala: Implications for sexual differentiation of the brain. *Brain Res. 213*: 422-426. (17)

Nitkin, R. M., Wallace, B. G., Spira, M. E., Godfrey, E. W. and McMahan, U. J. 1983. Molecular components of the synaptic basal lamina that direct differentiation of regenerating neuromuscular junctions. *Cold Spring Harbor Symp. Quant. Biol. 48*: 653-666. (2)

Noble, M., Fok-Seang, J. and Cohen, J. 1984. Glia are a unique substrate for the in vitro growth of central nervous system neurons. *J. Neurosci. 4*: 1892-1903. (11)

Nock, B. and Feder, H. 1984. Alpha-1 noradrenergic regulation of hypothalamic progestin receptors and guinea pig lordosis behavior. *Brain Res. 310*: 77-85. (18)

Nock, B., Blaustein, J. and Feder, H. 1981. Changes in noradrenergic transmission alter the concentration of cytoplasmic progestin receptors in hypothalamus. *Brain Res. 207*: 371-396. (18)

Noda, M., Furutani, Y., Takahashi, H., Toyosato, M., Tanabe, T. et al. 1983b. Cloning and sequence analysis of calf cDNA and human genetic DNA encoding alpha-subunit precursor of muscle acetylcholine receptor subunits. *Nature 302*: 818-823. (5)

Noda, M., Takahashi, H., Tanabe, T., Toyosato, M., Furutani, Y. et al. 1982. Primary structure of alpha-subunit precursor of *Torpedo californica* acetylcholine receptor deduced from cDNA sequence. *Nature 299*: 793-797. (5)

Noda, M., Takahashi, H., Tanabe, T., Toyosato, M., Kikyotani, S. et al. 1983. Primary structures of beta- and delta-subunit precursors of *Torpedo californica* acetylcholine receptor deduced from cDNA sequences. *Nature 301*: 251-255. (5)

Noden, D. M. 1980. The migration and cytodifferentiation of cranial neural crest cells. In *Current Research Trends in Prenatal Cranio-facial Development,* Pratt, R. M. and Christiansen, R. (eds.), Elsevier/North Holland, N. Y. (15)

Noden, D. M. 1983. The role of the neural crest in patterning of avian cranial skeletal, connective and muscle tissue. *Devel. Biol. 96*: 144-165. (15)

Noden, D. M. 1984. Neural crest development: New views on old problems. *Anat. Rec. 206*: 1-13. (15)

Nordeen, E. and Yahr, P. 1983. A regional analysis of estrogen binding to hypothalamic cell nuclei in relation to masculinization and defeminization. *J. Neurosci. 3*: 933-941. (18)

Norman, R. I., Mohraban, F., Barnard, E. A. and Dolly, J. O. 1982. Nicotinic acetylcholine receptor from chick optic lobe. *Proc. Natl. Acad. Sci. USA 79*: 1321-1325. (5)

Nottebohm, F. 1976. Sexual dimorphism in vocal control areas of the songbird brain. *Science 194*: 211-213. (VI)

Nottebohm, F. 1981. A brain for all seasons: Cyclical antomical changes in song control nuclei of the canary brain. *Science 214*: 1368-1370. (17)

Nottebohm, F. and Arnold, A. P. 1976. Sexual dimorphism in vocal control areas of the songbird brain. *Science 194*: 211-213. (17)

Numa, S., Noda, M., Takahashi, H., Tanabe, T., Toyosato, M. et al. 1983. Molecular structure of the nicotinic acetylcholine receptor. *Cold Spring Harbor Symp. Quant. Biol. 48*: 57-69. (2)

Nunez, E., Vallette, G., Benssayag, C. and Jayle, M-F. 1974. Comparative study on the binding of estrogens by human and rat serum proteins in development. *Biochem. Biophys. Res. Comm. 57*: 126-133. (18)

O'Connor, L. and Feder, H. 1984. Estradiol and progesterone influence a serotonin mediated behavioral syndrome (myoclonus) in female guinea pigs: Comparison with steroid effects on reproductive behavior. *Brain Res. 293*: 119-125. (18)

O'Connor, L. and Feder, H. 1985. Estradiol and progesterone influence α-5-hydroxytryptophan-induced (mycolonus in male guinea pigs). *Brain Res. 330*: 121-125. (18)

O'Connor, L. and Fischette, C. 1986. Hormone effects on serontonin-dependent behaviors. *Ann. N. Y. Acad. Sci. 474*: 437-444. (18)

O'Donovan, M. and Landmesser, L. In press. The development of alternation of flexor and extensor motor pools in the embryonic chick cord. *J. Neurosci.* (9)

O'Lague, P. H., Obata, K., Claude, P., Furshpan, E. J. and Potter, D. D. 1974. Evidence for cholinergic synapses between dissociated sympathetic neurons in cell culture. *Proc. Natl. Acad. Sci. USA 71*: 3602-3606. (4)

O'Shea, L., Saari, M., Pappas, B. A., Ings, R. and Stange, K. 1983. Neonatal 6-hydroxydopamine attenuates the neural and behavioral effects of enriched rearing in the rat. *Eur. J. Pharmacol. 92*: 43-47. (19)

Olmstead, C. E. and Villablanca, J. R. 1980. Development of behavioral audition in the kitten. *Physiol. and Behav. 24*: 705-712. (20)

Olsen, K. L. and Whalen, R. E. 1980. Sexual differentiation of the brain: Effects on mating behavior and [³H]estradiol binding by hypothalamic chromatin in rats. *Biol. Reprod. 22*: 1068-1072. (17)

Ono, R. D. 1983. Dual motor innervation in the axial musculature of fishes. *J. Fish Biol. 22*: 395-408. (8)

Oppenheim, R. 1981. Cell death of motoneurons in the chick embryonic spinal cord. Evidence on the role of cell death and neuromuscular function in the formation of specific connections. *J. Neurosci. 1*: 141-151. (9)

Oppenheim, R. W. 1981. Neuronal cell death and some related regressive phenomena during neurogenesis: A selective historical review and progress report. In *Studies in Developmental Neurobiology: Essays in Honor of Viktor Hamburger,* Cowan, W. M. (ed.). Oxford Univ. Press, New York. (8, 16)

Oppenheim, R. W., Chu-Wang, I.-Wu and Maderdrut, J. L. 1978. Cell death of motoneurons in the chick embryo spinal cord. III. The differentiation of motoneurons prior to their induced degeneration following limb-bud removal. *J. Comp. Neurol. 177*: 87-112. (16)

Oster-Granite, M. L. and Gearhart, L. 1981. Cell lineage analysis of cerebellar Purkinje cells in mouse chimeras. *Devel. Biol. 85*: 199-208. (16)

Oswald, R. E. and Freeman, J. A. 1980. Alpha-bungarotoxin binding and central nervous system nicotinic acetylcholine receptors. *Neurosci. 6*: 1-14. (5)

Owens, G. P., Chaudhari, N. and Hahn, W. E. 1985. Brain identifier sequence is not restricted to brain: Similar abundance in nuclear RNA of other organs. *Science 229*: 1263-1265. (6)

Paden, C., Gerlach, J. and McEwen, B. S. 1984. Estrogen and progestin receptors appear in transplanted fetal hypothalamus-preoptic area independently of the steroid environment. *Soc. Neurosci.* 5: 2374-2381. (18)

Palka, J. 1982. Genetic manipulation of sensory pathways in *Drosophila*. In *Neuronal Development,* Spitzer, N. C. (ed.). Plenum Press, New York. (8)

Palka, J. 1986. Neurogenesis and axonal pathfinding in invertebrates. *Trends Neurosci.* 9: 482-485. (III)

Palka, J., Schubiger, M. and Ellison, R. L. 1983. The polarity of axon growth in the wings of *Drosophila melanogaster*. *Devel. Biol.* 98: 481-492. (8)

Parsons, B., Rainbow, T. and McEwen, B. S. 1984. Organizational effects of testosterone via aromatization on feminine reproductive behavior and neural progestin receptors in rat brain. *Endocrinology 115:* 1412-1417. (18)

Patel, N. H., Snow, P. M. and Goodman, C. S. 1987. Characterization and cloning of fasciclin III: A glycoprotein expressed on a subset of neurons and axon pathways in *Drosophila*. *Cell 48:* 975-988. (7)

Patrick, J. and Stallcup, W. 1977a. Immunological distinction between acetylcholine receptor and the alpha-bungarotoxin-binding component on sympathetic neurons. *Proc. Natl. Acad. Sci. USA 74:* 4689. (5)

Patrick, J. and Stallcup, W. 1977b. Alpha-bungarotoxin binding and cholinergic receptor function on a rat sympathetic nerve line. *J. Biol. Chem. 252:* 8629. (5)

Patrick, J., McMillan, J., Wolfson, M. and O'Brien, J. C. 1977c. Acetylcholine receptor metabolism in a non-fusing muscle cell line. *J. Biol. Chem. 252:* 2143. (5)

Patterson, P. H. 1978. Environmental determination of autonomic neurotransmitter function. *Annu. Rev. Neurosci. 1:* 1-17. (15)

Patterson, P. H. and Chun, L. L. Y. 1977a. The induction of acetylcholine synthesis in primary cultures of dissociated rat sympathetic neurons. I. Effects of conditioned medium. *Devel. Biol. 56:* 263-280. (4)

Patterson, P. H. and Chun, L. L. Y. 1977b. The induction of acetylcholine synthesis in primary cultures of dissociated rat sympathetic neurons. II. Developmental aspects. *Devel. Biol. 60:* 473-481. (4)

Payette, R. F., Bennet, G. S. and Gershon, M. D. 1984. Neurofilament expression in vagal neural crest-derived precursors of enteric neurons. *Devel. Biol. 105:* 273-287. (15)

Payne, R. S. 1971. Acoustic location of prey by barn owls (*Tyto alba*). *J. Exp. Biol. 54:* 535-573. (20)

Peng, H. B. and Froehner, S. C. 1985. Association of the postsynaptic 43 k protein with newly formed acetylcholine receptor clusters in cultured muscle cells. *J. Cell Biol. 100:* 1698-1705. (2)

Peng, H. B. and Poo, M. 1986. Formation and dispersal of acetylcholine receptor clusters in muscle cells. *Trends Neurosci.* 9: 125-129. (12)

Perris, R. and Lofberg, J. 1986. Promotion of chromatophore differentiation in isolated premigratory neural crest cells by extracellular material explanted on microcarriers. *Devel. Biol. 113:* 327-342. (15)

Peterson, G. M. and Devine, J. V. 1963. Transfer of handedness in the rat resulting from small cortical lesions after limited forced practice. *J. Comp. Physiol. Psychol. 56:* 752-756. (19)

Pettigrew, A., Linderman, R. and Bennet, M. R. 1979. Development of the segmental innervation of the chick forelimb. *J. Embryol. Exp. Morphol. 49:* 115-137. (9)

Pfaff, D.W. 1966. Morphological changes in the brains of adult male rats after neonatal castration. *J. Endocrinol. 36:* 415-416. (17)

Pfeiffer, F., Graham, D. and Betz, H. 1982. Purification by affinity chromatography of the glycine receptor of rat spinal cord. *J. Biol. Chem. 257:* 7389-9393. (2)

Pfeiffer, F., Simler, R., Grenningloh, G. and Betz, H. 1984. Monoclonal antibodies and peptide mapping reveal structural similarities between the subunits of the glycine receptor of rat spinal cord. *Proc. Natl. Acad. Sci. USA 81:* 7224-7227. (2)

Phelps, C. and Sawyer, C. 1976. Postnatal thyroxine modifies effects of early androgen on lordosis. *Horm. Behav. 7:* 331-340. (18)

Phillips, R. J. S. 1960. *Lurcher,* a new gene in linkage group XI of the house mouse. *J. Genet. 57:* 35-42. (16)

Phoenix, C. H., Goy, R. W., Gerall, A. A. and Young, W. C. 1959. Organizing action of prenatally administered testosterone propionate on the tissues mediating mating behavior in the female guinea pig. *Endocrinology 65:* 369-382. (17)

Pilar, G. and Landmesser, L. T. 1976. Ultrastructural differences during embryonic cell death in normal and peripherally deprived ciliary ganglia. *J. Cell Biol. 68:* 339-356. (16)

Pilar, G., Landmesser, L. T. and Burstein, L. 1980. Competition for survival among developing ciliary ganglion cells. *J. Neurophys. 43:* 233-254. (16)

Pittman, R. and Oppenheim, R. 1979. Cell death of motoneurons in the chick embryo spinal cord. IV. Evidence that a functional neuromuscular interaction is involved in the regulation of naturally occurring cell death and the stabilization of synapses. *J. Comp. Neurol. 187:* 425-447. (9)

Plapinger, L. and McEwen, B. S. 1978. Gonadal steroid-brain interactions in sexual differentiation. In *Biological Determinants of Sexual Behavior,* Hutchinson, J. (ed.). John Wiley & Sons, New York. pp. 193-218. (18)

Pollerberg, E. G., Sadoul, R., Goridis, C. and Schachner, M. 1985. Selective expression of the 180-KD component of the neural cell-adhesion molecule N-CAM during development. *J. Cell Biol. 101:* 1921-1929. (1)

Ponder, B. A. J., Schmidt, G. H., Wilkinson, M. M., Wood, M. J., Monk, M. et al. 1985. Derivation of mouse intestinal crypts from single intestinal cells. *Nature 313:* 689-691. (16)

Popot, J.-L. and Changeux, J.-P. 1984. The nicotinic receptor of acetylcholine: Structure of an oligomeric integral membrane protein. *Physiol. Rev. 64:* 1162-1239. (5)

Porter, J. 1986. Relationship of age, sex and reproductive status to the quantity of tyrosine hydroxylase in the median eminence and superior cervical ganglion of the rat. *Endocrinology 118:* 1426-1432. (18)

Potter, D. D., Landis, S. C. and Furshpan, E. J. 1983. Adrenergic-cholinergic dual function in cultured sympathetic neurons of the rat. In *Development of the Autonomic Nervous System*. CIBA Foundation Symposium, Pitman Medical, London. (4)

Potter, D. D., Landis, S. C., Matsumoto, S. G. and Furshpan, E. J. 1986. Synaptic functions in rat sympathetic neurons in microcultures. II. Adrenergic/cholinergic dual status and plasticity. *J. Neurosci. 6*: 1080-1098. (4, 15)

Prestige, M. C. 1967. The control of cell number in the lumbar spinal ganglia during the development of *Xenopus laevis* tadpoles. *J. Exp. Morph. 17*: 453-471. (16)

Pruss, R. M., Mirsky, R., Raff, M. D., Thorpe, R., Dowding, A. J. et al. 1981. All classes of intermediate filaments share a common antigenic determinant defined by a monoclonal antibody. *Cell 27*: 419-428. (3)

Pukel, C. S., Lloyd, K. O., Travassos, L. R., Dippold, W. G., Oettgen, H. F. et al. 1982. GD3, a prominent ganglioside of human melanoma. *J. Exp. Med. 155*: 1133-1147. (15)

Purves, D. 1977. The formation and maintenance of synaptic connections. In *Function and Formation of Neural Systems*, Stent, G. S., ed., Dahlem Konferenzen, Berlin, pp. 21-49. (17)

Purves, D. 1980. Neuronal competition. *Nature 287*: 585-586. (8)

Purves, D. and Lichtman, J. W. 1980. Elimination of synapses in the developing nervous system. *Science 210*: 153-157. (5, 8, 13)

Purves, D. and Lichtman, J. W. 1985. *Principles of Neural Development*. Sinauer Associates, Sunderland, MA. (5, 12)

Purves, D., Thompson, W. and Yip, J. 1981. Re-innervation of ganglia transplanted to the neck from different levels of the guinea-pig sympathetic chain. *J. Physiol. 313*: 49-63. (9)

Pysh, J. J. and Weiss, G. M. 1979. Exercise during development induces an increase in Purkinje cell dendritic tree size. *Science 206*: 230-232. (19)

Raff, M. C., Fields, K. L., Hakomori, S., Mirsky, R., Pruss, R. M. et al. 1979. Cell-type-specific markers for distinguishing and studying neurons and the major classes of glial cells in culture. *Brain Res. 174*: 283-308. (11)

Rainbow, T., DeGroff, V., Luine, V. and McEwen, B. S. 1980. Estradiol 17B increases the number of muscarinic receptors in hypothalamic nuclei. *Brain Res. 198*: 239-243. (18)

Rainbow, T. C., Parsons, B. and McEwen, B. S. 1982. Sex differences in rat brain oestrogen and progestin receptors. *Nature 300*: 648-649. (17, 18)

Rainbow, T., Snyder, L., Berck, D. and McEwen, B. S. 1984. Correlation of muscarinic receptor induction in the ventromedial hypothalamic nucleus with the activation of feminine sexual behavior by estradiol. *Neuroendocrinology 39*: 476-480. (18)

Raisman, G. and Field, P. M. 1973. Sexual dimorphism in the neuropil of the preoptic area of the rat and its dependence on neonatal androgen. *Brain Res. 54*: 1-29. (17)

Rakic, P. 1972. Mode of cell migration to the superficial layers of fetal monkey cortex. *J. Comp. Neurol. 145*: 61-84. (11)

Rakic, P. and Riley, K. P. 1983a. Overproduction and elimination of retinal axons in the fetal rhesus monkey. *Science 219*: 1441-1444. (8, 10)

Rakic, P. and Riley, K. P. 1983b. Regulation of axon number in primate optic nerve by prenatal binocular competition. *Nature 305*: 135-137. (8)

Rakic, P. and Sidman, R. L. 1973. Sequence of developmental abnormalities leading to granule cell deficit in cerebellar cortex of Weaver mutant mice. *J. Comp. Neurol. 152*: 103-132. (11)

Rakic, P., Stensaas, L. J., Sayre, E. P. and Sidman, R. L. 1974. Computer-aided three dimensional reconstruction and quantitative analysis of cells from serial electron microscopic montages of foetal monkey brain. *Nature 250*: 31-34. (11)

Ramon y Cajal, S. 1893. [New findings about the histological structure of the central nervous system]. *Archiv fur Anatomie und Physiologie (Anatomie)* pp. 319-428. (19)

Ramon y Cajal, S. 1928. *Degeneration and Regeneration of the Nervous System*. Hafner, N.Y. (IV)

Ramon y Cajal, S. 1929. *Studies on Vertebrate Neurogenesis*. Translated by L. Guth, 1960. Charles Thomas, Springfield, IL. (13)

Rance, N., Wise, P., Selmanoff, M. and Barraclough, C. 1981. Catecholamine turnover rates in discrete hypothalamic areas and associated changes in median eminence luteinizing hormone-releasing hormone and serum gonadotropins on proestrus and diestrous day 1. *Endocrinology 108*: 1795-1802. (18)

Raper, J. A., Bastiani, M. and Goodman, C. S. 1983a. Pathfinding by neuronal growth cones in grasshopper embryos: I. Divergent choices made by the growth cones of sibling neurons. *J. Neurosci. 3*: 20-30. (7, 8)

Raper, J. A., Bastiani, M. J. and Goodman, C. S. 1983b. Pathfinding by neuronal growth cones in grasshopper embryos. II. Selective fasciculation onto specific axonal pathways. *J. Neurosci. 3*: 31-41. (7, 8)

Raper, J. A., Bastiani, M. J. and Goodman, C. S. 1983c. Guidance of neuronal growth cones: Selective fasciculation in the grasshopper embryo. *Cold Spring Harbor Symp. Quant. Biol. 48*: 587-598. (7, 10)

Raper, J. A., Bastiani, M. J. and Goodman, C. S. 1984. Pathfinding by neuronal growth cones in grasshopper embryos. IV. The effects of ablating the A and P axons upon the behavior of the G growth cone. *J. Neurosci. 4*: 2329-2345. (7)

Rasch, E., Swift, A., Riesen, H. and Chow, K. L. 1961. Altered structure and composition of retinal cells in dark-reared mammals. *Exp. Cell Res. 25*: 348-363. (19)

Ratnam, M. and Lindstrom, J. 1984. Structural features of the nicotinic acetylcholine receptor revealed by antibodies to synthetic peptides. *Biochem. Biophy. Res. Commun. 122*: 1225-1233. (5)

Ratnam, M., Nguyen, D. L., Rivier, J., Sargent, P. B. and Lindstrom, J. 1986. Transmembrane topography of nicotinic acetylcholine receptor: Immunochemical tests contradict theoretical predictions based on hydrophobicity profiles. *Biochem. 25*: 2633-2643. (5)

Ravdin, P. and Berg, D. K. 1979. Inhibition of neuronal acetylcholine sensitivity by α-toxins from *Bungarus multicinctus* venom. *Proc. Natl. Acad. Sci. USA* (5)

Reh, T. and Constantine-Paton, M. 1985. Eye-specific segregation requires neural activity in three-eyed *Rana pipiens. J. Neurosci. 5*: 1132-1143. (9)

Reh, T. and Kalil, K. 1981. Development of the pyramidal tract in the hamster: A light microscopic study. *J. Comp. Neurol. 200*: 55-67. (10)

Reh, T. and Kalil, K. 1982a. Development of the pyramidal tract in the hamster: An electron microscopic study. *J. Comp. Neurol. 205*: 77-88. (10)

Reh, T. and Kalil, K. 1982b. Functional role of regrowing pyramidal tract fibers. *J. Comp. Neurol. 211*: 276-283. (10)

Renner, K., Biegon, A. and Luine, V. 1985. Sex differences in long-term gonadectomized rats: Monoamines levels and [³H]nitroimipramine binding in brain nuclei. *Exp. Brain Res. 58*: 198-201. (18)

Rhoades, R. W. and Chalupa, L. M. 1980. Effects of neonatal enucleation on receptive-field properties of visual neurons in superior colliculus of the golden hamster. *J. Neurophysiol. 43*: 595-611. (13)

Rich, M. M. and Lichtman, J. W. 1986. Remodeling of endplate sites during muscle reinnervation in the living mouse. *Soc. Neurosci. Abst. 12*: 390. (12)

Richardson, G., Crossin, K. L., Chuong, C.-M. and Edelman, G. M. 1987. Expression of cell adhesion molecules in embryonic induction. III. Development of the otic placode. *Devel. Biol. 119*: 217-230. (1)

Richardson, P. M. and Issa, V. M. K. 1984. Peripheral nerve injury enhances central regeneration of primary sensory neurons. *Nature 309*: 791-793. (10)

Rieger, F., Grumet, M. and Edelman, G. M. 1985. N-CAM at the vertebrate neuromuscular junction. *J. Cell Biol. 101*: 285-293. (1)

Rieger, F., Daniloff, J. K., Pincon-Raymond, M., Crossin, K. L., Grumet, M. et al. 1986. Neuronal CAMs and cytotactin are colocalized at Nodes of Ranvier. *J. Cell Biol. 103*: 379-391. (1)

Rindler, M. J., Ivanov, I. E., Plesken, H. and Sabatini, D. D. 1985. Polarized delivery of viral glycoproteins to the apical and basolateral plasma membranes of Madin-Darby canine kidney cells infected with temperature-sensitive viruses. *J. Cell Biol. 100*: 136-151. (2)

Ritchie, J. M. and Rogart, R. B. 1977. Density of sodium channels in mammalian myelinated nerve fibers and nature of the axonal membrane under the myelin sheath. *Proc. Natl. Acad. Sci. USA 74*: 211-215. (2)

Roberts, A. and Clarke, J. D. W. 1982. The neuroanatomy of an amphibian embryo spinal cord. *Phil. Trans. Roy. Soc. Lond. B 296*: 195-212. (8)

Roberts, W. K. 1974. Use of benzoylated cellulose columns for the isolation of poly (adenylic acid) containing RNA and other polynucleotides with little secondary structure. *Biochem. 15*: 3516-3522. (4)

Robinson, T., Camp, D., Jacknow, D. and Becker, J. 1982. Sex differences and estrous cycle dependent variation in rotational behavior elicited by electrical stimulation of the mesostriatal dopamine system. *Behav. Brain Res. 6*: 273-287. (18)

Robins, L. N., Helzer, J. E., Weissman, M. M., Orvaschel, H., Gruenberg, E. et al. 1984. Lifetime prevalence of specific psychiatric disorders in three sites. *Arch. Gen. Psychiat. 41*: 949-958. (18)

Rodriguez-Sierra, J., Hagley, M. and Hendricks, S. 1986. Anxiolytic effects of progesterone are sexually dimorphic. *Life Sci. 38*: 1841-1845. (18)

Rogers, J. 1983. A straight LINE story. *Nature 306*: 113-114. (6)

Rogers, J. 1985a. Origins of repeated DNA. *Nature 317*: 765-766. (6)

Rogers, J. 1985b. The origin and evolution of retroposons. *Int. Rev. Cytol. 93*: 187-279. (6)

Rogers, S. L., Edson, K. J., Letourneau, P. C. and McLoon, S. C. 1986. Distribution of laminin in the developing peripheral nervous system of the chick. *Devel. Biol. 113*: 429-435. (9, 15)

Rohrer, H., Acheson, A. L., Thibault, J. and Thoenen, H. 1986. Developmental potential of quail dorsal root ganglion cells analyzed in vitro and in vivo. *J. Neurosci. 6*: 2616-2624. (15)

Rosenzweig, M. R. and Bennett, E. L. 1978. Experiential influences on brain anatomy and brain chemistry in rodents. In *Studies on the Development of Behavior and the Nervous System, Vol. 4: Early Influences,* Gottlieb, G. (ed.), Academic Press, New York, pp. 289-330. (19)

Rosenzweig, M. R., Bennett, E. L. and Diamond, M. C. 1972. Chemical and anatomical plasticity of brain: Replications and extensions. In *Macromolecules and Behavior,* 2nd ed., Gaito, J. (ed.), Appleton-Century-Crofts, New York, pp. 205-277. (19)

Ross, D., Johnson, M. I. and Bunge, R. D. 1977. Development of cholinergic characteristics in adrenergic neurones is age dependent. *Nature 267*: 536-539. (4)

Rossant, V. 1984. Somatic cell lineages in mammalian chimeras. In *Chimeras in Developmental Biology.* Le Douarin, N. and McLaren, A. (eds.). Academic Press, London, pp. 89-109. (16)

Rothblat, L. A. and Schwartz, M. 1979. The effect of monocular deprivation on dendritic spines in visual cortex of young and adult albino rats: Evidence for a sensitive period. *Brain Res. 161*: 156-161. (19)

Rovasio, R. A., Delouvee, A., Yamada, K. M., Timpl, R. and Thiery, J. P. 1983. Neural crest cell migration: Requirements for exogenous fibronectin and high cell density. *J. Cell Biol. 96*: 462-473. (15)

Runyan, R. B., Maxwell, G. D. and Shur, B. D. 1986. Evidence for a novel enzymatic mechanism of neural crest cell migration on extracellular glycoconjugate matrices. *J. Cell Biol. 102*: 432-441. (15)

Ruoslahti, E. and Engvall, E. 1978. Alpha-fetoprotein. *Scan. J. Immunol. 7*: 1-17. (18)

Ryffel, G. U. and McCarthy, B. J. 1975. Polyadenylated RNA complementary to repetitive DNA in mouse L-cells. *Biochemistry 14*: 1385-1389. (4)

Ryugo, D. K., Ryugo, R., Globus, A. and Killackey, H. P. 1975. Increased spine density in auditory cortex following visual or somatic deafferentation. *Brain Res. 90*: 143-146. (19)

Sabatini, D. D., Kreibich, G., Morimoto, T. and Adesnik, M. 1982. Mechanisms for incorporation of proteins in membranes and organelles. *J. Cell Biol. 92*: 1-22. (2)

Sah, D. W. Y. and Frank, E. 1984. Regeneration of sensory-motor synapses in the spinal cord of the bullfrog. *J. Neurosci. 4*: 2784-2791. (13)

Sakamoto, K. and Okada, N. 1985a. 5-methylcytidylic modification of in vitro transcript from the rat identifier sequence: Evidence that the transcript forms a tRNA-structure. *Nucleic Acids Res. 13*: 7195-7206. (6)

Sakamoto, K. and Okada, N. 1985b. Rodent type 2 Alu family, rat identifier sequence, rabbit C family, and bovine or goat 73-bp repeat may have evolved from tRNA genes. *Mol. Evol. 22*: 134-140. (6)

Salton, S. R. J., Richter-Landsburg, C., Greene, L. A. and Shelanski, M. L. 1983. Nerve growth factor-inducible large external (NILE) glycoprotein: Studies of a central and peripheral neuronal marker. *J. Neurosci. 3*: 441-454. (11)

Sanes, D. H. and Constantine-Paton, M. 1985. The sharpening of frequency tuning curves requires patterned activity during development of the mouse, *Mus musculus*. *J. Neurosci. 5*: 1152-1166. (9, 13)

Sanes, J. R., Marshall, L. M. and McMahan, U. J. 1978. Reinnervation of muscle fiber basal lamina after removal of myofibers. Differentiation of regenerating axons at original synaptic sites. *J. Cell Biol. 78*: 176-198. (12)

Sanyal, S. and Zeilmaker, G. H. 1977. Cell lineage in the retinal development of mice studied in experimental chimeras. *Nature 265*: 731-733. (16)

Sapienza, C. and St. Jacques, B. 1986. Brain-specific transcription and evolution of the identifier sequence. *Nature 319*: 418-420. (6)

Schlesinger, M. J., Ashburner, M. and Tissieres, A. (eds.). 1982. *Heat Shock: From Bacteria to Man*. Cold Spring Harbor Publications, Cold Spring Harbor, New York. (3)

Schmechel, D. E. and Rakic, P. 1979. A Golgi study of radial glial cells in developing monkey telencephalon: Morphogenesis and transformation into astrocytes. *Anat. Embryol. 156*: 115-152. (11)

Schmidt, G. H., Wilkinson, M. M. and Ponder, B. A. J. 1985. Cell migration pathway in the intestinal epithelium: An in situ marker system using mouse aggregation chimeras. *Cell 40*: 425-429. (16)

Schmidt, J. T. 1978. Retinal fibers alter tectal positional markers during the expansion of the half retinal projection in goldfish. *J. Comp. Neurol. 177*: 279-300. (IV)

Schmidt, J. and Edwards, D. 1983. Activity sharpens the map during regeneration of the retinotectal projection in goldfish. *Brain Res. 269*: 29-39. (9, 13)

Schmidt, J. T. and Eisely, I. E. 1985. Stroboscopic illumination and dark rearing block the sharpening of the regenerated retinotectal map in goldfish. *Neuroscience 14*: 535-546. (13)

Schmidt, J. T., Cicerone, C. M. and Easter, S. S. 1978. Expansion of the half-retinal projection to the tectum in goldfish: An electrophysiological and anatomical study. *J. Comp. Neurol. 177*: 257-277. (13)

Schoenheimer, R. 1942. *The Dynamic State of Body Constituents*. Harvard Univ. Press. (I)

Schubert, D., Heinemann, S., Carlisle, W., Tarikas, H., Kimes, B. et al. 1974. Clonal cell lines from rat nervous system. *Nature 249*: 224-227. (4)

Schuler, L. A., Weber, J. L. and Gorski, J. 1983. Polymorphism near the rat prolactin gene caused by insertion of an Alu-like element. *Nature 305*: 159-160. (6)

Schumacher, M. and Balthazart, J. 1986. Testosterone-induced brain aromatase is sexually dimorphic. *Brain Res. 270*: 285-293. (18)

Schweizer, G., Ayer-Le Lievre, C. and Le Douarin, N. M. 1983. Restrictions of developmental capacities in the dorsal root ganglia during the course of development. *Cell Differentiation 13*: 191-200. (15)

Scott, J. G. and Mendell, L. M. 1976. Individual EPSPs produced by single triceps surae Ia afferent fibers in homonymous and heteronymous motoneurons. *J. Neurophysiol. 39*: 679-692. (13)

Sealock, R., Wray, B. E. and Froehner, S. C. 1984. Ultrastructural localization of the M_r 43,000 protein and the acetylcholine receptor in *Torpedo* postsynaptic membranes using monoclonal antibodies. *J. Cell Biol. 98*: 2239-2244. (2)

Selmanoff, M. K., Brodkin, L. D., Weiner, R. I. and Siiteri, P. K. 1977. Aromatization and 5-α-reduction of androgens in discrete hypothalamic and limbic regions of the male and female rat. *Endocrinology 101*: 841-848. (17)

Sengel, P. 1976. *Morphogenesis of Skin*, Cambridge Univ. Press, Cambridge. (1)

Sessions, S. K. 1984. Cytogenetics and evolution in salamanders. Ph.D. dissertation, University of California, Berkeley. (14)

Shankland, M. 1984. Positional control of supernumerary blast cell death in the leech embryo. *Nature 307*: 541-543. (14)

Shankland, M. In press. Position-dependent cell interactions and commitments in the formation of the leech nervous system. In *Current Topics in Developmental Biology: Neurological Development IV*, Hunt, R. K. (ed.) (14)

Shankland, M., Bentley, D. and Goodman, C. S. 1982. Afferent innervation shapes the dendritic branching pattern of the medial giant interneuron in grasshopper embryos raised in culture. *Devel. Biol. 92*: 507-520. (8)

Shankland, S. M. and Stent, G. S. 1986. Cell lineage and cell interactions in the determination of developmental cell fates. In *Genes, Molecules and Evolution*, Gustafson, J. P., Stebbins, G. L. and Ayala, F. J. (eds.). Plenum Press, New York. (14)

Shankland, S. M. and Weisblat, D. A. 1984. Stepwise commitment of blast cell fates during the positional specification of the O and P cell fates during serial blast divisions in the leech embryo. *Devel. Biol. 106*: 326-342. (14)

Shatz, C. J. and Kliot, M. 1982. Prenatal misrouting of the retinogeniculate pathway in Siamese cats. *Nature 300*: 525-529. (8)

Shatz, C. J. and Stetavan, D. W. 1986. Interactions between retinal ganglion cells during the development of the mammalian visual system. *Annu. Rev. Neurosci. 9*: 171-207. (8)

Shojaeian, H., Delhaye-Bouchaud, N. and Mariani, J. 1985. Decreased number of cells in the inferior olivary nucleus of the developing *staggerer* mouse. *Devel. Brain Res. 21*: 141-146. (16)

Sicard, R. E. (ed.). 1985. *Regulation of Vertebrate Limb Regeneration*. Oxford Univ. Press. (I)

Sidman, R. L. 1968. Development of interneuronal connections in brains of mutant mice. In *Physiological and Biochemical Aspects of Nervous Integration*. Carlson, E. D. (ed.). Prentice Hall, Englewood Cliffs, N. J. pp. 163-193. (16)

Sidman, R. L. 1972. Cell interaction in the developing mammalian central nervous system. In *Cell Interaction* (Proc. 3rd Lepetit Colloquium). Silverstri, L. C. (ed.). North Holland Publishing Co., Amsterdam. pp. 1-13. (16)

Sidman, R. L. and Rakic, P. 1973. Neuronal migration, with special reference to developing human. *Brain Res. 62*: 1-35. (11)

Sidman, R. L., Green, M. C. and Appel, S. 1965. *Catalogue of the Neurological Mutants of Mice*. Harvard Univ. Press, Cambridge, MA. (16)

Sieber-Blum, M. 1984. Fibronectin-regulated methionine enkephalin-like and somatostatin-like immunoreactivity in quail neural crest cell cultures. *Neuropeptides 4*: 457-466. (15)

Sieber-Blum, M. and Cohen, A. M. 1978. Lectin binding to neural crest cells. *J. Cell Biol. 76*: 628-638. (15)

Sieber-Blum, M. and Cohen, A. M. 1980. Clonal analysis of quail neural crest cells: They are pluripotent and differentiate in vitro in the absence of noncrest cells. *Devel. Biol. 79*: 170-180. (15)

Sieber-Blum, M. and Patel, S. R. 1986. In vitro differentiation of quail neural crest cells into sensory neurons. *Prog. in Devel. Biol.* (Part B): 243-248. (15)

Sieber-Blum, M., Sieber, F. and Yamada, K. M. 1981. Cellular fibronectin promotes adrenergic differentiation of quail neural crest cells in vitro. *Exp. Cell Res. 133*: 285-295. (15)

Silver, J. 1978. Cell death during development of the nervous system. In *Handbook of Sensory Physiology*, Vol. IX, Development of Sensory Systems, Springer-Verlag, Berlin, pp. 419-436. (17)

Simerly, R. B., Swanson, L. W. and Gorski, R. A. 1984. Demonstration of a sexual dimorphism in the distribution of serotonin immunoreactive fibers in the medial preoptic nucleus of the rat. *J. Comp. Neurol. 225*: 151-166. (17)

Simerly, R. B., Swanson, L. W. and Gorski, R. A. 1985a. The distribution of monoaminergic cells and fibers in a periventricular preoptic nucleus involved in the control of gonadotropin release: Immunohistochemical evidence for a dopaminergic sexual dimorphism. *Brain Res. 330*: 55-64. (17, 18)

Simerly, R. B., Swanson, L. W., Handa, R. J. and Gorski, R. A. 1985b. Influence of perinatal androgen on the sexually dimorphic distribution of tyrosine-hydroxylase immunoreactive cells and fibers in the anteroventral periventricular nucleus of the rat. *Neuroendocrinology 40*: 501-510. (17, 18)

Simons, K. and Fuller, S. D. 1985. Cell surface polarity in epithelia. *Annu. Rev. Cell Biol. 1*: 243-288. (2)

Sine, S. M. and Steinbach, J. H. 1985. Activation of acetylcholine receptors on clonal mammalian BC_3H-I cells by low concentrations of agonist. *J. Physiol. 358*: 91-108. (5)

Sine, S. and Taylor, P. 1980. The relationship between agonist occupation and the permeability response of the cholinergic receptor revealed by bound cobra alpha-toxin. *Biol. Chem. 255*: 10144-10156. (5)

Singer, M. 1952. The influence of the nerve in regeneration of the amphibian extremity. *Quart. Rev. Biol. 27*: 169-200. (3)

Singer, M. 1965. A theory of the trophic nervous control of amphibian limb regeneration, including a reevaluation of quantitative nerve requirements. In *Regeneration in Animals and Related Problems*, Kiortsis, V. and Trampusch, H. A. L. (eds.). North-Holland Pub. Co., Amsterdam. (3)

Singer, M. 1974. Neurotrophic control of limb regeneration in the newt. *Ann. N. Y. Acad. Sci. 228*: 308-321. (3)

Singer, M., Nordlander, R. H. and Edgar, M. 1979. Axonal guidance during embryogenesis and regeneration in the spinal cord of the newt: The blueprint hypothesis of neuronal pathway patterning. *J. Comp. Neurol. 185*: 1-22. (III)

Sirevaag, A. M. and Greenough, W. T. 1985. Differential rearing effects on rat visual cortex synapses. II. Synaptic morphometry. *Devel. Brain Res. 19*: 215-226. (19)

Sirevaag, A. M. and Greenough, W. T. In press. Differential rearing effects on rat visual cortex synapses. III. Neuronal and glial nuclei, boutons, dendrites, and capillaries. *Brain Res.* (19)

Skene, J. H. P. and Willard, M. 1981a. Changes in axonally transported proteins during axon regeneration in toad retinal ganglion cells. *J. Cell Biol. 89*: 86-95. (10)

Skene, J. H. P. and Willard, M. 1981b. Axonally transported proteins associated with axon growth in rabbit central and peripheral nervous system. *J. Cell Biol. 89*: 96-103. (10)

Skene, J. H. P., Jacobson, R. D., Snipes, G. J., McGuire, C. B., Norden, J. J. et al. 1986. A protein induced during nerve regeneration (GAP-43) is a major component of growth cone membranes. *Science 233*: 783-786. (10)

Skillen, R., Thienes, C. and Strain, L. 1961. Brain 5HT, 5-hydroxytryptophan decarboxylase and monoamine oxidase of normal, thyroid fed and propyl-thiouracil-fed male and female rats. *Endocrinology 69*: 1099-1102. (18)

Slack, J. M. W. 1983. *From Egg to Embryo: Determinative Events in Early Development*. Cambridge Univ. Press, Cambridge. (14, V)

Slater, C. R. 1982. Postnatal maturation of nerve-muscle junctions in hindlimb muscles of the mouse. *Devel. Biol. 94*: 11-22. (12)

Slater, C. R., Allen, E. G. and Young, C. 1985. Acetylcholine receptor distribution on regenerating mammalian muscle fibers at sites of mature and developing nerve-muscle junctions. *J. Physiol.* (Paris) *80*: 238-246. (12)

Smith, C. B. 1984. Aging and changes in cerebral energy metabolism. *Trends Neurosci. 7*: 203-208. (19)

Smith, C. G. 1983. The development and postnatal organization primary afferent projections to the rat thoracic spinal cord. *J. Comp. Neurol. 220*: 29-43. (13)

Smith, C. and Frank, E. 1987. Peripheral specification of sensory neurons transplanted to novel locations along the neuraxis. *J. Neurosci. 7*: 1537-1549. (13)

Smith, C. and Frank, E. In press. Peripheral specification of sensory connections in the spinal cord. *Brain, Behav. and Evol.* (13)

Smith-Thomas, L. C., Davis, J. P. and Epstein, M. L. 1986. The gut supports neurogenic differentiation of periocular mesenchyme, a chondrogenic neural crest-derived cell population. *Devel. Biol. 115*: 293-300. (15)

So, K. F. and Aguayo, A. J. 1985. Lengthy regrowth of cut axons from ganglion cells after peripheral nerve transplantation into the retina of adult rats. *Brain Res. 328*: 349-354. (10)

Sobel, A., Weber, M. and Changeux, J.-P. 1977. Large-scale purification of the acetylcholine-receptor protein in its membrane-bound and detergent extracted forms from *Torpedo marmorata* electric organ. *Eur. J. Biochem. 80*: 215-224. (2)

Sotelo, C. 1975. Dendritic abnormalities of Purkinje cells in the cerebellum of neurologic mutant mice (Weaver and Staggerer). In *Advances in Neurology*. 12th ed. Kreutzberg, G. W. (ed.). Raven Press, New York, pp. 335-351. (16)

Sotelo, C. and Changeaux, J.-P. 1974. Bergmann fibers and granular cell migration in the cerebellum of homozygous *weaver* mutant mouse. *Brain Res. 77*: 484-491. (11)

Sotelo, C. and Changeux, J.-P. 1974. Transsynaptic degeneration "en cascade" in the cerebellar cortex of *staggerer* mutant mice. *Brain Res. 67*: 519-526. (16)

Sparenborg, S., Manley, D., Witzany, D. and Gabriel, M. 1986. Clonidine and scopolamine alter learning-related neuronal activity in rabbit cingulate cortex and AV thalamic nucleus. *Soc. Neurosci. Abst. 12*: 711. (19)

Spelsberg, T., Littlefield, B., Seelke, R., Dani, G., Toyoda, et al. 1983. Role of specific chromosomal proteins and DNA sequences in the nuclear binding sites for steroid receptors. *Rec. Prog. Horm. Res. 39*: 463-517. (18)

Sperry, R. W. 1943. Visuomotor coordination in the newt (*Triturus viridescens*) after regeneration of the optic nerve. *J. Comp. Neurol. 79*: 33-55. (IV)

Sperry, R. W. 1963. Chemoaffinity in the orderly growth of nerve fiber patterns and connections. *Proc. Natl. Acad. Sci. USA 50*: 703-710. (1, 9, IV, 13)

Sperry, R. W. and Miner, N. 1949. Formation within nucleus V of synaptic associations mediating cutaneous localization. *J. Comp. Neurol. 90*: 403-423. (13)

Spindel, E. R., Zilberberg, M. D., Habener, J. F. and Chin, W. W. 1986. Two prohormones for gastrin-releasing peptide are encoded by two mRNAs differing by 19 nucleotides. *Proc. Natl. Acad. Sci. USA 83*: 19-23. (4)

St. John, P. A., Froehner, S. C., Goodenough, D. A. and Cohen, J. B. 1982. Nicotinic postsynaptic membranes from *Torpedo*: Sideness, permeability to macromolecules, and topography of major polypeptides. *J. Cell Biol. 92*: 333-342. (2)

Stallcup, W. B. and Beasley, L. L. 1985. Involvement of the nerve growth factor-inducible large external glycoprotein (NILE) in neurite fasciculation in primary cultures of rat brain. *Proc. Natl. Acad. Sci. USA 82*: 1276-1280. (1)

Stallcup, W. B., Beasley, L. L. and Levine, J. M. 1985. Antibody against nerve growth factor-inducible large external (NILE) glycoprotein labels nerve fiber tracts in developing rat nervous system. *J. Neurosci. 5*: 1090-1101. (1)

Steen, T. P. and Thornton, C. S. 1963. Tissue interaction in amputated aneurogenic limbs of *Ambystoma* larvae. *J. Exp. Zool. 154*: 207-221. (3)

Steinbach, J. H. 1981. Developmental changes at acetylcholine receptor aggregates at rat skeletal neuromuscular junctions. *Devel. Biol. 84*: 267-276. (12)

Stent, G. S. 1985. The role of cell lineage in development. *Phil. Trans. Roy. Soc. Lond. B. 312*: 3-19. (14)

Stent, G. S. and Weisblat, D. A. 1985. Cell lineage in the development of invertebrate nervous systems. *Annu. Rev. Neurosci. 8*: 45-70. (14)

Stent, G. S., Weisblat, D. A., Blair, S. S. and Zackson, S. L. 1982. Cell lineage in the development of the leech nervous system. In *Neuronal Development*, Spitzer, N. (ed.). Plenum Press, New York. pp. 1-44. (14)

Sternberg, P. W. and Horvitz, H. R. 1986. Pattern formation during vulval development in *C. elegans. Cell 44*: 761-772. (14)

Stetavan, D. W. and Shatz, C. J. 1986a. Prenatal development of retinal ganglion cell axons: Segregation into eye-specific layers within the cat's lateral geniculate nucleus. *J. Neurosci. 6*: 234-251. (8)

Stetavan, D. W. and Shatz, C. J. 1986b. Prenatal development of cat retinogeniculate axon arbors in the absence of binocular interactions. *J. Neurosci. 6*: 990-1003. (8)

Steward, O. 1983. Polyribosomes at the base of dendritic spines of CNS neurons: Their possible role in synapse construction and modification. *Cold Spring Harbor Symp. Quant. Biol. 48*: 745-759. (19)

Steward, O. and Falk, P. M. 1986. Protein-synthetic machinery at postsynaptic sites during synaptogenesis: A quantitative study of the association between polyribosomes and developing synapses. *J. Neurosci. 6*: 412-423. (19)

Strader, C. D., Pickel, V. M., Joh, T. H., Strohsacker, M. W., Shorr, R. G. L. et al. 1984. Antibodies to the beta-adrenergic receptor: Attenuation of catecholamine-sensitive adenyl cyclase and demonstration of postsynaptic receptor localization in brain. *Proc. Natl. Acad. Sci. USA 80*: 1840-1844. (2)

Straznicky, C. 1983. The patterns of innervation and movements of ectopic hindlimb supplied by brachial spinal cord segments in the chick. *Anat. Embryol. 167*: 247-262. (III)

Stroud, R. M. and Finer-Moore, J. 1985. Acetylcholine receptor structure, function and evolution. *Annu. Rev. Cell Biol. 1*: 369-401. (5)

Stryker, M. P. and Harris, W. A. 1986. Binocular impulse blockade prevents the formation of ocular dominance columns in cat visual cortex. *J. Neurosci. 6*: 2117-2133. (9, 13)

Stuart, D. K., Blair, S. S. and Weisblat, D. A. 1987. Cell lineage, cell death, and the developmental origin of identified serotonin- and dopamine- containing neurons in the leech. *J. Neurosci. 7*: 1107-1122. (14)

Sulston, J. E. 1983. Neuronal lineages in the Nematode *Caenorhabditis elegans*. *Molecular Neurobiology*. Cold Spring Harbor Symp. Quant. Biol. 48: 443-452. (V)

Sulston, J. E. and Horvitz, H. R. 1977. Post-embryonic cell lineages of the nematode *Caenorhabiditis elegans*. *Devel. Biol. 56*: 110-156. (14)

Sulston, J. E., and White, J. G. 1980. Regulation and cell autonomy during postembryonic development of *Caenorhabditis elegans*. *Devel. Biol. 78*: 577-597. (14)

Sulston, J. E., Schierenberg, E., White, J. G. and Thomson, J. N. 1983. The embryonic cell lineage of the nematode *Caenorhabditis elegans*. *Devel. Biol. 100*: 64-119. (13, 14)

Sumikawa, K., Houghton, M., Smith, J. C., Bell, L., Richards, B. M. et al. 1982. The molecular cloning and characterization of cDNA coding for the alpha subunit of the acetylcholine receptor. *Nucleic Acids Res. 10*: 5809-5822. (5)

Sutcliffe, J. G., Milner, R. J., Bloom, F. E., and Lerner, R. A. 1982. Common 82-nucleotide sequence unique to brain RNA. *Proc. Natl. Acad. Sci. USA 79*: 4942-4946. (6)

Sutcliffe, J. G., Milner, R. J., Gottsefeld, J. M. and Lerner, R. A. 1984a. Identifier sequences are transcribed specifically in brain. *Nature 308*: 237-241. (6)

Sutcliffe, J. G., Milner, R. J., Gottsefeld, J. M. and Reynolds, W. 1984b. Control of neuronal gene expression. *Science 225*: 1308-1315. (6)

Sutcliffe, J. G., Milner, R. J., Shinnick, T. M. and Bloom, F. E. 1983. Identifying the protein products of brain-specific genes with antibodies to chemically synthesized peptides. *Cell 33*: 671-682. (6)

Swaab, C. F. and Fliers, E. 1985. A sexually dimorphic nucleus in the human brain. *Science 228*: 1112-1115. (17)

Taghert, P., Bastiani, M., Ho, R. K. and Goodman, C. S. 1982. Guidance of pioneer growth cones: Filopodial contacts and coupling revealed with an antibody to Lucifer Yellow. *Devel. Biol. 94*: 391-399. (8)

Tamarova, Z. A. 1977. Excitatory postsynaptic potentials induced in the frog lumbar motoneurones by muscle and cutaneous nerve stimulation. *Sechenov J. Physiol. U.S.S.R. 63*: 806-813. (13)

Tanaka, H. and Landmesser, L. 1986a. Interspecies selective motoneuron projection patterns in chick-quail chimeras. *J. Neurosci. 6*: 2880-2888. (9)

Tanaka, H. and Landmesser, L. T. 1986b. Cell death of lumbosacral motoneurons in chick, quail, and chick-quail chimera embryos: a test of the quantitative matching hypothesis of neuronal cell death. *J. Neurosci. 6*: 2889-2899. (16)

Tanaka, H. and Obata, K. 1984. Developmental changes in unique cell surface antigens of chick embryo spinal motoneurons and ganglion cells. *Devel. Biol. 106*: 26-37. (9)

Tanzi, E. 1893. I fatti e le induzioni nell'odierna istologia del sistema nervosa. *Riv. Sperim. freniatria medic. leg. 19*: 419-472. (19)

Tarkowski, A. K. 1959. Experiments on the development of isolated blastomeres of mouse eggs. *Nature 184*: 1286-1287. (V)

Tarkowski, A. K. 1961. Mouse chimeras developed from fused eggs. *Nature 190*: 357-360. (16)

Tassava, R. A. and Lloyd, R. M. 1977. Injury requirements for initiation of regeneration of newt limbs which have whole skin grafts. *Nature 268*: 49-50. (3)

Tate, K. and Westerman, R. A. 1973. Polyneuronal self-reinnervation of a slow-twitch muscle (soleus) in the cat. *Proc. Aust. Physiol. Pharm. Soc. 4*: 174-175. (12)

Taylor, A. C. and Kollros, J. J. 1946. Stages in the normal development of *Rana pipiens* larvae. *Anat. Rec. 94*: 7-23. (13)

Thanos, S. and Bonhoeffer, F. 1983. Investigations on the development and topographic order of retinotectal axons: Anterograde and retrograde staining of axons and perikarya with rhodamine in vivo. *J. Comp. Neurol. 219*: 420-430. (8)

Thiery, J.-P., Brackenbury, R., Rutishauser, U. and Edelman, G. M. 1977. Adhesion among neural cells of the chick embryo. II. Purification and characterization of a cell adhesion molecule from neural retina. *J. Biol. Chem. 252*: 6841-6845. (11)

Thiery, J.-P., Delouvee, A., Grumet, M. and Edelman, G. M. 1985a. Initial appearance and regional distribution of the neuron-glia cell adhesion molecule in the chick embryo. *J. Cell Biol. 100*: 442-456. (1)

Thiery, J.-P., Duband, J.-L. and Delouvee, A. 1985b. The role of cell adhesion in morphogenetic movements during early embryogenesis. In *The Cell in Contact: Adhesions and Junctions as Morphogenetic Determinants*, Edelman, G. M. and Thiery, J.-P. (eds.). John Wiley & Sons, New York. pp. 169-96. (1)

Thiery, J.-P., Duband, J.-L., Rutishauser, U. and Edelman, G. M. 1982. Cell adhesion molecules in early chick embryogenesis. *Proc. Natl. Acad. Sci. USA 79*: 6737-6741. (1)

Thoenen, H. and Edgar, D. 1985. Neurotrophic factors. *Science 229*: 238-242. (VI)

Thomas, J. B., Bastiani, M. J., Bate, C. M. and Goodman, C. S. 1984. From grasshopper to *Drosophila*: A common plea for neuronal development. *Nature 310*: 203-207. (7)

Thomas, J. O. and Thompson, R. J. 1977. Variation in chromatin structure in two cell types from the same tissue: A short DNA repeat length in cerebral cortex neurons. *Cell 10*: 633-640. (6)

Thompson, M. D., Woolley, D., Gietzen, D. and Conway, S. 1983. Catecholamine synthesis inhibitors acutely modulate [³H]estradiol binding by specific brain areas and pituitary in ovariectomized rats. *Endocrinology 113L*: 855-865. (18)

Thornton, C. S. 1938. The histogenesis of muscle in the regenerating fore limb of larval *Ambystoma punctatum*. *J. Morph. 62*: 17-47. (3)

Thornton, C. S. 1942. Studies on the origin of the regeneration blastema in *Triturus viridescens*. *J. Exp. Zool. 89*: 375-390. (3)

Thornton, C. S. and Steen, T. P. 1962. Eccentric blastema formation in aneurogenic limbs of *Ambystoma* larvae following epidermal cap deviation. *Devel. Biol. 5*: 328-353. (3)

Thornton, C. S. and Tassava, R. A. 1969. Regeneration and supernumerary limb formation under sparsely innervated conditions. *J. Morph. 127*: 225-232. (3)

Thornton, C. S. and Thornton, M. T. 1970. Recuperation of regeneration in denervated limbs of *Ambystoma* larvae. *J. Exp. Zool. 173*: 293-301. (3)

Tieman, S. B. 1984. Effects of monocular deprivation on geniculocortical synapses in the cat. *J. Comp. Neurol. 222*: 166-176. (19)

Tobet, S. A., Gallagher, C. A., Zahniser, D. J., Cohen, M. H. and Baum, M. J. 1983. Sexual dimorphism in the preoptic/anterior hypothalamic area of adult ferrets. *Endocrinology 112* (Suppl.): 240. (17)

Todd, J. T. 1823. On the process of reproduction of the members of the aquatic salamander. *Quart. J. Sci. Lit. Arts 16*: 84-96. (3)

Toran-Allerand, C. D. 1976. Sex steroids and the development of the newborn mouse hypothalamus and preoptic area in vitro: Implications for sexual differentiation. *Brain Res. 106*: 407-412. (18)

Toran-Allerand, C. D. 1978. Gonadal hormones and brain development: Cellular aspects of sexual differentiation. *Amer. Zool. 18*: 553-565. (18)

Toran-Allerand, C. D. 1980. Sex steroids and the development of the newborn mouse hypothalamus and preoptic area in vitro. II. Morphological correlates and hormonal specificity. *Brain Res. 189*: 413-427. (17, 18)

Toran-Allerand, C. D. 1984. On the genesis of sexual differentiation of the central nervous system: Morphogenetic consequences of steroid exposure and possible role of α-fetoprotein. In *Progress in Brain Research*, Vol. 61, *Sex Differences in the Brain*, DeBries, G. J., De Bruin, J. P. C., Uylings, H. B. M. and Corner, M. A. (eds.), Elsevier, Amsterdam, pp. 63-98. (17)

Toran-Allerand, C. D., Gerlach, J. L. and McEwen, B. S. 1980. Autoradiographic localization of [³H]estradiol related to steroid responsiveness in cultures of the hypothalamus and preoptic area. *Brain Res. 184*: 517-522. (17)

Torrealba, F., Guillery, R. W., Polley, E. H. and Mason, C. A. 1981. A demonstration of several independent, partially overlapping, retinotopic maps in the optic tract of the cat. *Brain Res. 219*: 428-432. (8)

Torrence, S. A. and Stuart, D. K. 1986. Gangliogenesis in leech embryos: Migration of neural precursor cells. *J. Neurosci. 6*: 2736-2746. (14)

Tosney, K. W. and Landmesser, L. 1984. Pattern and specificity of axonal outgrowth following varying degrees of chick limb ablation. *J. Neurosci. 4*: 2518-2527. (9)

Tosney, K. W. and Landmesser, L. T. 1985a. Specificity of motoneuron growth cone outgrowth in the chick limb. *J. Neurosci. 6*: 2336-2344. (8, 9)

Tosney, K. W. and Landmesser, L. T. 1985b. Growth cone morphology and trajectory in the lumbosacral region of the chick embryo. *J. Neurosci. 5*: 2345-2358. (8, 9, 10)

Tosney, K. W. and Landmesser, L. 1985c. Development of the major pathways for neurite outgrowth in the chick hindlimb. *Devel. Biol. 109*: 193-214. (9)

Tosney, K. W., Watanabe, M., Landmesser, L. and Rutishauser, U. 1986. The distribution of NCAM in the chick hindlimb during axon outgrowth and synaptogenesis. *Devel. Biol. 114*: 437-452. (9)

Triller, A., Cluzeaud, F., Pfeiffer, F., Betz, H. and Korn, H. 1985. Distribution of glycine receptors at central synapses: An immunoelectron microscopy study. *J. Cell Biol. 101*: 683-688. (5)

Tsou, A.-P., Lai, C., Danielson, P., Noonan, D. J. and Sutcliffe, J. G. 1986. Structural characterization of a heterogeneous family of rat brain mRNAs. *Molec. Cell Biol. 6*: 768-778. (4)

Tucker, R. P. and Erickson, C. A. 1984. Morphology and behavior of quail neural crest cells in artificial three-dimensional extracellular matrices. *Devel. Biol. 104*: 390-405. (15)

Turley, E. A., Erickson, C. A. and Tucker, R. P. 1985. The retention and ultrastructural appearances of various extracellular matrix molecules incorporated into three-dimensional hydrated collagen lattices. *Devel. Biol. 109*: 347-369. (15)

Turner, A. M. and Greenough, W. T. 1985. Differential rearing effects on rat visual cortex synapses. I. Synaptic and neuronal density and synapses per neuron. *Brain Res. 329*: 195-203. (19)

Ullu, E. and Weiner, A. M. 1985. Upstream sequences modulate the internal promotor of the human 7SL RNA gene. *Nature 318*: 371-374. (6)

Uylings, H. B., Kuypers, M. K., Diamond, M. C. and Veltman, W. A. M. 1978. Effects of differential environments on the plasticity of dendrites of pyramidal neurons in adult rats. *Exp. Neurol. 62*: 658-677. (19)

Vakaet, L. 1985. Morphogenetic movements and fate maps in the avian blastoderm. In *Molecular Determinants of Animal Form,* Edelman, G. M. (ed.). Alan R. Liss, Inc., New York. (1)

Van Der Loos, H. and Woolsey, T. A. 1973. Somatosensory cortex: Structural alterations following early injury to sense organs. *Science 179*: 395-398. (19)

Van Ness, J., Maxwell, I. H. and Hahn, W. E. 1979. Complex population of nonpolyadenylated messenger RNA in mouse brain. *Cell 18*: 1341-1350. (4)

van Raamsdonk, W., Van Veer, L., Veeken, K., Heyting, C. and Pool, C. W. 1982. Differentiation of muscle fiber types in the teleost *Branchydanio rerio,* the zebrafish. *Anat. Embryol. 164*: 51-62. (8)

van Raamsdonk, W., te Kronnie, G., Poole, C. W. and Van der Laarse, W. 1980. An immune histochemical and enzymatic characterization of the muscle fibers in myotomal muscle of the teleost *Brachydanio rerio,* Hamilton Buchanan. *Acta Histochem. 67*: 200-216. (8)

Varon, S. and Adler, R. 1980. Nerve growth factors and control of nerve growth. In *Current Topics in Developmental Biology,* Vol. 16, *Neuronal Development. II. Neural Development in Model Systems,* pp. 207-252. (17)

Vaughn, J. E. and Grieshaber, J. A. 1973. Morphological investigation of an early reflex pathway in developing rat spinal cord. *J. Comp. Neurol. 148*: 177-210. (13)

Vethamany-Globus, S., Globus, M. and Tomlinson, B. 1978. Neural and hormonal stimulation of DNA and protein synthesis in cultured regeneration blastemata, in the newt, *Notophthalmus viridescens. Devel. Biol. 65*: 183-192. (3)

Vincent, M. and Thiery, J.-P. 1984. A cell surface marker for neural crest and placodal cells: Further evolution in peripheral and central nervous system. *Devel. Biol. 103*: 468-481. (15)

Vito, C. and Fox, T. 1982. Androgen and estrogen receptors in embryonic and neonatal rat brain. *Devel. Brain Res. 2*: 97-110. (18)

Vogel, K. and Weston, J. A. 1985. Multiple subpopulations of avian neural crest cells arising in culture. *Cell Diff. 16*: 113S. (15)

Vogel, K. and Weston, J. A. 1986. Developmental changes in subpopulations of cultured neural crest cells. *J. Cell Biol. 103* (suppl.): 232.z. (15)

Vogel, M. W. and Herrup, K. 1986. Target related cell death in cerebellar granule cells: Are early generated cells at a competititive advantage? *Soc. Neurosci. Abst. 12*: 869. (16)

vom Saal, F. and Bronson, F. 1978. In utero proximity of female mouse fetuses to males: Effect on reproductive performance during later life. *Biol. Reprod. 19*: 842-853. (18)

Vrensen, G. and Cardozo, J. N. 1981. Changes in size and shape of synaptic connections after visual training: An ultrastructural approach of synaptic plasticity. *Brain Res. 218*: 79-98. (19)

Wake, D. B. 1966. Comparative osteology and evolution of the lungless salamanders, family Plethodontidae. *Memoirs, So. Calif. Acad. Sci. 4*: 1-111. (14)

Walicke, P. and Patterson, P. H. 1981. On the role of cyclic nucleotides in the transmitter choice made by cultured sympathetic neurons. *J. Neurosci. 1*: 333-342. (4)

Walicke, P., Campenot, R. and Patterson, P. 1977. Determination of transmitter function by neuronal activity. *Proc. Natl. Acad. Sci. USA 74*: 5767-5771. (4)

Wallace, C. S., Black, J. E., Hwang, H. M. and Greenough, W. T. 1986. Housing complexity affects body and adrenal body weight of adult rats within 10 days exposure. *Soc. Neurosci. Abst. 12*: 1283. (19)

Wallace, H. 1981. *Vertebrate Limb Regeneration.* John Wiley & Sons, New York. (3)

Wallis, I., Ellis, L., Suh, K. and Pfenninger, K. H. 1985. Immunolocalization of a neuronal growth-dependent membrane glycoprotein. *J. Cell Biol. 101*: 1990-1998. (1)

Walsh, F. S. and Moore, S. E. 1985. Expression of cell-adhesion molecule, N-CAM, in diseases of adult human skeletal-muscle. *Neurosci. Lett. 59*: 73-78. (1)

Walter, P. and Blobel, G. 1982. Signal recognition particle contains a 7S RNA essential for protein translocation across the endoplasmic reticulum. *Nature 299*: 691-698. (6)

Ward, I. L. 1984. The prenatal stress syndrome: Current status. *Psychoneuroendocrinology 9*: 3-11. (18)

Watanabe-Nagasu, N., Itoh, Y., Tani, T., Okano, K., Koga, N. et al. 1983. Structural analysis of gene loci for rat U1 small nuclear RNA. *Nucleic Acid Res. 11*: 1791-1801. (6)

Waterhouse, B. D. and Woodward, D. J. 1980. Interaction of nerepinephrine with cerebrocortical activity evoked by stimulation of somatosensory afferent pathways in the rat. *Exp. Brain Res. 67*: 11-34. (19)

Watson, J. B. and Sutcliffe, J. G. In preparation. A monkey brain-specific Pol III transcript of the Alu repeat family. (6)

Weibel, E. R. 1979. *Stereological Methods.* Vol. 1: *Practical Methods for Biological Morphometry.* Academic Press, New York. (19)

Weiner, A. M., Deininger, P. L. and Efstratiadis, A. 1986. Nonviral retroposons: Genes, pseudogenes, and transposable elements generated by the reverse flow of genetic information. *Annu. Rev. Biochem. 55*: 631-661. (6)

Weisblat, D. A. and Blair, S. S. 1984. Developmental indeterminacy in embryos of the leech *Helobdella triserialis. Devel. Biol. 101*: 326-335. (14)

Weisblat, D. A. and Kristan, Jr., W. B. 1985. The development of serotonin containing neurons in the leech. In *Model Neural Networks and Behavior,* Selverston, A. I. (ed.). Plenum Press, New York. pp. 175-190. (14)

Weisblat, D. A. and Shankland, M. 1985. Cell lineage and segmentation in the leech. *Phil. Trans. Roy. Soc. Lond. B. 312*: 39-56. (14)

Weisblat, D. A., Harper, G., Stent, G. S. and Sawyer, R. T. 1980b. Embryonic cell lineage in the nervous system of the glossiphoniid leech *Helobdella triserialis. Devel. Biol. 76*: 58-78. (14)

Weisblat, D. A., Kim, S. Y. and Stent, G. S. 1984. Embryonic origins of cells in the leech *Helobdella triserialis. Devel. Biol. 104*: 65-85. (14)

Weisblat, D. A., Sawyer, R. T. and Stent, G. S. 1978. Cell lineage analysis by intracellular injection of a tracer enzyme. *Science 202*: 1295-1298. (14)

Weisblat, D. A., Zackson, S. L., Blair, S. S. and Young, J. D. 1980a. Cell lineage analysis by intracellular injection of fluorescent tracers. *Science 209*: 1538-1541. (14)

Weiss, P. 1942. Lid closure reflex from eyes transplanted to atypical locations in *Triturus torosus*: Evidence of a peripheral origin of sensory specificity. *J. Comp. Neurol. 77*: 131-169. (13)

Wennogle, L. P. and Changeux, J.-P. 1980. Transmembrane orientation of proteins present in acetylcholine receptor-rich membranes from *Torpedo* studied by selective proteolysis. *Eur. J. Biochem. 106*: 381-393. (2)

Werle, M. J. and Herrera, A. A. 1986. The elimination of excess polyneuronal innervation in reinnervated adult frog muscles. *Soc. Neurosci. Abst. 12*: 547. (12)

Wernig, A., Carmody, J. J., Anzil, A. P., Hansert, E., Marciniak, M. and Zucker, H. 1984. Persistence of nerve sprouting with features of synapse remodelling in soleus muscles of adult mice. *Neuroscience 11*: 241-253. (12)

Wesa, J. M., Chang, F.-L. F., Greenough, W. T. and West, R. W. 1982. Synaptic contact curvature: Effects of differential rearing on rat occipital cortex. *Dev. Brain Res. 4*: 253-257. (19)

West, J. D. 1978. Analysis of clonal growth using chimeras and mosaics. In *Development in Mammals,* Johnson, M. H. (ed.), Elsevier/North Holland Biomedical Press, New York. pp. 413-460. (16)

West, R. W. and Greenough, W. T. 1972. Effect of environmental complexity on cortical synapses of rats: Preliminary results. *Behav. Biol. 7*: 279-284. (19)

Westerfield, M. and Eisen, J. S. 1985. The growth of motor axons in the spinal cord of *Xenopus* embryos. *Devel. Biol. 109*: 96-101. (8)

Westerfield, M., McMurray, J. V. and Eisen, J. S. 1986. Identified motoneurons and their innervation of axial muscles in the zebrafish. *J. Neurosci. 6*: 2267-2277. (8)

Weston, J. A. 1970. The migration and differentiation of neural crest cells. *Adv. Morphogen. 8*: 41-114. (15)

Weston, J. A. 1981. The regulation of normal and abnormal neural crest cell development. *Adv. Neurol., Vol. 19: Neurofibromatosis (von Recklinghausen Disease),* Riccardi, V. M. and Mulvihill, J. J. (eds.), Raven Press, New York, pp. 77-95. (15)

Weston, J. A. 1982. Motile and social behavior of neural crest cells. In *Cell Behaviour,* Bellairs, R., Curtis, A. and Dunn, G. (eds.), Cambridge Univ. Press, Cambridge, pp. 429-470. (15)

Weston, J. A. 1983. Regulation of neural crest cell migration and differentiation. In Yamada, K. M. (ed.), *Cell Interactions and Development: Molecular Mechanisms,* John Wiley & Sons, New York, pp. 153-184. (15)

Weston, J. A. 1986. Phenotypic diversification in neural crest-derived cells: The time and stability of commitment during early development. *Curr. Topics in Devel. Biol. 20*: 195-210. (15)

Weston, J. A. and Butler, S. 1966. Temporal factors affecting localization of trunk neural crest cells in the chicken embryo. *Devel. Biol. 14*: 246-266. (15)

Weston, J. A., Ciment, G. and Girdlestone, J. 1984. The role of extracellular matrix in neural crest development: A reevaluation. In *The Role of Extracelluar Matrix in Development,* Trelstad, R. (ed.), Alan R. Liss, New York, pp. 433-460. (15)

Weston, J. A., Derby, M. A. and Pintar, J. E. 1978. Changes in the extracellular environment of neural crest cells during their early migration. *Zoon 6*: 103-113. (15)

Wetts, R. and Herrup, K. 1982a. Interaction of granule, Purkinje and inferior olivary neurons in Lurcher chimeric mice. I. Qualitative studies. *J.E.E.M. 68*: 87-98. (16)

Wetts, R. and Herrup, K. 1982b. Interaction of granule, Purkinje and inferior olivary neurons in Lurcher chimeric mice. II. Granule cell death. *Brain Res. 250*: 358-363. (16)

Wetts, R. and Herrup, K. 1983a. Direct correlation between Purkinje and granule cell number in the cerebella of Lurcher chimeras and wild-type mice. *Dev. Brain Res. 10*: 41-47. (16)

Wetts, R. and Herrup, K. 1983b. Cerebellar Purkinje cells are descended from a small number of progenitors committed during early development: Quantitative analysis of *lurcher* chimeric mice. *J. Neurosci. 2*: 1494-1498. (16)

Whalen, R. and Massicci, J. 1975. Subcellular analysis of the accumulation of estrogen by the brain of male and female rats. *Brain Res. 89*: 255-264. (18)

Whalen, R. E. and Rezek, D. L. 1974. Inhibition of lordosis in female rats by subcutaneous implants of testosterone, androstenedione or dihydrotestosterone in infancy. *Horm. Behav. 5*: 125-128. (17)

White, B. A. and Bancroft, F. C. 1982. Cytoplasmic dot hybridization. *J. Biol. Chem. 257*: 8569-8572. (4)

White, E. L. 1978. Identified neurons in mouse SmI cortex which are postsynaptic to thalamocortical axon terminals: A combined Golgi-electron microscopic and degeneration study. *J. Comp. Neurol. 181*: 627-662. (19)

White, J. G., Southgate, E., Thomson, J. N. and Brenner, S. 1983. Factors that determine connectivity in the nervous system of *Caenorhabditis elegans. Cold Spring Harbor Symp. Quant. Biol. 43*: 633-640. (8)

Whiting, P. J. and Lindstrom, J. 1986. Purification and characterization of a nicotinic acetylcholine receptor from chick brain. *Biochem. 25*: 2082-2093. (5)

Whitman, C. O. 1878. The embryology of *Clepsine. Quart. J. Microscop. Sci. 18*: 215-315. (14)

Whitman, C. O. 1892. The metamerism of *Clepsine. Festschrift zum 70. Geburtstage R. Leuckarts,* pp. 385-395. (14)

Whitney, F. R. and Fuarano, A. V. 1984. Highly repeated DNA families in the rat. *J. Biol. Chem. 259*: 10481-10492. (6)

Wiegand, S. and Terasawa, E. 1982. Discrete lesions reveal functional heterogeneity of suprachiasmatic structures in the regulation of gonadotropin secretion in the female rat. *Neuroendocrinology 34*: 395-404. (17)

Wiesel, T. N. and Hubel, D. H. 1963. Single cell responses in striate cortex of kittens deprived of vision in one eye. *J. Neurophysiol. 26*: 1003-1017. (5, 8)

Wiesel, T. N. and Hubel, D. H. 1965. Comparison of effects of unilateral and bilateral eye closure on cortical unit responses in kittens. *J. Neurophysiol. 28*: 1029-1040. (8)

Wigston, D. and Sanes, J. 1985. Selective reinnervation of intercostal muscles transplanted from different segmental levels to a common site. *J. Neurosci. 5*: 1208-1224. (9)

Wilkinson, R. S. and Lichtman, J. W. 1985. Regular alternation of fiber types in the transversus abdominis muscle of the garter snake. *J. Neurosci. 5*: 2979-2988. (12)

Williams, R. K., Gordidis, C. and Akeson, R. 1985. Individual neural cell-types express immunologically distinct N-CAM forms. *J. Cell Biol. 101*: 36-42. (1)

Williams, R. W. and Rakic, P. R. 1985. Dispersion of growing axons within the optic nerve of the embryonic monkey. *Proc. Natl. Acad. Sci. USA 82*: 3906-3910. (10)

Williams, R. W., Bastiani, M. J., Lia, B. and Chalupa, L. M. 1986. Growth cones, dying axons, and developmental fluctuations in the fiber population of the cat's optic nerve. *J. Comp. Neurol. 246*: 32-69. (10)

Wilson, D. B. 1981. The cell cycle during closure of the neural folds in the C57B1 mouse. *Dev. Brain Res. 2*: 420-424. (16)

Wilson, J. D., Griffin, J. E., Leshin, M. and George, F. W. 1981. Role of gonadal hormones in development of the sexual phenotypes. *Hum. Genet. 58*: 78-84. (17)

Wilson, M. and Roy, E. 1986. Pharmacokinetics of od imipramine are affected by age and sex in rats. *Life Sci. 38*: 711-718. (18)

Windle, W. F. 1934. Correlation between the development of local reflexes and reflex arcs in the spinal cord of cat embryos. *J. Comp. Neurol. 59*: 487-505. (13)

Wise, P., Rance, N. and Barraclough, C. 1981. Effects of estradiol and progesterone on catecholamine turnover rates in discrete hypothalamic regions in ovariectomized rats. *Endocrinology 108*: 2186-2193. (18)

Wise, S. and Jones, E. G. 1976. The organization and postnatal development of the commissural projection of the rat somatic sensory cortex. *J. Comp. Neurol. 168*: 313-344. (10)

Wold, B. J., Klein, W. H., Hough-Evans, B. R., Britten, R. J. and Davidson, E. H. 1978. Sea urchin embryo mRNA sequences expressed in the nuclear RNA of adult tissues. *Cell 14*: 941-950. (4)

Wolinsky, E. and Patterson, P. H. 1983. Tyrosine hydroxylase activity decreases with induction of cholinergic properties in cultured synaptic neurons. *J. Neurosci. 3*: 1495-1500. (4)

Woodruff, M. L., Theriot, J. and Burden, S. J. 1987. 300-kd subsynaptic protein copurifies with acetylchoine receptor-rich membranes and is concentrated at neuromuscular synapses. *J. Cell Biol. 104*: 939-946. (2)

Woolsey, T. A. and Van der Loos, H. 1970. The structural organization of layer IV in the somatosensory region 9SI) of the mouse cerebral cortex: The description of a cortical field composed of discrete cytoarchitectonic units. *Brain Res. 17*: 205-242. (19)

Woolsey, T. A., Dierker, M. L. and Wann, D. F. 1975. Mouse SmI cortex: Qualitative and quantitative classification of Golgi-impregnated barrel neurons. *Proc. Natl. Acad. Sci. USA 72*: 2165-2169. (19)

Wyman, R. J. 1973. Somatotopic connectivity or species recognition connectivity? In *Control of posture and locomotion,* Stein, R.B. (ed.), Plenum Press, New York. pp. 45-53. (13)

Xue, Z. G., Smith, J. and Le Douarin, N. M. 1985. Differentiation of catecholaminergic cells in cultures of embryonic avian sensory ganglia. *Proc. Natl. Acad. Sci. USA 82*: 8800-8804. (15)

Yamamoto, K. 1985. Steroid receptor regulated transcription of specified genes and gene networks. *Annu. Rev. Gen. 19*: 209-252. (18)

Yntema, C. L. 1959. Regeneration of sparsely innervated aneurogenic forelimbs of *Ambystoma* larvae. *J. Exp. Zool. 140*: 101-123. (3)

Yoon, C. H. 1972. Developmental mechanisms for changes in cerebellum of *staggerer* mouse, a neurological mutant of genetic origin. *Neurology 22*: 743-754. (16)

Yoshikawa, K. and Kong, J. 1983. Sex-related difference in substance P level in rat anterior pituitary: A model of neonatal imprinting by testosterone. *Brain Res. 273*: 362-365. (18)

Yoshikami, D. and Okun, L. M. 1984. Staining of living presynaptic nerve terminals with selective fluorescent dyes. *Nature 310*: 53-56. (12)

Young, B. D., Birnie, G. D., and Paul, J. 1976. Complexity and specificity of polysomal poly (A⁺) RNA in mouse tissues. *Biochemistry 15*: 2823-2829. (4)

Young, E. F., Ralston, E., Blake, J., Ramachandran, J., Hall, Z. W. et al. 1985. Topological mapping of acetylcholine receptor: Evidence for a model with five transmembrane segments and a cytoplasmic COOH-terminal peptide. *Proc. Natl. Acad. Sci. USA 82*: 262-630. (5)

Yu, W. A. and Yu, M. C. 1983. Acceleration of the regeneration of the crushed hypoglossal nerve by testosterone. *Exp. Neurol. 80*: 349-360. (17)

Zackson, S. L. 1982. Cell clones and segmentation in leech development. *Cell 31*: 761-770. (14)

Zackson, S. L. 1984. Cell lineage, cell: Cell interaction, and segment formation in the ectoderm of a glossiphoniid leech embryo. *Devel. Biol. 104*: 143-160. (14, 16)

Zengel, J. E., Sypert, G. W. and Munson, J. B. 1983. Relation of single fiber EPSP amplitude in spinal moto-neurons to location of afferent entry. *Soc. Neurosci. Abst. 9*: 862. (13)

Ziller, C. E., Dupin, E., Brazeau, P., Paulin, D. and LeDouarin, N. M. 1983. Early segregation of a neuronal precursor cell line in the neural crest as revealed by culture in a chemically defined medium. *Cell 32*: 627-638. (15)

Ziskind-Conhaim, L., Geffen, I. and Hall, Z. W. 1984. Redistribution of acetylcholine receptors on developing rat myotubes. *J. Neurosci. 4*: 2346-2349. (2)

Zolman, J. F. and Morimoto, H. 1965. Cerebral changes related to duration of environmental complexity and locomotor activity. *J. Comp. Physiol. Psychol. 60*: 382-387. (19)

Zolovick, A., Pearse, R., Boehlke, K. and Elefthriou, B. 1966. Monoamine oxidase activity in various parts of the rat brain during the estrous cycle. *Science 154*: 649. (18)

Zurn, A. D. 1982. Identification of glycolipid binding sites for SBA and differences in the surface glycolipids of cultured adrenergic and cholinergic sympathetic neurons. *Devel. Biol. 94*: 483-498. (15)

Index

ABOUT THE BOOK

This book was set in Itek Times Roman at Ampersand Inc., and manufactured at R. R. Donnelley & Sons. The book was edited by Joanne Fraser. Joseph Vesely supervised production.